The Handbook of Privacy Studies

The Handbook of Privacy Studies

An Interdisciplinary Introduction

Edited by
Bart van der Sloot & Aviva de Groot

First published in 2018 by Amsterdam University Press Ltd.

Published 2025 by Routledge
4 Park Square, Milton Park, Abingdon, Oxon OX14 4RN
605 Third Avenue, New York, NY 10158

Routledge is an imprint of the Taylor & Francis Group, an informa business

ISBN: 9789462988095 (pbk)
ISBN: 9781003706434 (ebk)
NUR 740

Cover design: Moker Ontwerp

DOI: 10.5117/9789462988095

Contents

Introduction

The background of this book

It was the beginning of 2010 when I (Bart) first joined the University of Amsterdam as a privacy researcher. What struck me immediately were two things: how interdisciplinary the topic of privacy was and how limited the interaction was between the different fields, disciplines, and researchers from the various faculties at the university. Out of personal interest and a desire to map the field of privacy, I decided to invite for coffee and formally interview over 50 colleagues about their privacy research. It brought me to the fields of medicine, anthropology, economy, political science, informatics, philosophy, law, sociology, communication science, psychology, and a couple more.

I think I failed miserably at grasping and properly describing everyone's research in the small report I made on the basis of the interviews. Although privacy was certainly the central theme, the role it played in the various disciplines, the methodology they used, and the types of questions scholars were trying to answer varied widely, not to mention the jargon. Some philosophers tried to define the universal value of privacy, while anthropologists and sociologists stressed its contextual and cultural nature; while political scientists viewed data as means of power and control, lawyers tended to see privacy as a right to be safeguarded from intrusions; in communication science and economy, personal information was seen primarily as an asset, while informaticians focused mainly on building secure and confidential information systems without any data leakage.

What became clear from the interviews was that each researcher felt that in order to properly discuss and answer research questions within his own field of research, he needed to have insight into aspects from other disciplines. People working at the informatics department, for example, built information systems in health care environments, and sought a better grasp of informational secrecy and doctor-patient confidentiality in order to properly design infrastructure. People at the medical department of the university called for more knowledge about the legal protection of patient data and the exception in law for using their data for scientific (medical) research. Faced with the different approaches in different countries and regions in the world, lawyers wanted to have more insights from the fields of sociology and anthropology. And people working within the latter disciplines often were more than interested in the ethical debates about values and principles underlying the right to privacy.

At the same time, people stressed that their knowledge of other fields and disciplines was limited. So I started organizing bimonthly research meetings, each time with two or three speakers from different backgrounds and disciplines, to discuss their research and get feedback from the audience and learn from each other. Gradually, we became more formal and structurally connected and Nico van Eijk (Faculty of Law), Guido van 't Noordende (Faculty of Science), until he left the university, Beate Roessler (Faculty of Humanities), Edo Roos Lindgreen (Faculty of Economics and Business), and I formed the spearhead leading the initiative now officially coined the Amsterdam Platform for Privacy Research (APPR).

We decided to organize public seminars and meetings, aimed at a broader audience. Although right now – at least in Europe – privacy is high on the political agenda, 2010 and 2011 were the years that 'I've got nothing to hide' and 'privacy is dead' dominated as slogans. We felt that it was necessary to explain in what ways privacy plays a role in many aspects in work and life. Doing so, APPR grew to be the organisation it is today – a network of more than 70 scholars at the University of Amsterdam that do research on aspects related to privacy.

We decided to expand and organized the Amsterdam Privacy Conference 2012 and the Amsterdam Privacy Conference 2015, which aimed to be a truly interdisciplinary conference, going beyond the many law and tech (sometimes with the inclusion of ethics) seminars and workshops already taking place. The Amsterdam Privacy Conference 2018 is being organized as we write this introduction. Finally, we felt that not only our own research community and the international research community would benefit from the interdisciplinary approach to privacy research, but students as well. I started the interdisciplinary privacy course, which after two years evolved into the minor Privacy Studies, which was attended by Aviva de Groot (the second editor of this book), who then took over the coordination of the minor programme.

Although my personal interest in privacy was already well developed during my earlier career in the film industry, I (Aviva) first engaged with privacy professionally when I entered legal practice. At that time, the European Commission had just issued its first communication on the reform of data protection law, and the focus on informational privacy had become predominant. The Amsterdam Minor Privacy Studies provided a timely programme promoting a broader and deeper understanding of a concept that was increasingly being discussed as nearing extinction.

The rich notions of privacy that interdisciplinary study offers easily resonate with lay conceptions that students develop earlier on in life. In my case, the early 70's represented the late age of a 'social conventions battlefield', in a country where some families had already learned to shed their identities after the Second World War. My contemporaries and I explored the ruinous landscapes with both curiosity and vigilance. Just like confinement, we observed that exposure can both be life-ruining and life-threatening. Later, academic literature on privacy and the broad relevance of the principles that it addressed deepened my retrospective understanding. The feminist critiques and debates were of special relevance in their insights into the politics of privacy.

The first edition of the minor programme already took place in the 'I have nothing to hide' era. Students' traditional starting assignment was – and still is – to define and argue their individual notion of privacy. When I took over the coordination of the minor from Bart, I saw students surprised to find the solemn voice of law echoing some of their heartfelt notions. They were intrigued and curious about the law's lacunas, and its paternalistic potential. They were relieved to find there is no need to resort to law to sustain any argument, and that supposed dichotomies (like that of privacy versus security) could also be seen as interdependent relations. However, characterizations of privacy as (either/or) rights, freedoms, values, defined by breaches or by context, by ethical or cultural norms, narrowed down to intimate aspects or tradable data, often made for confusing discussions. In the afterhours of many a lecture, students confessed to being overwhelmed. The different vocabularies, academic cultures and methods of the disciplines, in addition to those of the students themselves, posed challenges to the conception of a cross-disciplinary, comprehensive understanding that they wanted to develop.

These discussions and other interdisciplinary teaching experiences partly informed the design of this book. And although it is called '*The* Handbook of Privacy Studies,' we aim to do justice to the diversity not only between, but also within the disciplines, which is reflected in the chapter's titles that present *a*, rather than *the* perspective from their discipline. We hope this book sustains the analysis of common understandings and differences, so that these can be taught in a meaningful way.

Why this book?

The reason to initiate and edit this book was to promote the interdisciplinary line of privacy research we had built over the past few years at the University of Amsterdam. It is also the combined result of our continued search for an interdisciplinary *understanding* of privacy, and a way to share these insights with students of different disciplines. Since the idea for this book was developed during the discussions at APPR meetings and the minor programme, many of the authors are based in Amsterdam. Others have their roots there. Bart and Aviva now work at the Tilburg University, Willemijn Aerdts and Gilliam Valk decided to move from the University of Amsterdam to Leiden University, and Matthijs Koot started working for a security firm. Still other authors were either teachers for the minor programme, such as Jo Pierson, students of the minor programme, such as Ine van Zeeland, or keynote speakers at the Amsterdam Privacy Conferences, such as Sandra Petronio, Deirdre Mulligan, Viktor Mayer-Schöberger, Anita Allen, and Amitai Etzioni. Some new faces also appear, such as Robin Pierce, Miko Hypponen, and Cas Sunstein.

This book is intended for three types of audiences:

- It is written for privacy researchers who are interested in other fields of research. Suppose you are a lawyer and are faced in your research with aspects of ethics and informatics – this book will provide you with a basic understanding of those disciplines and suggest further readings on specific topics that may be of interest to you. The chapters are written so that a researcher from every scientific background should be able to understand the disciplinary approach to privacy from every other academic discipline.
- It is written for students who are interested in privacy from a multidisciplinary background. It can be used as a basic textbook in interdisciplinary educational programmes such as the minor Privacy Studies. It can also be used for disciplinary courses of which privacy is one of the aspects. A chapter may be used to explain to a student what, among other themes and topics, the role of privacy is from the perspective of the discipline covered in that chapter.
- It is also written for a general audience interested in privacy. Privacy is in the news almost every day – Facebook and Cambridge Analytica, hacks and political profiling, medical research and big data technology, the General Data Protection Regulation, and mass surveillance by intelligence agencies, etc. This book will provide you with more background information about these developments and how to understand and properly evaluate them.

Privacy itself is a multidisciplinary phenomenon. A common 'playground' and language needs to be instated for researchers to present the role of privacy within their discipline, and the interdisciplinary value of their knowledge to a common understanding of privacy. When Aviva and Ine presented the minor and its challenges at the National Interdisciplinary Education Conference, the need for something that could be conceptualized as 'a handbook' was shared widely. Calls included 'a red thread', 'oversight', 'tables with aspects', 'a reader for teachers', 'creative examinations', and 'strictly structured lecture schemes'.

This book caters to some of these needs. It is the first book that makes an earnest attempt at bringing together some of the most important disciplinary approaches in the field of privacy in a comprehensive way. Nevertheless, it is only a first scan and a selection of relevant disciplines. We already envisage a second edition that includes fields that were also part of the minor and that we are eager to incorporate, such as psychology, sociology, architecture, internet studies, and political science. This first edition contains chapters on history, law, ethics, economy, informatics, intelligence studies, archival studies, medicine, media studies, communication studies, and anthropology. We asked each author to provide the reader with an introduction to her field of research, the role privacy plays within that discipline, to introduce the reader to the classic texts that have helped shape that discipline, and to map the debates and schools that have been dominant over the past few decades. Finally, we have asked them to list a number of questions on their current research agenda (or that of their peers) – what are the difficult challenges, what burning dilemmas are provoked by new technological developments, and what unresolved issues remain to be addressed by scholars? Each chapter concludes with a few suggestions for further reading.

Between those chapters, introducing the disciplinary approaches to privacy, we have added small snippets and reflections by famous authors and defining intellectuals in the field of privacy. We are honoured to have a star line-up of Priscilla Regan, Beate Roessler, Cass Sunstein, Miko Hyponen, Charles Raab, Amitai Etizoni, Robin Pierce, Kenneth Bamberger, Deirdre Mulligan, Viktor Mayer-Schönberger, and Anita Allen. Like soloists to the orchestra, these voices lead but also resonate with the score produced through the combined effort of the book's authors.

After all these years, our interdisciplinary privacy research meetings frequently result in discussions about what a person precisely means to say, or why certain research questions are valid at all. Although some researchers have seized the opportunity to work together and expand to multi- or interdisciplinary research, most are still clearly centred in their own field of research. In part, this stems from the perception that what is considered to be essential research, ground-breaking research, works that attract funding and positions, are still mostly disciplinary. We were therefore quite surprised to be faced with quite the opposite problem upon receiving the first draft chapters. We had specifically asked authors to keep to their own discipline. To introduce it, to explain what role privacy plays within it, what debates about privacy exist in their discipline and in their specific research. Almost every author took an interdisciplinary or at least multidisciplinary approach. Had they finally come to, had we asked them to revive old habits? We were especially surprised that many authors chose to discuss legal aspects – either laws, codes of conducts, case law, or specifically legal authors such as Warren & Brandeis. Perhaps it reflects the character of present-day privacy research.

Content of this book

The first chapter is on history, written by Ronald Kroeze and Sjoerd Keulen. They argue that the history of privacy shows that privacy is an ever-changing and context-dependent phenomenon. As such, opportunities for and threats to privacy are highly related to broader societal developments. Several of these broader developments are been distinguished and discussed and we briefly sum them up here. First, changing morals, cultural and religious ideas about the individual, family, household, and 'natural' relationships have had an effect on individual privacy. Second, privacy has been influenced throughout history by political changes, especially the rise of the idea of private individual rights and the acceptance of an individual sphere that the state, society, and legal system should respect and protect. Furthermore, the development of liberal-democracies – with individual freedom and the non-interference principle as its core values – and the internationalization of human rights in the past decades, have had a big impact on the politics and history of privacy. Finally, as the first but certainly not the only chapter to address the fact that technological changes, especially in the field of infrastructure, media, and communication, have had and will have great impact on privacy matters. For those that started the book (and the chapter) expecting a definite overview of the history of privacy, the chapter may serve as a 'training phase': rather than provide accounts and definitions, the chapters of this book afford insight into the disciplinary lives of privacy, and how each discipline takes care of the subject.

A snippet by professor Priscilla Regan introduces her seminal text *Legislating Privacy: Technology, Social Values, and Public Policy* (1995), in which she argued that privacy is not only of value to the individual but to society in general as well. She also suggested three bases for the social importance of privacy: its common value, its public value, and its collective value. Her thinking about privacy as a social value was informed both by the philosophical and legal writing at the time, as well as by legislative politics and processes in the United States that sought to protect a 'right to privacy'. She concluded that the individualistic conception of privacy, popular in the 1960s and 1970s, did not provide a fruitful basis for the formulation of privacy protective policy. When privacy is regarded as being of social importance, she argued, different policy discourse and interest alignments are likely to follow. Regan's text provides a natural bridge from broader social/societal understanding of privacy to commencing to learn about the subject in more legal detail.

The second chapter is written by Bart van der Sloot. He explains that rather recently, privacy has been incorporated into human rights instruments such as the Universal Declaration on Human Rights and the European Convention on Human Rights. The European Court of Human Rights has granted the right to privacy, provided under Article 8 ECHR, a very broad scope, covering almost every aspect of a person's life. The EU Charter of Fundamental Rights contains a right to data protection, in addition to a right to privacy. Data protection is regulated in the EU by the General Data Protection Regulation. The GDPR provides detailed rules on how and when data controllers may legitimately process personal data of citizens. Famous rules include the purpose limitation, data minimalization and storage limitation principle, the right to be forgotten, the right to resist profiling, and the obligation to perform data protection impact assessments. We have put this chapter in the front row for various reasons. One is the earlier mentioned fact that the legal discourse is particularly big, broad and growing. Another is that several authors expressed the wish to refer to this chapter directly, to avoid conceptualizing legislative aspects in theirs. That is why this chapter is substantially bigger than the other ones; it provides a point of reference for the other disciplines.

The next snippet is written by Beate Roessler, who discusses her widely cited book *The Value of Privacy*. In this book, she discusses three dimensions of privacy: locational privacy, informational privacy, and decisional privacy. She argues that conceptions of privacy based upon a concept of autonomy or individual freedom provide the most interesting and forward-looking possibilities for a conceptualization of the term. The three dimensions – not realms, not spaces – of privacy serve to protect, facilitate, and effectuate individual liberties in a variety of respects. Freedom-oriented theories of privacy are to be found within the whole range of theories of privacy, from those that deal with the privacy of (intimate) actions to those concerned with informational privacy or the privacy of the household.

The third chapter, dealing with the ethical perspective, logically follows. It is written by Marijn Sax. As both society and technology are constantly developing and changing, he argues, we are also confronted with a constant reconfiguration of norms that regulate what we may know of each other, what we may see of each other, what places we may enter, what information we may share, and what private decisions we may (try to) influence. Many of the theories discussed are an attempt to (1) make sense of these shifting norms, and (2) suggest how we *should*, ideally, understand and enforce privacy norms. Marijn Sax explains, inter alia, the difference in ethics between access-based and control-based approaches to privacy.

At this stage in the book, where privacy-as-autonomy has been properly introduced, a following snippet is presented by Cass Sunstein, who offers a general introduction to the idea of 'nudging', the theory of manipulating people's choices to serve their own well being. A list of the most important 'nudges' illustrates the practice. Nudging was made famous by the book *Nudge* he wrote together with Richard Thaler. The snippet also provides a short discussion of the question whether to create some kind of separate 'behavioural insights unit', capable of conducting its own research, or instead to rely on existing institutions. The snippet is followed by a chapter on a discipline that addresses the costs and benefits to privacy of the actors on either side of the nudging (and other) business, and takes an economical perspective.

This fourth chapter is written by Edo Roos Lindgreen. The chapter explores an economical approach to privacy. Roos Lindgreen identifies and clarifies various factors of influence on the economics of privacy in the digital age. As it turns out, it is relatively easy to identify positive economic factors (benefits) and negative economic factors (costs) of privacy for individuals, organizations, and society at large. For individuals, controlling the disclosure of personal data has significant direct benefits, but also leads to opportunity costs: the indirect costs of not being able to enjoy other benefits. For private and public organizations, collecting and using personal data leads to significant economic benefits; prohibiting them from doing so will erode their competitive advantage and incur opportunity costs. For society at large, however, the situation is quite unclear.

The next snippet by IT-expert Miko Hypponen takes a leap into the architecture beneath the applications and techniques at play in the former chapter. Hypponen argues that the Internet wasn't built for security or privacy. We built it first and have had to play catch-up to secure it afterwards and are still working on that, all the time. Unfortunately, the Internet of Things was not 'built' for security either. But it's not too late. We need to take the Internet of Things' security seriously, and do it now, before the problems caused by neglecting it become too difficult to handle. By now, some of these problems are (un)fortunately foreseeable.

The former snippet introduced an important subject, and maybe the hardest one to present in a way that serves all envisaged categories of readers: the informatics perspective. Matthijs Kootand Cees de Laat, working at the University of Amsterdam, have taken it upon themselves to author the chapter. They explain how ICT poses privacy challenges, and how privacy poses ICT challenges. Selected topics relating to both perspectives are discussed. From a technical perspective, cryptography, PETs, and access controls are building blocks for privacy and data protection. They discuss the various challenges of building secure and safe systems and networks. The chapter is salient in a time where governments are intensively exploring the use of these techniques, directly and indirectly funding developments, and taking sides in the ensuing public discourse on privacy that in the process frequently narrows down to data protection. Which takes us naturally to the subjects of politics and intelligence studies.

Charles Raab activates the appropriate mental muscles in his snippet, where he argues that contributions to the study of information privacy issues can be grounded in empirical research and analytical approaches derived from the discipline of political science. Moreover, research and commentary on other dimensions of privacy besides the informational one serve to broaden the field and constructively blur the boundary that has developed between information privacy and other domains of privacy: e.g. the body, public and private space, thoughts and movement. Governance and regulatory regimes (including the law) and policy activity for these other objects of study could also be investigated as part of the analysis.

The fifth chapter logically follows with the intelligence perspective on privacy, written by Willemijn Aerdts and Gilliam Valk. They suggest that next to the rather 'technical' debate about the degree to which intelligence- and security services are allowed to invade personal space and infringe upon the right of privacy, there is a debate on how services are to actively protect civilians and their personal rights. Data mining is an important instrument of intelligence- and security services. Being able to collect, process, and analyse big data and the search for suspicious correlations seem to be indispensable to avert threats. Henceforth, an adequate oversight is of utmost importance. As shown in this chapter, this relates to the position of services and their special power in the democratic legal order (proportionality), the confidence and trust society has in the services, and the prevention of the abuse of special powers.

Amitai Etzioni has given these matters ample thought and has provided analyses throughout his career. In this snippet, he shows that in order to maintain privacy in the cyber age, boundaries on information that may be used by the government should be considered along three major dimensions: (1) the level of sensitivity of the information; (2) the volume of information collected; and (3) the extent of cybernation. These considerations guide one to find the lowest level of intrusiveness while holding the level of common good constant. A society ought to tolerate more intrusiveness if there are valid reasons to hold that the threat to the public has significantly increased (e.g. there is a pandemic), and reassert a lower level of intrusiveness when such a threat has subsided.

From the field of security, the book turns to two other disciplines in which the need to limit privacy is a central element: archival sciences and medicine.

The sixth chapter is written by Tjeerd Schiphof, who discusses the relationship between archival studies and privacy. He explains how Privacy issues are salient in the archival field. For example, individuals might experience harm because of the fact that certain materials will be stored for the long term, and so can be accessed during their lifetime. The archival institutions, private and governmental, and individual archivists have considerable responsibilities in this respect, especially at certain stages in the archival process. Schiphof explains how archivists need to navigate a sometimes complex field of law, professional ethics and national and international standards, and how these are challenged by the affordances of new technologies.

This is followed by the introduction to another field where professional ethics play an important role, and where much is asked from individual practitioners and of the field as a whole. In her snippet, Robin Pierce discusses privacy from a medical perspective. She stresses the importance of intersecting normative strands of medical privacy, derived from different sources, to form a set of norms designed to protect a bundle of interests that is essential to the maintenance of an effective healthcare system that encourages and protects appropriate care-seeking and treatment. Whether and how technological changes in the collection, storage, and processing of data affect the construct of medical privacy is a pressing question. Just as a bell cannot be unrung, erosion of the sphere of medical privacy is unlikely to be restored. The eager embrace of technological innovation such as big data, machine learning, AI, eHealth, data sharing, essentially forming a virtual explosion of connectedness is likely to present challenges to the construct of medical privacy. She argues for the teleological basis for medical privacy and suggests that at least one aspect of evaluating and potentially remedying instances of erosion is assessing the impact on the ability of the current construct of medical privacy to achieve its objectives.

In the seventh chapter, Wouter Koelewijn channels this focus of medical privacy to explore in depth the data protection norms and regulations at play in healthcare relationships. He underlines the high importance of privacy and data protection in this sector, and addresses the challenges of bridging legal complications and contradictions that entail the right to privacy ands doctor-patient confidentiality, especially in light of the development of electronic information systems for the storage of medical data – and in those of e-Health, big data, and artificial intelligence in healthcare. Changes in the perceptions of patients and physicians vis-à-vis each other and adaptations of the data-protection concepts seem inevitable.

In an interesting follow-up after discussing the interplay of law and professional ethics, Kenneth Bamberger and Deirdre Mulligan suggest in their snippet that for too long, scholarship and advocacy around privacy regulation has focused almost entirely on law 'on the books' – legal texts enacted by legislatures or promulgated by agencies. By contrast, the debate has surprisingly ignored privacy 'on the ground' – the ways in which those who collect and control data in different countries have (or have not) operationalized privacy protection in the light of divergent formal laws, decisions made by local administrative agencies, and other jurisdiction-specific social, cultural, and legal forces. They introduce their influential book, *Privacy on the ground.*

Having made the shift from the books and to the ground, we continue to focus on human interaction, and how this is increasingly mediated, influencing many privacy aspects. The eighth chapter, written by Jo Pierson and Ine van Zeeland, discusses the field of media studies. They argue that given the transition from social media to online platforms, the media-studies perspective generates a uniquely interdisciplinary insight into how these digital media and society mutually articulate each other. This is particularly relevant as these media and technologies are penetrating all fibres of society, from social communication to domains like health, education, mobility, urban life, and smart cities. Consequently, the need to investigate and address fundamental public values like privacy and data protection from a media and communications perspective will only increase. Media are thereby interpreted in a broad sense, namely as technological tools that mediate the interaction between people. After this chapter, it is high time to re-visit the players that make these mediated communications possible, and how they do it.

Viktor Mayer-Schönberger, famous for many books such as *Big Data* and *Delete*, argues in his snippet that today's data-rich markets are mostly online (because of the very low transaction cost of information flows online) and run by private companies. Amazon operates a data-rich market, and so do Google, Apple, Facebook, Alibaba, etc., but also current niche players such as Airbnb or Spotify. Consumers prefer such marketplaces because of the superior matching experience compared to most conventional markets, enabling them to share information with their peers, the other market 'customers'. But for this matching to happen, they also have to share their information with the market providers Market providers are the central conduit and know everything about everyone on the market that can be gleaned online. This is a tremendous (and potentially troubling) concentration of power.

A large part of this market consists of (or incorporates elements from) what is known as 'social media' or 'social networks' – platforms for people to exchange information. Where they communicate. The ninth chapter is written by Sandra Petronio, who writes about privacy in communication sciences. She stresses that the nature of privacy has long been a part of the human condition, yet, our attention to this important aspect of life, where individuals need both privacy and the ability to be social with others is in constant need of new discoveries. A mission of communication privacy management theory is to bring new insights into this phenomenon. The mission is to push these ideas further and help others to advance their interests in privacy inquiries. Petronio has developed the Communication Privacy Management Theory, which helps to understand privacy challenges and provides for teaching tools and devising ways to translate research into meaningful practice to help others.

This dual manifestation of privacy in 'freedom from' and 'freedom to' has been extensively analysed by Anita Allen. the author of the tenth and final snippet. The title refers to her ground-breaking book: *Uneasy Access: Privacy for Women in a Free Society.* This was not only the first book-length treatment of privacy by a philosopher to focus on women, it was the first book-length treatment by an academic philosopher to focus on *any* aspect of privacy. The work was a response both to the academic debates about the meaning and value of privacy found in analytic-style philosophy journals; and to feminist critiques of privacy emanating from many disciplines. While conceding that women have historically lived their lives as ancillaries and inferiors, Allen argued in *Uneasy Access* that they have had 'too much of the wrong kind of privacy'. After this, we zoom out for the last time to explore other(s) social practices and to what extent these are being studied and analysed as privacy practices – or from a privacy perspective to begin with.

The tenth and final chapter is written by Sjaak van der Geest and engages with an anthropological perspective on privacy. He argues that the old definitions and concepts of privacy still provide fruitful starting points for the exploration of meanings and experiences around privacy, in varying social and cultural settings. The chapter shows privacy as a dynamic process of having control over what one wants to share with selected others, and what not. Importantly, he points out that there is a relative neglect of privacy described as such by anthropologists, although working in other cultures and living closely with their interlocutors confronted them with striking differences in local managements and experiences of privacy. Observations about this however remained largely implicit in their ethnographic work. These indirect allusions to privacy can be found in debates about shame, social manners, witchcraft, family life, stigmatization (HIV/AIDS), gossip, secrets, lying, and disgust.

We hope that this book will help researchers around the globe to understand each other's disciplines and inspire interdisciplinary privacy research. We hope that students will find in the Handbook of Privacy Studies a reliable and intelligible introduction in to the enormous world of privacy research, and that it enables them to use the knowledge it contains in their careers. Finally, we hope that this book will help anyone interested in the subject, to gain a better grasp of privacy, to critically reflect on its role in current society. We hope you enjoy reading *The Handbook of Privacy Studies*!

Bart van der Sloot & Aviva de Groot

1. Privacy from a Historical Perspective

Sjoerd Keulen & Ronald Kroeze

1.1 Introduction

Privacy has never been a major topic for historians. After the first publication of a study on the concept of privacy in colonial history in 1972, it took another 44 years before David Vincent published the first monograph on the history of privacy. However, over the last twenty years privacy has received more attention of historians, especially in an attempt to historicize growing concerns about modern surveillance techniques. This has indeed provided new insights into contemporary challenges as well as the history of privacy, for example that privacy has had different meanings and as an ideal came into existence under specific historical circumstances. Moreover, over the last 30 years concerns of privacy and privacy regulations have influenced the profession of historians.

Here it is important to stress that historians have their own research methods. They focus on continuity and change over time and pay ample attention to the context in which certain ideas and practices have developed.[1] The historical discipline's main concern is therefore to understand the past on its own terms. The methodology historians use assumes that the past can only be made accessible through source criticism, the interpretation of sources and literature, and the construction of a historical narrative. Historical narratives may change when new sources are discovered, old sources are restudied with the help of new (digital) methods or when a new generation of historians asks new questions about the past informed by contemporary challenges.[2] This explains why historians make a distinction between the past as such and historical narratives about the past. The latter, the history of history writing, is called historiography. Studying the historiographical trends in general and the historiography of the topic under scrutiny more precisely, is essential for historians. It provides insight into how historians have dealt with the past, the methods they have used, and the different interpretations of the same past that can (co)exist and the debates this variety has caused among historians.[3] Understanding and

1 Tosh 2010; Lorenz 2006.

2 Ankersmit 1985, 15; Ankersmit 2001.

3 Iggers 1997.

accepting these aspects is what might be called 'historical awareness', as the historiographer and methods historian John Tosh has stressed. It also includes being sceptical towards nostalgia (the past was better) and progress (the present and future are better than the past) as well as anachronism ('the unthinking assumption that people in the past behaved and thought as we do', as Tosh puts it).[4]

We are inclined to this understanding of history writing and offer a historical interpretation of privacy in this chapter. We touch upon some of the most important topics in Western(-European) history and historiography when it comes to the history of privacy. Other historians using a different geographical scope or other sources and methods may want to stress different developments.

In this chapter, we first look at the history of privacy by using a long-term perspective and by focusing on the broader context. Thereafter we discuss several classic texts, which provide a good entrance to understanding the turning points in the history of privacy. These classic texts can be viewed as essential sources for understanding various past meanings of privacy. In the third section we introduce the historiography of privacy. Here we discuss the main texts of historians on privacy as well as the different historical methods and historical schools and how they have contributed to different (and sometimes conflicting) understandings of privacy in history. As privacy is not only an object of study, we will discuss the challenges privacy holds for the (future) profession of historians in the fifth section. We will finish with some concluding remarks.

1.2 The meaning and function of privacy

Privacy is not a clear-cut concept. Neither today, nor in history. As present-day dictionaries, such as the Merriam-Webster or Oxford Dictionary, already show, privacy can be defined as freedom from unauthorized intrusion or one's right to privacy, but also as (a place of) seclusion, secrecy, a private matter, and the state of being free from public attention. But as history shows, these interpretations have not always been around and were developed in specific historical circumstances. In this chapter we give a historical overview of how privacy has been understood throughout history. By using a long-term perspective and focusing on the broader context we illustrate that the concept of privacy was never fixed, and that the discussions and

4 Tosh 2010.

discourses on privacy reflect the larger societal changes. We use the most common periodization in Western-European historiography. As we illustrate, the history of privacy can be traced back to Ancient Times but the rise of more modern and contemporary interpretations of privacy have been related to the premodern period (ca. 1500-1789) which includes the Renaissance, the Reformation, and the Enlightenment. The third period deals with The Long 19th Century, the era from the French Revolution until the First World War (1789-1914). The fourth period covers more or less the Short Twentieth Century (1914-1991), which includes two World Wars and the Cold War, and its aftermath. However, we also made an intervention in the periodization. Because the latest changes of privacy are very much influenced by the historical impact of new information technology we divided the twentieth century in a pre- and a post computer age, the latter starting in the 1970s.[5] To illustrate that the borders of the periodizations for privacy are not as strict as in for example political periodizations, we used round numbers.

As subthemes in every historical period we touch upon the most emergent changes in those time periods. Those changes come mostly in the form of discussions and anxieties about sociopolitical and technological change. These changes have similarities but also differ for every period, which is one of the explanations that the concept of privacy was both characterized by recurrent features and debates as well as by fluidity in time. We do not focus on the judicial and legal aspects of privacy which are covered in the legal chapter of this handbook.

1.2.1 Until 1500: Privacy before the Middle Ages

Scholars have traced the history of privacy back to ancient civilizations. The sociologist of totalitarian regimes Barrington Moore wrote a social and cultural history of privacy in the ancient world. He emphasized that 'totalitarian' regimes throughout history have been trying to control their subjects' lives by either denying them privacy or through surveillance. Moore, for example, looked at the Chinese Qin dynasty (221-206 BC) and the Indian Maruya Empire (322-187 BC), and stressed how they were unsuccessful in controlling privacy as they lacked modern equipment like phone tapping or CCTV for surveillance.[6]

5 As an introduction: Jordheim 2012.

6 Moore 1984.

Aristotle (384-322 BC) is another common starting point for a historical review of privacy.[7] Many scholars of privacy consider the distinction Aristotle made between the private domestic sphere of the family, the *oikos*, and the public sphere of politics and political activity, the *polis*, as the first classical reference to the existence of a distinctive private domain. Both Aristotle's *Politics* and *Ethics* cover these subjects. The political philosopher Hannah Arendt (1906-1975) made this distinction famous when she argued that this split also separated the world of women and children (oikos) from that of men (polis), and that this distinction has continued to exist into the modern era.[8] By using these references, historical reviews of privacy, suggest that in over 2200 years of history privacy was mainly understood in the same way.

Several historians have stressed that this view on privacy as an unchanged concept is problematic as can be illustrated by the example of the Greek oikos and polis. From historical research we know that the oikos differed much from our modern nuclear family house(hold) aimed at consumption. The ancient household was foremost a place for production, a farm, a catering of a much larger family (and their slaves), through which the oikos as a group – and not the individuals that made up the oikos – had access to the polis. The oikos was the place where traditions of the polis were taught, making the oikos a political phenomenon. The role of women was also more complex. Religion was pivotal in the polis and women played a central, sometimes even decisive, role in religious ceremonies and festivals. This makes the (political) influence of women in the polis considerable.[9] Since the organization of society was made out of groups and people who foremost identified themselves as a group member, there was only a limited notion of individuality if we use a contemporary Western perspective. This makes a research that starts from the idealized modern notion of privacy as an aspired and equal individual right historically problematic.

1.2.2 Privacy from the Renaissance till the French Revolution (ca.1500-1800)

1.2.2.1 The importance of a middling sector

Amongst historians the position now commonly held, is that, in the words of Harvard historian Jill Lepore, 'the history of privacy is bounded; privacy, as an aspiration, didn't really exist before the rise of individualism, and it

7 For example: DeCew 2018.

8 Arendt 1998.

9 Nagle 2006.

got good and going only with the emergence of a middle class'.[10] Privacy as a concept is essentially linked to the emergence of individualism and a middling sector in society that had both the time to take up intellectual labour and – unlike the rulers and the lower strata of society – the liberty to choose their own living space.[11]

We can see the emergence of such a middling group in the period of the Renaissance and the Reformation (c. 1450-1650). Merchants, scholars, and clergy had the luxury and time to reflect and to write to fellow souls about their inner feelings. After the invention of the printing press (c. 1440) books and letters were quickly dispersed throughout Europe. When private letters are compared to public outlets, one sees how individuals created a distinction between the private and the public persona. This is typical for the Renaissance. For example, by analysing the work of Thomas More (1478-1538) Renaissance scholar Stephen Greenblatt shows how More purposely draws a 'calculated distance between his public persona and his inner self. (...) His whole identity depended upon the existence of a private retreat'. More also built such a retreat in a literal sense, in the form of his house. His inner feelings and needs sharply contrasted with More's most famous work, the ironic *Utopia*. In this antonym work the private (*privatus*) is identified as the root to all social injustice and the prime hindrance to the public interest. The urge for retreat is a characteristic of the time of Renaissance, which can be seen both in monastic and in civic life. With priests seeking voluntary periods of seclusion. 'As the public, civic world made increasing claims on men's lives, so, correspondingly, men turned themselves, sought privacy, withdrew for privileged moments from urban pressures'. This was one of the driving forces that generated individuality, which is one of the key characteristics of the Renaissance.[12]

The diary became a place of definition and management of the self and thus a place of privacy. According to historian Philippe Ariès, England at the end of the fifteenth century was 'the birthplace of privacy', since diaries were widely kept there. Private letters, diaries, and autobiographies, but also closets and the study got popular.[13] However, privacy was not a clear positive thing for contemporaries. The linguist and cultural historian Cecile Jagodzinski shows that privacy in the days and works of Shakespeare (1564-1616) was mainly discussed in a negative manner. In plays like *Love's*

10 Lepore 2007.

11 Webb 2007.

12 Greenblatt 2005, 45, 46.

13 Phillipe Ariès 2003, 5.

Labours Lost or *The Tempest* privacy is portrayed as negative. Solitude and the contemplative private are treated as suspicious. They are the 'instigators of vice and political conspiracy' which are trying to create chaos, and disrupt the stability of the natural state in which kings have the divine right to rule.[14]

1.2.2.2 An emerging individuality

An emerging individuality had a profound effect on society. The Reformation (1517-1648) can be viewed as a struggle between collective readership by a traditional church authority and hierarchy of the Catholic Church versus the authority of the individual believer and his interpretations of private reading of the scripture. Jagodzinski shows how the concept of privacy changed in the seventeenth century in a context of rising popularity of reading. The number of printed books increased, as well as their circulation. Readers started to acquire 'a new sense of personal autonomy, a new consciousness of the self'. This helped to shape the concept of privacy to become a personal right and the core of individuality. According to Jagodzinski, continuing religious struggles in post-Reformation England 'eventually ratified the right to individual autonomy in all things (including the religion): and that the catalyst for these changes lay in the practice of private spiritual reading'. This was not a revolutionary process but a steadily evolving one.[15] *Two Treatises of Government* (1690) of the protoliberal and philosopher John Locke (1632-1704) are symbolic for this new understanding of privacy as personal autonomy and individuality. In his contract theory he argues that cooperation in and stability of a political society is the result of the legitimate aim of rational individuals to protect their private life, liberty, and property.[16]

Changes in the understanding of privacy also changed family life and housing. In his book on the history of childhood Phillipe Ariès proposes that the formation of the modern nuclear family was a result of 'a desire for privacy and also a craving for identity: the members of the family were united by feeling, habit and their way of life'.[17] This was very much a middle-class affair, both the higher and the lower classes still lived in larger groups. In the eighteenth century 'the family began to hold society at a distance, to push it back beyond a steadily extending zone of private life'. The layout of houses began to change to accommodate the urge for privacy, most strikingly by the introduction of a corridor on which rooms opened. Rooms also got

14 Jagodzinski 1999, 1-25.

15 Jagodzinski 1999, 1-25.

16 Locke 1988.

17 Ariès 1962, 413.

distinct functions and beds that used to be all over the house ended up exclusively in a bedroom. Servants were kept at more distance by installing bells, while the introduction of the first post services were used for making appointments to visit – instead of just dropping by. 'The rearrangement of the house and the reform of manners left more room for private life; and this was taken up by a family reduced to parents and children, a family from which servants, clients and friends were excluded', as Ariès states.[18]

In his book *The Secret History of Domesticity* the cultural scholar Michael McKeon shows how the modern notion of the public-private relation emerged in the seventeenth and eighteenth centuries in England. He describes this development throughout the whole private-public spectrum. At the private side of the spectrum this is visible in developments like the privatization of the family and marriage. McKeon also stresses the political impact of this development which becomes apparent at the 'public extreme' in the rise of contractual thinking, the devolution of absolutism and the shaping of a civil society separated from the state.[19]

The rise of a public sphere in the eighteenth century also had an impact on privacy. In Georgian England (1714-1830), printing was deregulated which lead to a spectacular rise in periodicals and newspapers. The establishment of the private persona became the fundament of citizenship. Those elements were combined in the increasing fascination of newspapers, biographers, and gossipers for the individual. Those stories circulated in a larger public sphere of coffeehouses, clubs, pubs, and playhouses. The effects of this shift were clearly visible in how a new class of entertainment professionals, the eighteenth-century London 'celebrities', protected their good reputation and their private feelings. As the cultural historian Stella Tillyard famously wrote: 'Celebrity was born at the moment private life became a tradable public commodity'.[20] For the 'celebrity' stage workers, for those who lived in and from their life in the public eye, controlling their self-representation became very important.[21] This relationship between privacy and new communication technology (newspapers), which became apparent in eighteenth century London, took off in a spectacular way after 1800 and influenced the whole of society. From 1800 onwards, the relationship between privacy and technology thickens and becomes a recurrent theme in history.[22]

18 Ariès 1962, 398, 399.
19 McKeon 2007.
20 Tillyard 2005, 64.
21 Fawcett 2016, 1-22.
22 Lepore 2013.

1.2.3 Privacy in an age of modern urbanization, communication, and state-formation (ca. 1800-1900)

1.2.3.1 Privacy threatened, privacy as an ideal?

It has been argued that after 1800 two interpretations of privacy emerged, that have kept their relevance until today. First, this period gave birth to the modern 'surveillance state' and the concept of the 'all-seeing eye' which threatened privacy and will eventually lead to privacy's death.[23] In the late 1780s, Jeremy Bentham developed the idea of the panopticon, a (prison) design with guards watching everything without prisoners/ citizens knowing when and how. The panopticon is often taken as the starting point of this modern rationale.[24]

This metaphor can only be understood against the background of an emerging second interpretation: privacy as an ideal and aspiration for every citizen. Legal historians have stressed that the democratic revolutions of around 1800 played an important role in the shaping of this ideal. The American Revolution was a defence against the right of not being insulted by the government. The Bill of Rights (1791) explicitly stated the 'right of people to be secure in their persons, houses, papers, and effects'.[25] The French Revolution gave birth to the Universal Declaration on the Rights of Man in 1789. Georges Duby, in volume IV of *A History of the Private Life*, claims that 'the nineteenth century was the golden age of private life, a time when the vocabulary and reality of private life took shape'.[26] David Vincent in *Privacy. A Short History* also stresses the importance of the rise of the modern household: the members of the household were free and secure, behind the front door they could read their books and have intimate relationships without interference, here modern privacy could flourish.[27]

1.2.3.2 Crowded places and new technologies

The rise of two paradoxical views on privacy were a result of the same developments. First, they were a reaction to extreme population growth which raised the question how to control society as well as maintain individual space. When we take the British example we clearly see the opportunities and challenges. The British population doubled between 1801

23 Froomkin 2000, 1463.
24 Vincent 2016, 53.
25 Solove 2006, 4, 5.
26 Ariès, Duby and Veyne 1987.
27 Vincent 2016, 63.

and 1851, and had doubled again by 1911, a process that went hand in hand with urbanization: up to 80% lived in a city around 1900. As cities grew, they became places of strangers in which it was impossible to know every person, street, or event. Gaslights were introduced in cities (in London in 1807) to create more visibility and safety for individuals in the night. Traffic rules were drafted to separate pedestrians from horses and, later, from cars, and social rules developed how to keep physical distance in crowded places such as train cabins. Separating people and their different tasks, became central in Victorian housing design. 'The family must have privacy', one could read in books on planning. Study, living, kitchen, and dining room were separated, servants and family were not expected to share rooms and gardens were fenced to offer privacy, seclusion, and intimacy. Of course, only the middle and higher echelons of society could afford a house that met these conditions but privacy became the ideal for all.[28]

Secondly new (communication) technology had its impact.[29] Written correspondence was not new, but new was the well-organized postal system that became increasingly reliable, easy, and cheap. In the nineteenth century low standard prices were introduced and postmen stopped in every town. Together with state investments in schools, the number of people in Western Europe that could write and read, and send letters, increased dramatically. Innovations such as the telegraph and telephone offered extra communication possibilities.[30] Journalism flourished in the nineteenth century and in the final quarter of this century, what has been called *New Journalism* developed: the emergence of the 'modern' committed, well-informed, and respectable journalist who wrote columns or tried to find out what 'really' happened. But New Journalism also refers to the emergence of American-style *boulevardism* or mass media newspapers focusing on gossip, scandal, and celebrity life.[31] Issues of immorality such as political corruption or 'unnatural' sexual affairs (adultery, homosexuality) were covered. Royals turned to the law to prevent privacy insults. A much-cited ruling of Prince Albert v. Strange in 1849 prevented that stolen etchings of Prince Albert were published. A main argument for the decision was that there existed 'the abstraction of one attribute of property, which was often its most valuable quality, namely, privacy'.[32] In a mediatized society, privacy literally became

28 Vincent 2016, 54-61.

29 Lepore 2013.

30 Van der Woud 2013; Henkin 2007; Wenzlhuemer 2015.

31 Wijfjes and Voerman 2009; Wijfjes 2004.

32 Mitchell and Mitchell 2012.

valuable. There are many examples of nineteenth-century elite men and royals that in return for money prevented publications of their 'lapses'. In the Netherlands king William II (1840-1849) was blackmailed for supposed homosexual relationships, sums of money and lucrative positions prevented his enemies from publication.[33] Especially the fear of losing their honour and reputation made people willing to pay. Newly drafted formal-legal rules on adultery, homosexuality, and divorce – another breeding ground for scandal – could quite easily turn someone's private affairs into newsworthy public stories.

1.2.3.3 Modern information collecting techniques

Changes of the state and how it was governed had an impact on privacy as well. The emerging modern bureaucratic nation-state was clearly represented by the establishment of post offices and the postman in the street, who worked on schedule and followed standardized procedures.[34] The postal system connected the nation and its inhabitants and was, together with the security forces like the police and the army, a clear representative of the modern state. But the modern state was a paradoxical thing when it comes to privacy. On the one hand the government took measures to protect privacy, on the other hand it infringed further in private life through data collection. For example, it actively engaged in the prohibition of certain stories or in forcing newspapers to destroy complete issues when the privacy of high-placed persons was threatened. At the same time the government structurally collected more and more information. The Census and the collecting of Government Records were 'threats' to privacy in the nineteenth century according to privacy law professor Daniel Solove. In the US the number of questions asked during the census dramatically increased from only four in 1790 to 142 in 1860.[35] In England the General Register Office collected and archived information on marriage and childbirth since 1801 but officials steadily collected more sensitive information on economic status, languages spoken, and illnesses for 'security' reasons.[36]

Not surprisingly, in such a context privacy scandals could emerge. Such as the one in 1844, when it became known that with permission of sir James Graham, Secretary of the Home Department, the post of Italian freedom fighter Giuseppe Mazzini living in exile in London was opened on request

33 Van Zanten 2014. But only for a while, in the end several anecdotes reached out to the public.

34 Bayly 2003.

35 Solove 2006, 6.

36 Levitan 2011.

of the Austrian government. In Parliament Graham denied his actions because state security was not a topic to be discussed publicly. It showed that state security could and would be used as an argument to intrude on privacy.[37] Moreover, if and when private correspondence was a matter of public concern was a source of scandal throughout the nineteenth century.[38]

1.2.3.4 The paradox of the liberal state

Although, the nineteenth century is widely regarded as an era of liberalism,[39] one sees how liberal reforms such as freedom of opinion in post, speech, and in the press, more room for private entrepreneurship in the media sector and new laws to protect individual rights were in reality both an opportunity as well as a challenge to privacy. On the one hand the liberal emphasis on private space and individual rights that need to be guaranteed by the law and the state was supportive towards the development of privacy as an individual right. On the other hand, even in an era of liberal reform, citizens would only enjoy their privacy when the state granted it to them. As the historian of privacy David Vincent puts it: 'Liberal governmentality derived its authority from a deliberate act of withdrawal from the private sphere'.[40] In other words, the liberal state gave privacy to its citizens on certain conditions. The emergence of the modern state made people, therefore, rethink their individual privacy and possible threats.

This is clearly visible in the work of the eminent liberal scholar John Stuart Mill (1806-1873) who dedicated much of his work to the dangers of the 'overgrown state' for private individuals. He stressed that in a liberal democracy, the freedom of private individuals should not be limited by a bureaucratic state or other unnecessary forms of state control; interference in one's private life should be only allowed when an individual harms someone else.[41]

From important scandals and debates from this period, we can also derive how the emergence of liberal rights in combination with the technological and communication developments we discussed above, informed a new debate about privacy. The struggle to accommodate new communication devices which could expose the private life to ever-larger audiences in often novel ways played a crucial role in these debates. Besides the secret post

37 Vincent 2008.
38 Kroeze 2008.
39 Kahan 2003.
40 Vincent 2016, 75, 76 and 118. Based on Barry, Osborne, and Rose, 1996.
41 Held 2016; Mill 1869.

example of 1844 and the case of Prince Albert vs. Strange (1849), 'The Right to Privacy' article of Samuel Warren and Louis Brandeis of 1890 is a crucial text of this period. It was a reaction to the intrusion of *boulevardism* on the private life of the first author, whose daughter's marriage was without consent covered in the media.[42] The article was a plea for a 'right to be let alone'. This challenged the idea that privacy was a relational thing and only to be found in the context of the family and domestic home. In short, the text can be seen as one of the first pleas for private 'isolation', for a desire to control personal image and information and for a legal system that would protect these rights, an interpretation that would become dominant in the twentieth century.[43]

1.2.4 Privacy in an era of international conflict and the emergence of the welfare state (ca. 1900-1970)

1.2.4.1 Extending individual rights

Warren and Brandeis contributed to a more radical interpretation of privacy and urged for legal protection but their desire to better protect the private individual fit well a broader development of protecting human rights. In the twentieth century privacy became a more fundamental and international desire, a development which was a reaction to experiences with racism in a colonial context and the atrocities and disrespect for private life and dignity during the Second World War (1939-1945). For those reasons initiatives to strengthen the formal-legal protection of individual rights on the international level were widely supported. The United Nations were founded in 1945 and article 12 of the UN Declaration of Human Rights of 1948 stressed that 'no one shall be subjected to arbitrary interference with his privacy'. The 1950 European Convention on Human Rights issued that 'Everyone has the right to respect for his private and family life, his home and his correspondence'.[44]

Still, some other important changes took place on a national scale in relation with the emergence of the welfare state. From the beginning of the twentieth century, in different Western countries, new laws were established that protected vulnerable individuals and their individuality

42 'The Right to Be Let Alone', 1890.

43 Vincent 2016, 77 and 78.

44 UN Declaration of Human Rights, see http://www.ohchr.org/EN/UDHR/Documents/UDHR_Translations/eng.pdf; European Convention on Human Rights, see https://www.echr.coe.int/Documents/Convention_ENG.pdf. See also Stuurman 2017, Chapter 9 'The Age of Human Rights'.

such as children, women, and the elderly.[45] Acts that promoted children's rights (in England the 1908 Children's Act, the Punishment of Incest Act of 1908 and the Maternity and Child Welfare Act of 1918; internationally also the UN Declaration of the rights of the child 1959/1989 could be mentioned) allowed the state to intervene in family life when the child was neglected.[46] Women's rights were strengthened as well. Women were more and more recognized as autonomous citizens with an individuality that did not depend on their relationship with a man and on their position in the household. Very important in this respect was the universal right to vote that was established in many countries in the first half of the twentieth century. But it was a long, and still-lasting struggle. Not only did women lack the right to have their own bank account or to work after marriage in countries such as the Netherlands in the 1950s and 1960s, a 'modern' country such as Switzerland established full women's suffrage only in 1971, to name but a few examples.[47]

These changes were clearly related to the welfare state, which cautiously emerged in the years around the First World War (1914-1918) and was embraced by most political groups in the West in the decades after 1945.[48] Besides laws on women and child rights, the welfare state established new town planning acts and set basic standards for housing (in Great Britain in 1918 and 1919 and in the Netherlands with the Housing Law of 1901 and the Rental Law of 1950). These acts prescribed that new houses, especially in the social housing sector, should have a separate kitchen, an indoor toilet,[49] and preferably three bedrooms so that parents, sisters, and brothers could sleep in their own room and have their privacy. Housing acts however also contained basic rules about how families were supposed to use their house and under what conditions welfare workers were allowed to intervene. In the 1950s in the Netherlands, public officials who selected farm helpers for the new Noordoost-Polder selected on how housewives made beds and were dressed in unannounced house visits.[50] So, the welfare state provided a basis for home, security, literacy, income, and health but those collective claims always went hand in hand with the right of the state to interfere.[51]

45 Renwick 2017. In Germany this process started even earlier: Grimmer-Solem 2003.

46 Vincent 2016, 80.

47 Adams 2016.

48 Judt 2007; Keulen 2014.

49 Vincent 2016, 81: Large groups – 20-30% – had no fixed bath and no water closet. In 1951 in Manchester 40% of the homes did not have an exclusive use of a bath. Near-universal availability of basic sanitation was achieved after 1975. Across Europe we find comparable figures.

50 Vriend 2014.

51 Young and Willmott 2011; Vincent 2016, 127.

New communication and entertainment technology had, again, another impact on privacy. The telephone, a nineteenth-century invention, displaced the letter as the most important means of communication by 1970. Radio and television were new for the twentieth century and were readily adopted in the new homes. They were consumer products but also created new forms of fear about the harmful effects of too much privacy as authorities became suspicious about the moral impact of the television on private and family life.[52]

1.2.4.2 New fears of the surveillance state

The twentieth century also added another chapter to the fear of the emergence of the surveillance state and its impact on privacy. Although, statistics and surveillance had started in the nineteenth century, as did the debate on the surveillance state, the twentieth century made it more of a reality. Because of the rise of the welfare state, more files of individuals were created and kept. If people wanted social housing, a pension, or unemployment benefits they had to register and apply for support and often had to accept inspection at home to determine both the financial need and the decency and skill set of the prospective recipients. Surveillance, therefore, changed from being controlled and supervised by one's neighbourhood and family to an anonymous and systematic control by the state and social welfare organizations.[53] Other forms of registration were introduced as well. Almost nobody used to register for a passport, but from around 1900 a passport was needed to travel abroad and the document became universal.[54]

As part of the surveillance state police, security, and intelligence services advanced as well. Criminal organizations were infiltrated more often by police, and they started to use phone taps. In Britain, in 1957 an inquiry committee chaired by judge Lord Birkett, investigating the tapping of the phone of a barrister of a London gangster, stated:

> There is no doubt that the interception of communications, whether by the opening or reading of letters or telegrams, or by listening to and recording telephone conversations, is regarded with general disfavour. (…) [They are] an invasion of privacy and an interference with the liberty of the individual in his right to be 'let alone when lawfully engaged in his own affairs.'[55]

52 Vincent 2016, 91, 93, and 94.

53 For Germany: Lutz 2017.

54 Bayly 2003.

55 As cited in: Vincent 2016, 105

The committee expressed reservations for phone tapping for national security and thought it best to continue these activities and to not be transparent about whom or what was being monitored. In addition, without real parliamentary consultation, security organizations extended their activities in the period around World War Two and during the Cold War. For example, in many Western democracies communists and communist organizations were monitored and spied upon in these decades.[56]

Interestingly enough, at the same time privacy became perceived and presented as a core value of liberal democracy during the Cold War. Famous books like Hannah Arendt's *The Origins of Totalitarianism* of 1951 emphasized how totalitarian governments could only exist because of their destruction of 'the public realm of life' and by the isolation of every individual – it 'destroys private life as well'.[57] George Orwell illustrated the dangers of an illiberal state in his novel *Nineteen Eighty-Four* of 1949. Here, he presented a world without private life in which the 'Thought Police' controlled everything by permanent surveillance.[58]

1.2.5 Privacy in the computer age (1970-present)

1.2.5.1 The digitalization of privacy

The rise of the computer (1960s), Internet (1983), and World Wide Web (1993) in the past few decades has brought the impact of technological change on privacy issues at the centre of public debate. Information gathering and archiving were central for the modern state since the nineteenth century but the introduction of the computer started a whole new debate about data collecting and privacy threats. In 1969, Jerry Rosenberg wrote *The Death of Privacy* in which he argued that computers were in use with complete access to personal data.[59] Arthur R. Miller wrote in 1971 that computers would create a 'surveillance system that will turn society into a transparent world in which our homes, our finances, and our associations will be bared to a wide range of casual observers'. The growing concerns about state interference can also be derived from the renewed attention for Burke's panopticon concept, for example in Foucault's *Discipline and Punish*.[60]

56 Vincent 2008, 116-128; Hooper 1987, 29-31, 104.
57 Deborah 2002; Müller 2013.
58 Orwell 2008, 165.
59 Rosenberg 1969.
60 Foucault 1991.

Civil unrest urged politicians to take measures. In the Netherlands and Sweden in the 1970s, civilians protested against the census and the storage of the census data in the new mainframe computers. This led to the introduction of a real Privacy Law in Sweden in 1973, the adoption of the Convention for the Protection of Individuals with regard to Automatic Processing of Personal Data by the Council of Europe in 1981, and to the first national data protection law in the Netherlands in 1989.[61] In Britain in 1972, the government issued a committee headed by Kenneth Younger to consider legislation on privacy and the United States adopted their Privacy Act in 1974. Attempts to add fluoride to drinking water as a public health measure was annulled by the Dutch High Court in 1973 because the Court thought that such far-reaching measures needed a basis in law.[62] Thus, interference in private life by the government had been acceptable in the welfare state of the 1950s but no longer in the 1970s when these forms of interference in personal life needed a clear judicial foundation.

But not all contemporaries discussed digitalization as a threat. Some saw it as democratization. The computer would destroy the privacy of the typical bourgeois family and end the privilege of elites to control their private life, property, and information. Thus, in the 1970s privacy was redefined: it was used to emphasize the autonomy of the individual rather than the family and it concentrated on (the end of) information privacy.[63]

1.2.5.2 Spread of progressive values?

What by the 1970s was called progressivism further strengthened the idea of privacy as an individual and legal right. Clearly the 'traditional' marriage went into decline in Western society and single life, living together, and other forms of non-traditional relationships increased providing more options for individuals how to live and where to find their privacy. In recent years the number of single-person households has even risen to a European average of 30% of the population. Widespread availability of new and modern houses accommodated these personal choices. Legal changes, such as those that ended the criminalization of homosexuality or widened the possibility for divorce also had a huge impact on individual opportunities.[64]

61 Vincent 2016, 111 and 112; Overkleeft-Verburg 1995; Council of Europe, *Treaty 108*, 1981. This rise of literature on the end of privacy has continued up until today. See for example, David Holtzman 2006: 'Our privacy is shrinking quicker than the polar ice gap'.

62 Edeler 2009; HR 22-06-1973, NJ 1973, 386 Fluoridering.

63 Vincent 2016, 113-115.

64 Vincent 2016, 118-129, 212.

There are even signs that progressive privacy interpretations have become global aspirations. Western and non-Western ideas about privacy may still differ greatly but have also converged as privacy, at least on paper, has become a global aspired human right. The establishment of the earlier mentioned UN declaration on Human Rights and the European Convention on Human Rights has also supported this change, as well as the fact that the European Court of Human Rights has the right to rule on alleged claims of interference. Same-sex marriage was first introduced in the Netherlands in 2001, by 2018 almost 30 countries in all continents have adopted it.[65]

In sum, although orthodox religious groups and other conservative forces may have never accepted these changes and in some Western countries have retained their influence, in countries where these liberal-progressive values and laws have been established they have remained in place and put constraints on societal and state interference with private lives of citizens.

1.2.5.3 The impact of 9/11 and anti-terrorism

In the most recent period, the terrorist attacks of 9/11 in 2001 and the antiterrorism laws that were issued in reaction to it, have made privacy a more complex and disputed issue. In 2007, Julian Assange's Wikileaks revealed documents about the impact of antiterrorism actions, which stirred up emotions on privacy issues. Assange justified his actions with the slogan: 'Privacy for the weak and transparency for the powerful'. According to him we stand at a crossroads because of the rise of '[I]nternet that transfers power over entire populations to an unaccountable complex of spy agencies and their transnational corporate allies'.[66] In 2013, as a public warning Edward Snowden published classified documents about what the government had been collecting, including private information, under the umbrella of counterterrorism.

Not unlike the era of the Cold War, intelligence agencies are little transparent about their actions, and politicians are hardly asking them to be. The British Intelligence and Security Committee of Parliament's Privacy and Security report of 2015 stated: 'While the Committee has been provided with the exact figures relating the number of authorisations and warrants held by the Agencies, we have agreed that publishing that level of detail would be damaging to national security'. In the Netherlands, the parliamentary subcommittee on intelligence and security issues is even

65 For example the General Data Protection Regulation (GDPR) of the European Union which will be enforced in all the EU member states from 25 May 2018.
66 Assange et al. 2016.

called the secretive committee (commissie Stiekem).[67] Hence, the main line of defence of different Western governments has been in line with what we have seen throughout history: whenever infiltrations are reported, the government, with support of parliament, neither confirms nor denies accusations, all for the sake of security and with reference to the argument that those who have nothing to hide, will not be harmed.[68]

In recent history, different voices can be heard in the debate on privacy. Edward Snowden is one of the critical voices when it comes to the 'nothing-to-hide-argument': 'Arguing that you don't care about the right to privacy because you have nothing to hide is no different than saying you don't care about free speech because you have nothing to say'. He added that individuals do not have to justify the right to be let alone, on the contrary, governments should convincingly explain why they collect personal data in the first place. There is also a growing number of, mainly legal, experts who have analysed the 'nothing-to-hide argument' and came to the conclusion that it is a dangerous, ill-convincing, and false representation of how these laws work.[69] The larger public seems concerned as well. In 2017 in the Netherlands, the Law on the Intelligence and Security Services passed parliament, but a popular comedian launched a successful campaign to rally popular support to hold a referendum in March 2018 on this 'Big-Data-Trawl Law' (Sleepwet). The turnout showed that a (small) majority did not support the law, which forced the government to make changes.[70] The debate is hot-tempered because 'not only privacy is at stake but above all democracy', as privacy sociologist Jan Holvast has claimed.[71]

On the other hand, there are experts who have nuanced these recent fears. The historian of privacy Vincent has stressed that throughout modern history there has always been a tendency to overestimate the possibilities and techniques, and therefore the dangers, of the surveillance state.[72] He claims that misreading of the history of privacy contributes to recent fears. And unlike critical voices like to claim, there is no historical evidence that supports the claim that people were more in control of their personal image and private information in the past. The examples of the annoyed Warren or the fear of the London celebrities in the nineteenth century illustrate this. And although social media may have blurred existing lines too and Facebook

67 Versteegh 2017.
68 Vincent 2016, 131.
69 Solove 2011.
70 Lonkhuyzen 2017.
71 Holvast 2009.
72 Vincent 2016, 132-134.

CEO Mark Zuckerberg may claim that privacy is no longer a 'social norm',[73] face-to-face communication is still highly important and many social media messages only have relevance for a small group of users. Therefore, some scholars have stressed to look at privacy more as a contextual value instead of only an individual and absolute principle. The philosopher of technology Helen Nissenbaum has stressed the importance of 'contextual integrity': privacy is about rules and expectations between you and the environment.[74] Clearly, in the contemporary period these rules and expectations are being reformulated, as they were in the past, and this explains ongoing debates on privacy in society, politics, and science.

1.2.6 Conclusion of the meaning and function of privacy in history

To sum up, from a long-term perspective privacy should not be understood as a linear development from less to an ever more complete set of individual rights. Nor is the context in which privacy has been discussed fixed in time. In addition, privacy in history was not always valued as something very important, nor always as a positive value. Debates about its relevance should be understood against the background of the great changes in history such as the rise of individualism, the Protestant Reformation, liberalism, and the emergence of individual rights, as well as ongoing changes in technology and communication. In the early modern era of the Renaissance and the Reformation privacy became attached to the individual but this was mainly in the context of having a private place in your home for and within the household and family life, for example to read or to pray in seclusion. Literacy and the rise of the printing press, which improved people's ability to read and communicate, contributed to privacy as an information issue as well. In a world of emerging liberalism and the modern state in the eighteenth and nineteenth century, privacy became more and more associated with protection by the state and the law, also against foreign threats. Paradoxically, the state and its security forces were also viewed as a danger to privacy, especially its interference in personal life or the gathering of personal information. The Second World War and the Cold War contributed to a belief that individual human rights, of which privacy was one, were the essential elements of a modern democracy which required more legal protection, also on the international level. Changes in modern

73 Johnson 2010. Already in 1999 Sun Microsystems CEO Scott McNealy at the introduction claimed: 'You already have zero privacy. Get over it!'
74 Nissenbaum 2010.

communication techniques, from the printing press and telephone to the computer and Internet, have had a great impact on the way privacy was understood as well. All these changes have made privacy a slippery concept that is difficult to grasp in general terms. Yes, it can be said that privacy is a form of seclusion, a right, and about the protection of private life and personal information, but in what way specifically requires that one delves into the social, political, economic, and international circumstances of the historical period one is interested in. We provided an introduction to these issues in the text above.

1.3 Classic texts and authors

In this section we will turn to four historical sources on privacy which highlight important shifts and developments in the history of privacy. Although in the texts the word 'privacy' was not always used, or not very often, they are about issues that are clearly part of the broader history of privacy. Moreover, the sources provide an entrance to how privacy in a certain period was understood. We chose Thomas More's *Utopia* (1516) because his text highlights the relationship between privacy and the rise of individualism against the background of the Reformation in the Renaissance and early modern era. Thereafter we discuss John Locke's *Second Treatise on Government* (1690) for his text is a clear example of the importance of the rise of liberalism for the acceptance of private individual rights in the seventeenth and eighteenth century. Then we discuss Jeremy Bentham's *Panopticon* (1791), for his text provides a good introduction into modern efforts, and obsessions, to control society and his idea of a panopticon has become a metaphor when it comes to discussions about the surveillance state up until the contemporary era. Finally, we chose Samuel Warren and Louis Brandeis' 'The Right to Privacy' (1890) for this text is a clear example of how in the industrial era individual privacy became defined as the right to be let alone, worthy of protection by law. The text can also be read as a clear example of individual's reactions to the growing modern communication techniques and growing role of the media on private life in the nineteenth century.

1.3.1 Thomas More, Utopia (1516)

Thomas More (1478-1535) was a leading Renaissance humanist. He was a chancellor to the English king Henry VIII but against the Reformation and opposed to the views of his patron to split the Church of England from the Catholic Church of Rome. More corresponded with many fellow humanists, such as Erasmus of Rotterdam. From his correspondence, we know that More purposely tried to shield his private life off from his public persona. This combination makes More a symbol for the emerging idea of individuality that needs privacy, which is one of the key characteristics of the Renaissance era. His views on privacy are clearly visible in More's most famous work: the novel *Utopia* from 1511. It was More who coined the term utopia. Historian Quentin Skinner has argued that More wrote *Utopia* as an ironic satire to prove that a perfect society could not exist without private property. This interpretation is now widely accepted but is an idea that started to emerge in this period. In *Utopia* More sketches a just society in the form of the Island Utopia. On this island there is no private property, but also, or therefore, no privacy. Privacy in Utopia is not viewed as a freedom; on the contrary, privacy is viewed as highly suspicious.[75] To keep its inhabitants in view full, in order to make sure that they behave well, there are no private spaces. Utopians eat in public halls and do not have a private home. The citizens rotate between the houses every year and the houses do not have a lock. Even the individual body is not private. In Utopia it is custom to make the private parts public to the partner before marriage.[76]

Thomas More wrote the book in Latin. More smartly used the Latin rendering of his name, Morus, which is similar to the Greek word for fool. He used this as a device to distance his personal self from the views in the text, while at the same time making it clear that the island Utopia is not real. Thus, the text shows how Renaissance thinkers created a distance and a distinction between their public persona and the inner self which is symbolic for the emergence of individualism in society. Secondly, because Utopia is an antonym, the ironic function helps to get a clear picture on the Renaissance thoughts on privacy. The book remains influential until today. For example, it ranks as text number 51 in the collection of one million curricula of English-language colleges and universities, while libraries over the world today hold over 700 different forms and (language) editions of this text, outranking by far any other text with utopia in its title.[77]

75 Skinner 1987.

76 More 1985.

77 Search in the Open Syllabus Project, via: http://explorer.opensyllabusproject.org/, worldcat.org.

1.3.2 John Locke, Second Treatise on Government (1690)

John Locke (1632-1704) is a founder of liberalism and a philosopher who is famous for his social contract theory. Locke published his *Second Treatise* anonymously in 1690 as part of his book *Two Treatises of Government*. The *Second Treatise* was a defence of the Glorious Revolution (1688) in which the absolute Catholic King James II was overthrown by Parliamentarians in favour of the protestant King William III. The *Second Treatise* can be seen as a counterargument to Thomas Hobbes (1588-1679) *Leviathan* (1651) in which Hobbes promotes an absolutist government as the solution to protect the people from civil war ('a war of all against all'), which he views as the state of nature. Locke had a different view on the state of nature. His state of nature is that of law and reason, which would prevent people 'to harm another in his life, liberty and or property'. But since there is no impartial authority to judge, the state of nature is neither stable nor safe for individual humans.

> This makes him [man] willing to quit a condition, which, however free, is full of fears and continual dangers: and it is not without reason, that he seeks out, and is willing to join in society with others, who are already united, or have a mind to unite, for the mutual preservation of their lives, liberties and estates, which I call by the general name, property.

Thus in order to protect private life, liberty, and/or property men is willing to unite in a society under a social contract. Since the protection of these liberties is the main reason for collaboration, a ruler of this society should not infringe on those liberties. To make certain that the ruler's sole purpose is to protect those private rights, he is tied to the social contract. When he breaks it, the people are entitled to revolt and overthrow the government.[78] So Locke argues that the state has to protect private life and individual rights, and has no right to harm them, or only on those conditions agreed under a social contract. This is a crucial principle of liberalism as well as liberal democracy. From the mid-eighteenth century the thoughts of Locke gained new popularity. Most significantly was the adoption of his thinking on private individual rights ('unalienable rights [...] Life, Liberty and the pursuit of Happiness') in the American Declaration of Independence in 1776 (see also section two).[79] Thereafter Locke's writings also became influential

78 Locke 1988.
79 Glenn 2003, 17, 18.

in the rising debates on the abolishment of the slave trade and up until the contemporary era his work is a point of reference when it comes to discussions about individual rights, including privacy.

1.3.3 Jeremy Bentham's Panopticon (1791)

Bentham's *Panopticon* from 1791 is a classic text for it has served since its publication as a metaphor for what will happen when privacy is disrespected. In the twentieth century it became the symbol for modern state's obsession with control, total oversight, and social engineering. His text is the original source for contemporary references to the panopticon and the surveillance state.

What was the panopticon? The philosopher, utilitarianist, and social reformer Jeremy Bentham (1748-1831) presented the panopticon as a proposal for social reform. The panopticon is a circular institutional building for constant surveillance, most famously in the form of a prison. The name panopticon refers to *Panoptes*, the giant watchman with hundred eyes from Greek mythology. The basic idea is that a group of people, such as prisoners, could be (cost) effectively supervised by a single watchman from a watchtower in the middle. The watchtower should be built in such a way that prisoners could not see if the guard was actually looking at them, but a rightly designed tower guaranteed that they *could* be watched at every moment. In the words of Bentham: 'I mean, the apparent omnipresence of the inspector (...) combined with the extreme facility of his real presence.'[80] Since it would be impossible for prisoners to verify if the watchman was watching them, Bentham predicted that all prisoners would act as if they were being watched constantly. This was 'a new mode of obtaining power of mind over mind, in a quantity hitherto without example'. Bentham had high hopes for his new inspection model: 'Morals reformed – health preserved – industry invigorated – instruction diffused – public burdens lightened – Economy seated, as it were, upon a rock – the Gordian Knot of the poor-law not cut, but untied – all by a simple idea in Architecture!'[81] The panopticon is perhaps most famous as an architectural design for a prison. Not least because Bentham ordered sketches and unfruitfully tried to persuade the British government for years to build a prison according to his plans. But Bentham saw the panopticon foremost as a tool of management for any institution. His brother would build a panoptical factory, and Bentham

80 Bentham 2011; Vincent 2016, 53.

81 Bentham 2011.

saw its surveillance capacities fitting for schools, hospitals, mad-houses, and the like.[82]

Bentham's description of continuous surveillance has been very influential and shaped the thinking of later scholars. It is clearly visible in the constant surveillance through telescreens by the totalitarian state in George Orwell's *Nineteen Eighty-Four* (1949).[83] In 1975 the idea of the panopticon gained influence once again thanks to the work of the French Philosopher Michel Foucault. In his book *Discipline and Punish: the Birth of the Prison* (1975) he used 'panopticism' as a metaphor for modern disciplinary societies.[84] According to Foucault the panopticon principle is not only used for prisons, but the mechanism of constant surveillance is a mechanism that controls modern social life. Power structures need docile bodies which are ideal to work in factories, create order in military regiments, or strengthen discipline in schools. In order to instil docility, the constant threat of surveillance is needed to discipline society to behave by its rules and norms. This requires a particular structure, that of the panopticon. More recently, for example during the Edward Snowden-affair on the global surveillance programmes of the National Security Agency, the panopticon was often referred to in order to emphasize how in today's digital age oversight and monitoring are organized.[85]

1.3.4 Samuel Warren and Louis Brandeis, 'The Right to Privacy' (1890)

Samuel Warren and Louis Brandeis' 'The Right to Privacy' (1890) shows a change. 'Publicity which had meant the opposite of secrecy', for men like Jeremy Bentham a century ago, 'had come to mean the attention of the press (the opposite of privacy)', as Jill Lepore argues.[86] Moreover, the text is a modern plea why there should be a right to be let alone, worthy of protection by law. 'The Right to Privacy' article of Samuel Warren and Louis Brandeis is therefore a classic.[87] The article has been called 'the single most influential article on privacy' and 'the most profound development in privacy law'.[88] They clearly responded to the changes of their time. Explicitly Warren and Brandeis referred to the 'recent inventions and business methods', such as new communication technology and mass media – the circulation of newspapers rose by about

82 Vincent 2016, 52-54.
83 Orwell 2008.
84 Foucault 1991.
85 For example: Rule 2013, A27; Simpson 2013; Julian Sanchez 2014.
86 Lepore 2013, 10.
87 Warren and Brandeis 1890, 193.
88 Solove 2006, 10; Vincent 2016, 76.

1000% between 1850 and 1890[89] – which threatened personal privacy. Warren, through his family fortune a member of the Boston commercial elite, was furious when he found out that in his view intimate details of his family were publicly shared without his consent: the *Boston Saturday Evening Gazette* had infiltrated into the wedding breakfast of Warren's daughter and published about it.[90] 'The press is overstepping in every direction the obvious bounds of propriety and decency' and 'Gossip had become trade', the authors wrote.[91] Warren and Brandeis largely build their argument on Prince Albert v. Strange (1849).[92] They wanted to protect 'the sacred precincts of private and domestic life'. But the Warren and Brandeis article also reflects a change in how privacy should be understood. At issue was a family occasion but their plea was a rejection of *any* form of personal infiltration without clear consent and a legal basis, as the article held a plea for 'the right to be let alone'. Moreover, it was a response to the modern world in which 'solitude and privacy have become more essential to the individual'.[93] Privacy, especially the right to be let alone, was not a universal right, but necessary in a modern era of mass media, and so was its legal protection, they argued. The influence of the article of Warren and Brandeis is further discussed in the chapter on privacy and law.

1.4 Traditional debates and dominant schools

Although privacy has never been a major theme in the work of historians, when we analyse historiography (the history of history writing) we can distinguish several influential works and three significant methodological streams of history writing on privacy: the history of law, social history, and cultural history.

1.4.1 The first wave: History of Law (legal history)

Privacy was first explored by historians of law. This field is mainly practised in faculties of law for the purpose of the development and interpretation of the law.[94] Due to the nature of common law, this discipline is less well established or developed in continental Europe. One should keep in mind that law history has a different purpose than much of the work of mainstream

89 Solove 2006, 10.
90 Glancy 1979.
91 Warren and Brandeis 1890, 193.
92 Post 1991, 647.
93 Vincent 2016, 77, 78.
94 For an oversight of the methods and historiography of the history of law: Ibbetson, 2003.

historians. The latter tend to work in faculties of arts or of humanities. As a result of working separatedly, there is not much cooperation or interaction between the mainstream historians and legal historians. The field of history of law is however a productive field. Newer work on the history of privacy can for example be found in David Garrow's monumental work on the historic roots of the judicial struggle for abortion rights which were concluded in Roe v. Wade (1973).[95] Another subdomain of this discipline is less interested in the jurisprudence, but focuses more on the context in which law or interpretations came about. A good example is the article of Dorothy Glancy on 'the invention of privacy law' in which she researches the context of *boulevardism* to explain why Warren and Brandeis wrote their article.[96]

1.4.2 The second wave: Social History (1960s)

Privacy as a field of study found its way into the academic discipline of mainstream history through the field of social history. It was David Flaherty who became a professor of law and history at the University of Western Ontario and wrote the first monograph which had the history of privacy as its main subject. His *Privacy in Colonial New England* (1972) can be seen as a bridge between the fields of the history of law and social history. The book originated from a subsidy of the Association of the Bar of New York City to assess the growing concern about privacy at the end of the 1960s. Flaherty's book starts from his belief that privacy is not a modern notion but a basic law of biology and ecology. He tried to prove this by turning to puritanism in colonial New England, because Puritans in the 1960s also had an ambivalent attitude towards privacy. He showed how individual New Englanders valued privacy and how with the growth of the colony and its economy privacy became more valued as houses could grow larger and settlements got more scattered. Moreover, he stressed that the control and authority over the personal life waned by the eighteenth century.

The enthusiasm to study the history of the daily life of ordinary people which were heretofore underrepresented in history, is typical for social history which became the main discipline of history writing in the 1970s. The rise of this type of social history can be understood as a democratization process within history writing, which mirrored the democratization process in society. The discipline used a wide range of methods, from microhistory focusing on small examples to the *annales* approach focusing on long-term

95 Garrow 1994.
96 Glancy 1979.

changes in mentalities. Perhaps the best-known example of an *annales* historian who wrote on the private life and on privacy is Phillipe Ariès from France. He wrote *Centuries of Childhood: A Social History on Family Life* (1962), in which privacy is discussed as one of the explanations for changes in the treatment of children as children.[97] He was also one of the editors of the five volume-series *A History of Private Life* (1985-1987).[98] In this history on daily life from antiquity to the present, the emergence of privacy is one of the themes. Diana Webbs' history of privacy and solitude in the Middle Ages is a recent example of the *annales* school.[99]

1.4.3 The third wave: New Cultural History (1990s-present)

David Vincent wrote several books on the history of different aspects of privacy, such as secrecy and the public discourse on privacy in the 19th century.[100] His *Privacy: a Short History* is the only monograph that covers the history of privacy from the Middle Ages up until the present era.[101] Although it is not a world history as it focuses primarily on the history of privacy in Great Britain, his approach and use of sources is exemplary for a cultural history approach of privacy. Starting from accounts of medieval court cases on watching windows of neighbours, Vincent leads us through the history of privacy. Vincent's main argument is that history of privacy is not linear. Notions of privacy have differed throughout history. Changes in daily life and the development of the house and bedrooms as private places are a central theme of his book.

Vincent is a social historian by training but his work is clearly influenced by New Cultural History. The New Cultural History approach emphasizes the importance of studying language and other social and cultural utterances traditionally neglected by historians, with the help of (insights from) language, narrative, and discourse theory. The influence of New Cultural History is very visible in David's book on the history of privacy: *I Hope I Don't Intrude*. The book discusses the changing concept of privacy by studying nineteenth-century plays. The book title is the catch phrase of Paul Pry, the main character and eponymous of a very popular play of the time. Moreover, the work of Fawcett on celebrity and privacy in the eighteenth century fits this category.[102] Cultural-history studies on privacy primarily look at privacy

97 Ariès 1962.
98 Especially in the third volume: Aries 2003.
99 Webb 2007.
100 Vincent 2008; Vincent 2015.
101 Vincent 2016.
102 Fawcett 2016.

in terms of reputation and domestic life. In the last years cultural histories used artefacts or personal letters as main sources to study privacy in the early modern era (ca. 1500-1750), such as is the case in Cecile M. Jagodzinski *Privacy and Print: Reading and Writing in Seventeenth-century England* (1999) or Lena Cowen Orlin's *Locating Privacy in Tudor London* (2010).[103]

Cultural history has also become the main method for political historians. This becomes visible in the recent works on privacy, modernity, and the development of the modern state. Examples are Higgs', Moran's, and Frost's work on secrecy and the state, focusing on the endeavours of the British and the United States Government in keeping official secrets secret.[104] Kathrin Levitans *A Cultural History of the Census* shows how society responded to the introduction and use of census data. The book is a good example of how cultural history has entered the field of the history of privacy in relation with policy history.[105] Not only privacy policies and the 'politics of privacy' are now more commonly researched, but also the private aspects of elites and their struggle to maintain their privacy. Examples are the recent autobiography of Jeroen van Zanten of the Dutch King William II or popular histories on the private aspects of royalty such as Michael Paterson's *A Brief History of the Private Life of Elizabeth II.*[106]

To sum up, the historiography of privacy has broadened in recent decades. It changed from a purely legal history into something to be understood in the context of social, political, and technological change that has had an effect on both elites and common people as social and cultural historians have stressed. Moreover, in the recent period there is a tendency to not only see privacy as a history of emerging individualism, Protestantism and liberalism, like in historical studies on the Renaissance. Privacy is now more often researched in relation with housing, modern state formation, globalization, and technological and communication innovation, for example in the recent book of David Vincent. This has led to the result that by now privacy is treated as a more complex and paradoxical phenomenon, worthy of studying on its own terms. In the section below, we will further elaborate on how changes in the field of history writing have affected how historians understand and deal with privacy.

103 Jagodzinski 1999; 2010.

104 Higgs 2003; Moran 2013; Frost 2017.

105 Levitan 2011.

106 Van Zanten 2014; Paterson 2012.

1.5 New challenges and topical discussions

In 1980, David Flaherty was one of the first to draw attention to the responsibilities of the historian for the privacy of his research objects. In contrast to neighbouring fields such as the social sciences and law, historians were late to give attention to privacy of sources. The main reason is that historians were long occupied with writing about people who no longer lived.[107] This is clearly related to the professionalization of history since the nineteenth century. Central in this professionalization process was the belief that historians could best study histories of people, events, or cultures that had come to an end. This assertion, often summarized in Hegel's quote: 'the owl of Minerva spreads its wings only with the falling of dusk' was a guiding principle for historians. This has changed since the 1970s, through the emergence of the field of contemporary history or *Zeitgeschichte*. As a result, historians started researching and writing about processes that still last and about people still alive.[108] Moreover, especially in the American context, historians became more conscious of privacy because they increasingly made use of the Freedom of Information Act to retrieve sensitive government information for historical research.[109]

The emergence of the relatively new field of oral history has had an impact as well. One of the goals of oral history was (and is) to give voice to the voiceless in history, by interviewing people in length about their daily lives or about traumatic experiences.[110] Unlike much of the ethnographic research in the social sciences, oral history interviews are typically not anonymous and they are being collected to be archived and thus are being kept publicly available for further research.[111] Asking for consent has become part of professional oral history research. The adoption of consent forms started in the United States where oral history has a stronger developed tradition of interviewing elitist groups who were concerned with controlling their views. By 1994, the Oral History Association had adopted ethical guidelines in which the interviewee got options to put restrictions on the accessibility of the information, to restrict access to the archives, or to request for anonymity and confidentiality.[112]

107 Flaherty 1980.
108 Palmowski and Spohr Readman 2011.
109 Flaherty 1980, 421.
110 Keulen and Kroeze 2012; Thompson 2000.
111 Thompson 2000.
112 Boschma and Mychajlunow 2003.

Changing copyright laws have influenced the fields of (oral) history and archiving in recent decades. Up until 1989 it was typical to have an informal understanding about consent in the social sciences and oral history research in the United Kingdom. Access to archives was generally an informal issue between researcher and archivist. Nowadays archives have to ensure that copyright is transferred to the archive or a licence is needed which approves broad public access while leaving the copyright with the producer of the archived material or interviewee.[113] Consent forms and copyrights may be an official solution to make consent and privacy more transparent, but this is not the end of the matter. These legal solutions lead to new dilemmas and problems for historians, as is discussed in the edited volume *Doing Recent History*.[114] Laura Clark Brown and Nancy Kaiser describe how archives struggle with interpreting privacy laws in the archival context. At first archives attempted to develop policies for sensitive materials but this proved to be unworkable as every new set of material brought its own unprecedented challenges. Now archives are inclined to turning to 'legal loopholes' in order to work around highly specialized privacy laws if they attain school records or hospital archives.[115] More information on privacy and archives can be found in the chapter on archival studies in this book.

In the same volume Gail Drakes sheds light on privacy laws and intellectual property rights. She argues how the expansion of copyright laws in the United States since the Copyright Act of 1976 has hindered historians to use newsreels or TV programmes as their content is privately owned or stored behind pay walls. Another example is the use of copyright and the 'right to publicity' laws to maintain, protect, or polish the image of a family member posthumously. The use of these laws has restricted the access to historical information on certain individuals, even after their death.[116]

The rise of women's history and the subject of privilege in the field of history in recent decades has had another impact on historians working on privacy-related topics. Feminists have pointed at the politics behind private-public distinctions and have criticized dominant notions of non-interference and privacy. What is considered private and privacy by someone, may be an urgent public matter for another. They also have made historians aware of power relations in interviewing. The historian Joan Sangster for example has argued that it is impossible for an interviewer to be detached

113 Thompson 2000.
114 Potter and Romano 2012.
115 Brown and Kaiser 2012.
116 Drakes 2012, 85.

and objective about the (interview) subject. She has also questioned the democratic assumptions of oral history by pointing out that differences in status, background, gender, or class between interviewer and interviewee could lead to 'unequal, intrusive and potentially exploitative relationships'.[117] For example, who decides what is 'true' when the interviewee, referring to his or her memory, and the historian, referring to historical knowledge, clash on the meaning of a subject? This debate has since widened and plays an eminent role in Afro-American History, postcolonial history, and the history of underprivileged groups. For example, was the collection of human remains by physio-anthropologists, the production of photos of naked indigenous people by Westerners and their exhibition in colonial museums, even up until today, a breach of privacy? And how can it be redressed?[118]

These considerations have also influenced archiving. In a recent publication, Michelle Moravec, a scholar on women's history and digital history, asks herself the question how we should treat ethics, consent and privacy of interviewees in paper magazines with small circulations amongst likeminded readers, which are now being digitized and made freely accessible to the world.[119] One recent reaction to this debate is that archives are starting to adopt restrictions to full access for the general public. They grant only full access to specific communities to 'their' materials.[120]

The lack of structural archiving of online information is one of the most important recent challenges. Whereas primary sources, printed newspapers, books, and many oral history collections are collected and categorized by national and local archives or libraries, websites are typically not. The Dutch situation is exemplary and not an exception. Here, every online published article of the largest news organization of the Netherlands, the publicly financed NOS, from before 2010 has disappeared. Hundreds of thousands of online articles from the largest Dutch newspaper *De Telegraaf* and the complete online archive of free newspapers (*Spits, De Pers, DAG.nl*) suffered the same fate.[121] The problem is related to continuous updates of digital online search, storage, and visual tools that will also continue in the future. Adobe has already announced to stop supporting the video tool *Flash* by 2020, threatening the accessibility of millions of online movie clips from individuals and organizations. The same is true for Data Management

117 Sangster 1994.
118 Sysling 2017.
119 Moravec 2017, 186.
120 Brown 2016.
121 Sedee 2018.

Systems (DMS) of organizations including the government. By updating or replacing software older DMS versions cannot be read, making digital (governmental) archives completely inaccessible, which hinders the democratic control and legitimation of decision-making.[122]

The lack of a structural approach and the unavailability of past online data to the larger public makes online data accessibility highly dependent on arbitrary decision-making and to those who have the means and interests to dig up lost information. From an academic and democratic perspective this is not desirable. For the near future historians and archivists have to rethink this dilemma, also in relation to the 'right to be forgotten' adopted in the EU.[123] The newly proclaimed 'right to refuse to be researched' which questions whether 'overstudied others' – such as native communities, ghettoized and orientalized communities – benefit themselves from the ethics and usefulness from social science research,[124] will cause further complications but nonetheless makes debating those issues inevitable.

1.6 Conclusion

The history of privacy shows that privacy has been understood as and in relation to seclusion, individual rights and protection of personal information which requires protection from the law and the government. Secondly, the history of privacy shows that debates on privacy can be understood as fears about the impact of new information technology, government interference in personal life and the rise of the so-called surveillance state. Moreover, to explain and understand how privacy was understood in specific time periods, the treatment of privacy as a context-dependent phenomenon is needed.

As the meaning of privacy is context-dependent, opportunities for and threats towards privacy are highly related to broader societal developments. Of these broader developments, several have been distinguished and discussed in this chapter but we briefly sum them up here. First, changing morals, cultural and religious ideas about the individual, family, household, and 'natural' relationships have had an effect on privacy. Second, privacy has been influenced throughout history by political changes on the national

122 For a Dutch example: paragraph 7.2.2: Kamerstuk II 2014/15 33 606, nr. 4. Hoofdrapport Parlementair Enquête Woningcorporaties.

123 EU, Judgment of the Court in Case C131/12 'the right to be forgotten' (13 May 2014).

124 Tuck and Yang 2014.

and international level, especially the rise of the idea of individual rights, including privacy, and the acceptance of an individual sphere which the state, society, and legal system should respect and protect from internal and external oppression. The development of liberal-democracy – with individual freedom and the non-interference principle as its core values – and the internationalization of human rights in the past decades have had a big impact on the politics of privacy, and the history of privacy. Finally, technological change, especially in the field of infrastructure, media, and communication, from the printing press up to Internet, have had a great impact on privacy matters and will continue to do so.

These changing technological, political, cultural and judicial shifts are not only worthy of historical research but have had an impact on the profession and the ethics of the historian and historical research as well. With the development of digital databases and online sources new technical possibilities have emerged but these have given rise to new debates on how to deal with privacy and accessibility. Debates about the essence of privacy will continue and thus make privacy a fruitful object of study for historians but also a matter of ethical reflection for citizens, politicians, and historians alike. Clearly, privacy is not only a contextual and relational issue but also a paradoxical one.

Further reading

David Vincent's *Privacy: a Short History* (2016) is the only available monograph on the history of privacy and provides an introduction to privacy in history. Other suggestions for further reading are mentioned in Chapter 4 and throughout the text. Alternatively, one could check the references.

References

Adams, Jad. (2016). *Women and the Vote: a World History.* City: Publisher.

Ankersmit, F.R. (2001) *Historical Representation.* Stanford: Stanford University Press.

Ankersmit, F.R.(1985). *Narrative Logic.* Groningen: Martinus Nijhof.

Arendt, Hannah and M. Canovan. (1998). *The Human Condition.* Chicago: University of Chicago Press.

Ariès, Philippe, Georges Duby, and Paul Veyne. (1987). *A History of Private Life. Volume 4: From the Fires of Revolution to the Great War.* Cambridge, MA: Harvard University Press.

Ariès, Philippe, Georges Duby, Roger Chartier, and Goldhammer, Arthur. (2003). *A History of Private Life. Volume III.* Cambridge, MA: Harvard University Press.

Ariès, Philippe and R. Baldick. (1962). *Centuries of Childhood: a Social History of Family Life.* New York: Random House.

Assange, Julian, Jacob Appelbaum, Andy Müller-Maguhn, and, Jérémie Zimmermann. (2016). *Cypherpunks: Freedom and the Future of the Internet.* City: Publisher.

Barry, A., T. Osborne and N. Rose. (eds). (1996). 'Introduction: Foucault and Political Reason' in *Foucault and Political Reason: Liberalism, Neo-Liberalism and Rationalities of Government.* City: Publisher.

Bayly. C.A. (2003). *The Birth of the Modern World, 1780-1914.* City: Wiley-Blackwell.

Bentham, Jeremy and Miran Božovič. (2011). *The Panopticon Writings.* London/New York: Verso.

Boschma, G., O. Yonge, and L. Mychajlunow. (2003). 'Consent in Oral History Interviews: Unique Challenges', *Qual? Health Res?* 1, p. 129-135.

Brown, K.L. (2016). 'On the Participatory Archive: The Formation of the Eastern Kentucky African American Migration Project', *Southern Cultures* 1, p. 113-127.

Deborah, N. (2002). 'Pursuing Privacy in Cold War America', Journal? Volume? Issue? Pages?.

DeCew, J. (2018). 'Privacy', in Edward N. Zalta (ed.), *The Stanford Encyclopedia of Philosophy,* https://plato.stanford.edu/archives/spr2018/entries/privacy/.

Edeler, H.A. (2009). *De drinkwaterfluoridering: tandartsen, staat en volksgezondheid in Nederland, 1946-1976.* Houten: Bohn, Stafleu, van Loghum.

Fawcett, J.H. (2016). *Spectacular Disappearances. City:* University of Michigan Press.

Flaherty, D.H. (1980-1983). 'Privacy and Confidentiality: The Responsibilities of Historians', *Reviews in American History,* p. 419-429.

Foucault, M. (1991). *Discipline and Punish: the Birth of the Prison.* London: Penguin Books.

Froomkin, A.M. (2000). 'The Death of Privacy?', *Stanford Law Review,* p. 1461-1543.

Frost, D.B. (2017). *Classified.* City: McFarland & Company, Inc.

Garrow, David J. (1994). *Liberty and Sexuality: the Right to Privacy and the Making of Roe v. Wade.* City: Publisher.

Glancy, D.J. (1979). 'The Invention of the Right to Privacy', *Arizona Law Rev*iew, p. 1.

Glenn, R.A. (2003). *The Right to Privacy: Rights and Liberties under the Law.* Santa Barbara: Publisher.

Greenblatt, Stephen. (2005). *Renaissance Self-fashioning: from More to Shakespeare.* Chicago/London: The University of Chicago Press.

Grimmer-Solem, E. (2005). *The Rise of Historical Economics and Social Reform in Germany, 1864-1894.* Oxford: Clarendon Press.

Held, D. (2016). *Models of Democracy.* Cambridge/Malden, MA: Polity.

Henkin, D.M. (2007). *The Postal Age: the Emergence of Modern Communications in Nineteenth-century America.* Chicago/Bristol: University of Chicago Press.

Higgs, E. (2003). *The Information State in England: the Central Collection of Information on Citizens, 1500-2000.* Basingstoke: Palgrave Macmillan.

Hooper, D. (1987). *Official Secrets: the Use and Abuse of the Act.* City: Harvill Secker.

Ibbetson, David. (2003). 'Historical Research in Law' in Mark Tushnet and Peter Cane (eds.), *The Oxford Handbook of Legal Studies.* Oxford: Oxford University Press.

Iggers, David (1997). *Historiography in the Twentieth Century.* Middletown: Weseylyan University Press.

Jagodzinski, Cecile M. (1999). *Privacy and Print: Reading and Writing in Seventeenth-century England.* Charlottesville/London: University Press of Virginia.

Jordheim, Helge. (2012) 'Against Periodization: Kosselleck's theory of Multiple Temporalities', *Theory and History* 51(2), 151-171.

Judt, T. (2007). *Postwar: a History of Europe since 1945.* London: Pimlico.

Kahan, A. (2003). *Liberalism in Nineteenth-century Europe: The Political Culture of Limited Suffrage*. City: Springer.

Keulen, S. (2014). *Monumenten van beleid. De wisselwerking tussen Nederlands overheidsbeleid, sociale wetenschappen en politieke cultuur, 1945-2002*. Hilversum: Verloren.

Keulen, S. and R. Kroeze. (2012). 'Back to Business: A Next Step in the Field of Oral History: the Usefulness of Oral History for Leadership and Organizational Research', *The Oral History Review* 1, p. 15-36.

Kroeze, R. (2008). 'Dutch Liberal Politics between Private and Public: the Letters-affaire of 1865', *Public Voices* 2, p. 25-43.

Lepore, J. (2013). 'The Prism: Privacy in an Age of Publicity', *The New Yorker* 24(June).

Levitan, Kathrin. (2011). *A Cultural History of the British Census: Envisioning the Multitude in the Nineteenth Century*. New York: Palgrave Macmillan.

Locke, J. (1988). *Two Treatises of Government*. Cambridge: Cambridge University Press.

Lorenz, Chris. (2006). *De constructie van het verleden. Een inleiding in de theorie van de geschiedenis*. City: Publisher.

McKeon, Michael. (2007). *The Secret History of Domesticity: Public, Private, and the Division of Knowledge*. Baltimore: Johns Hopkins University Press.

Mill, J.S. (1869). *On Liberty*. City: Longmans, Green, Reader, and Dyer.

Mitchell, C.C.J. and P. Mitchell (2012). *Landmark Cases in Equity*. Oxford: Hart.

Moore, B.J. (1984). *Privacy: Studies in Social and Cultural History*. Armonk: Publisher.

Moran, Christopher R. (2013). *Classified: Secrecy and the State in Modern Britain*. Cambridge: Cambridge University Press.

Moravec, M. (2017). 'Feminist Research Practices and Digital Archives', *Australian Feminist Studies* 91(92), p. 186-201.

More, Thomas. (1985). *Utopia*. Harmondsworth: Penguin Books.

Müller, Jan-Werner. (2013). *Contesting Democracy: Political Ideas in Twentieth-century Europe*. New Haven: Yale University Press.

Nagle, D.B. (2006). *The Household as the Foundation of Aristotle's Poli.*, Cambridge: Cambridge University Press.

Orlin, L.C. (2008). *Locating Privacy in Tudor London*. Oxford: Oxford University Press.

Orwell, George. (2008). *Nineteen Eighty-four*. London: Penguin Books.

Overkleeft-Verburg, G. (1995). *De Wet persoonsregistraties: norm, toepassing en evaluatie*. Zwolle: W.E.J. Tjeenk Willink.

Palmowski, J. and K. Spohr Readman. (2011). *Speaking Truth to Power: Contemporary History in the Twenty-First Century*. City: Publisher.

Paterson, Michael. (2012). The Private Life of Elizabeth II. London: Robinson.

Post, R.C. (year). *Rereading Warren and Brandeis: Privacy, Property, and Appropriation Symposium: The Right to Privacy One Hundred Years Later*. City: Publisher.

Potter, Claire, Romano Bond, and Christine Renee. (2012). *Doing Recent History: on Privacy, Copyright, Video Games, Institutional Review Boards, Activist Scholarship and History that Talks Back*. Athens/London: University of Georgia Press.

Lutz, Raphael. (2017). *Poverty and Welfare in Modern German History*. New York: Berghahn.

Renwick, C. (2017). *Bread for All: theOorigins of the Welfare State*. City: Publisher.

Rosenberg, J.M. (1969). *The Death of Privacy*. New York: Random House.

Sangster, J. (1994). 'Telling Our Stories: Feminist Debates and the Use of Oral History', *Women's History Review* 1, p. 5-28.

Skinner, Q. (1987). 'Sir Thomas More's Utopia and the Language of Renaissance Humanism', *The Languages of Political Theory in Early-Modern Europe*, Volume(issue),p. 123-157.

Solove, D.J. (2011). *Nothing to Hide: the False Tradeoff between Privacy and Security.* City: Yale University Press.

Solove, D.J. (2006). 'A Brief History of Information Privacy Law', *Journal* Volume(issue, pages).

Stuurman, S. (2017). *The Invention of Humanity. Equality and Cultural Difference in World History.* City: Harvard University Press.

Sysling, F. (2017). 'Skulls, Restitution and Internal Colonialism in the Netherlands', *Contemporanea* 1, p. 140-146.

Thompson, P.R. (2000). *The Voice of the Past: Oral History.* Oxford: Oxford University Press.

Tillyard, S. (2005). *'Paths of Glory': Fame and the Public in Eighteenth-Century London.* City: Publisher.

Tosh, J. (2010). *The Pursuit of History: Aims, Methods, and New Directions in the Study of Modern History.* New York: Longman/Pearson.

Tuck E. and K.W. Yang. (2014). 'R-Words: Refusing Research', *Humanizing Research: Decolonizing Qualitative Inquiry with Youth and Communities* Volume(issue), p. 223-248.

Vincent, David. (2016). *Privacy: a Short History.* Cambridge: Polity.

Vincent, David (2015). *I Hope I Don't Intrude: Privacy and Its Dilemmas in Nineteenth-century Britain. City: Publisher.*

Vincent, David. (2008). *The Culture of Secrecy: Britain, 1832-1998.* Oxford: Oxford University Press.

Vinis, P. (2016). *Raymond Wacks, Privacy: a Very Short Introduction.* City: Publisher.

Vriend, Eva. (2014). *Het nieuwe land. Het verhaal van een polder die perfect moest zijn.* Amsterdam: Balans.

Warren, S.D. and L.D. Brandeis. (1890). 'The Right to Privacy', *Harvard Law Review* 5, p. 193-220.

Webb, D. (2007). *Privacy and Solitude in the Middle Ages.* London: Hambledon Continuum.

Wenzlhuemer, R. (2015). *Connecting the Nineteenth-century World: the Telegraph and Globalization.* Cambridge: Cambridge University Press.

Wijfjes, Huub and G. Voerman. (2009). *Mediatization of Politics in History.* Leuven/Walpole: Peeters.

Wijfjes, W. (2004). *Journalistiek in Nederland, 1850-2000: Beroep, cultuur en organisatie,* Amsterdam: Boom.

Woud, Auke van der. (2013). *Een nieuwe wereld: Het ontstaan van het moderne Nederland.* City: Publisher.

Young, Michael and Peter Willmott. (2011). *Family and Kinship in East London.* London: Routledge.

Zanten, Jeroen van. (2014). *Koning Willem II: 1792-1849.* Amsterdam: Boom.

Legislating Privacy: Technology, Social Values, and Public Policy

Priscilla Regan

In *Legislating Privacy: Technology, Social Values, and Public Policy* (1995), I argued that privacy is not only of value to the individual but also to society in general and I suggested three bases for the social importance of privacy: its common value, its public value, and its collective value. My thinking about privacy as a social value was informed both by the philosophical and legal writing at the time, and also by the legislative politics and processes in the United States that sought to protect a 'right to privacy'. I concluded that the individualistic conception of privacy, popular in the 1960s and 1970s, did not provide a fruitful basis for the formulation of policy to protect privacy. I argued that if privacy is also regarded as being of social importance, different policy discourse and interest alignments are likely to follow.

As society moved into the 20th century, thinking about the importance of privacy was largely shaped by Warren and Brandeis' 1890 *Harvard Law Review* article defining privacy as the 'right to be let alone'. Alan Westin in his seminal book *Privacy and Freedom* adopted this individual rights view of privacy, defining it as the right 'of the individual to control information about himself' (1967). This focus on the individual right and the emphasis on individual control dominated much of liberal, legal, and philosophical thinking about privacy during the late 1960s and through the 1980s – a time when information and communication technologies transformed the ways that businesses, governments, and individuals collected, retained, analysed, and transferred information about individuals.

Starting in the 1970s, a group of philosophical thinkers also began to consider a broader social value of privacy. In a compendium of essays (Pennock and Chapman 1971), Carl Friedrich and Arnold Simmel both acknowledged that privacy has some broader social importance. Friedrich wrote that he was 'not concerned… with the private aspect of this privacy, individualistic and libertarian, but with the political interest that may be involved' (115). Simmel argued that privacy is 'part and parcel of the system of values that regulates action in society' (71). In a series of articles in *Philosophy and Public Affairs* in 1975, Judith Jarvis Thomson, Thomas Scanlon, and James Rachels each considered how to broaden the interest in privacy beyond traditional liberal thinking in order to expand and revitalize its importance. In an anthology on privacy (1984) and a later book (1992), Ferdinand Schoeman began a more serious and broader scholarly discussion about the social importance of privacy. Increasingly, privacy scholars at that time recognized that a narrow individual rights justification for and basis of privacy was inadequate to the actual importance of privacy in modern life. Spiros Simitis argued that privacy should not be regarded as a 'tolerant contradiction' but a 'constitutive element of a democratic society' (1987, 732).

In examining the dynamics of congressional policymaking for three issues – information privacy, communication privacy, and psychological privacy – my analysis revealed that the conception of privacy as an individual right contributed to limited congressional support for legislation. All three issues were placed on the congressional agenda in response to technological changes perceived as threatening privacy and all remained on the agenda for years, if not decades, before weakened legislation was passed. Although the idea of privacy was a good symbol with rather broad if somewhat vague public appeal, in each case protecting privacy involved costs to fairly defined interests. These interests were successful in redefining the issue to something seemingly more concrete and more in the public interest such as efficiency, crime control, or honesty and productivity in the workplace. The definition of privacy as an individual right hampered policy formulation both because policy discussion often became dominated by lawyers debating the relevance of certain legal precedents and because it entailed the balancing of an individual right to privacy against other competing rights and values that were more clearly seen as of broad social importance.

Drawing upon philosophical and legal writings and my analysis of the difficulties of legislating privacy protections in the US, I developed arguments for the common, public, and collective value of privacy. My thinking about the common value of privacy was based on the belief that all individuals value some degree of privacy and have some common perceptions about privacy. Although individuals may indeed have different definitions of privacy and may draw dissimilar boundaries about what they regard as private and public, all recognize privacy as important. I drew upon both theoretical and empirical arguments to support privacy as a common value. Theoretically, my analogy was to freedom of conscience – individuals may believe in different religions or no religion, but they similarly acknowledge the importance of freedom of conscience. In the same way that one need not agree on the particulars of religious beliefs, one need not agree on the particulars of privacy beliefs to accept that privacy is essential to one's individual and social existence. Drawing on the thinking of John Stuart Mill (1859) and Ruth Gavison (1980), I argued that privacy is important for the development of a type of individual that forms the basis for the contours of society that we have in common. Mill's concern was echoed by John Dewey in his claim that the perception of the 'public' arises from the perception of broader consequences – 'concern on the part of each in the joint action and in the contribution of each of its members to it' (1927, 181).

Empirically, I turned to public opinion data for support of common perceptions. Public opinion surveys from the 1970s to the 1990s provided support that people were concerned about their privacy, that they shared such concern in rather large numbers, and that their perceptions of privacy issues were quite similar. The data supported the notion that people had a shared meaning regarding the value, importance, and meaning of privacy – even if they applied that meaning somewhat differently in their own lives. Respondents to a series of Louis Harris and Alan Westin surveys during this time, as well as a 1994 ACLU survey, reported that they did care about privacy in a number of social, political, and economic contexts and that generally they supported more government action to protect privacy (See Regan 1995, 50-68).

I based my original thinking about the public value of privacy on the argument that privacy was important to the democratic political system and the workings of the democratic political process. In most of the legal and constitutional writing about privacy and democracy in the US literature, privacy is seen as an instrumental right particularly important in two respects: furthering the exercise of First Amendment rights and providing constraints on the use of government power, especially in Fourth Amendment terms. I argued that privacy was also independently important to the democratic process as the development of commonality, essential to the construction of a 'public' or Arendt's 'community of one's peers', required privacy so that people were not over-differentiated (Regan, 226-227). I claimed that the use of personal information for targeting political messages, for example, violates the integrity of the electoral process because they fragment the body politic.

Regarding the collective value of privacy, I advocated that technology and market forces were making it harder for any one person to have privacy without all persons having a similar minimum level of privacy. I argued that privacy was in effect a 'collective or public good', as used in economics (Coase, 1974), for three reasons. First, I maintained that privacy was not a 'private good' in that one could not effectively buy back or establish a desired level of privacy because of the non-voluntary nature of many record-keeping relationships. Second, I contended that the market will not produce an optimal supply of the good. As with clean air and national defence, the market is an inefficient mechanism for supplying privacy. And third, I held that the complexity and interrelatedness of the computer and communication infrastructure make it more difficult to divide privacy. This claim that privacy is a collective value may be seen as counterintuitive so I will briefly review each of my reasons.

It is somewhat difficult to regard privacy as a 'good' in economic terms but even in 1995 it was fairly obvious that it was difficult to disengage from societal relationships that might impinge on one's privacy. The list of record-generating relationships that were necessary components of modern life – including for example banking, credit, and healthcare – was growing. If individuals exited these relationships in order to protect their privacy not only would they make their own lives more complicated to live, they would also make the functioning of a modern economy and society more complicated and less efficient. These developments arguably make privacy less of a 'private good', where one could buy back or establish a desired level of privacy, and more of a 'collective good', where one's level of privacy affects not only others' level of privacy but also the functioning of the institutions whose activities might implicate privacy.

The contention that the market will produce a suboptimal supply of privacy is an easier one to understand. It is widely recognized, and borne out by experience, that the calculus of any organization is to collect as much information as possible about individuals in order to reduce any risk of decision-making about that individual. An organization will rationally be privacy invasive in its information gathering and use. But for individuals, the rational calculus is often to not see the privacy implications of their decisions. Privacy choices are often hidden transaction costs; the individual is focused on the purchase or service being negotiated – not focused on the opportunity or need to make a decision about privacy. Both the organizational calculus and the individual calculus thus result in less privacy – a suboptimal supply both because the quality of the information flowing within the system may be degraded and because trust in the system may be compromised. Left to its own devices, privacy invasions are the result of market failures.

The idea that the complexity and interrelatedness of the communication infrastructure made it more difficult to divide privacy was supported by the acknowledgment that the design of an overall system determines what is possible. For example, in communication systems hardware and software determine the level of privacy possible. Somewhat similarly, it was also difficult to isolate one record from a system of records and give that record a particular level of privacy.

I concluded that viewing privacy from the broader perspective of its common, public, and collective value would change the definition of privacy policy problems, the terms of policy discourse, and the patterns of interest group and legislative activity. Acknowledging that privacy is a common and public value would weaken the criticism that privacy is a negative value. Aligning privacy with societal interests would remove some of the difficult philosophical and policy issues involved in reconciling the balance between individual and society. Recognition that privacy has some features of a public or collective good would make clearer the institutional or organizational interests in personal information and the weaknesses of a market solution in providing privacy protection. Since the publication of my book in 1995, several scholars – including Julie Cohen, Helen Nissenbaum, Paul Ohm, Beate Rossler, Paul Schwartz, Daniel Solove, and Valerie Steeves – have continued to develop, refine, and strengthen arguments about privacy's social importance. More work remains in terms of incorporating these arguments into effective privacy policy.

References

Dewey, John. 1927. *The Public and its Problems*. Chicago: Swallow Press.

Friedrich, Carl J. 1971. 'Secrecy versus Privacy', in J. Roland Pennock and John W. Chapman (eds.), *Privacy*, Nomos Series 13, Yearbook of the American Society for Politica and Legal Philosophy. New York: Atherton Press.

Gavison, Ruth. 1980. 'Privacy and the Limits of the Law', *Yale Law Journal* 89(3): 347.

Mill, John Stuart. 1859. 'On Liberty', in *The English Philosophers from Bacon to Mill*. New York: Modern American Library, 1939), p. 998.

Pennock, J. Roland and John W. Chapman (eds.). 1971. *Privacy*, Nomos Series 13, Yearbook of the American Society for Political and Legal Philosophy. New York: Atherton Press.

Rachels, James. 1975. 'Why Privacy is Important', *Philosophy and Public Affairs* 4(4): 323.

Regan, Priscilla M. 1995. *Legislating Privacy: Technology, Social Values, and Public Policy*. Chapel Hill: University of North Carolina Press.

Scanlon, Thomas. 1975. 'Thomson on Privacy', *Philosophy and Public Affairs* 4(4): 315.

Schoeman, Ferdinand. 1984. 'Privacy and Intimate Information', in David Schoeman (ed.), *Philosophical Dimensions of Privacy*. Cambridge: Cambridge University Press.

Simitis, Spiros. 1987. 'Reviewing Privacy in an Information Society', *University of Pennsylvania Law Review* 135 (March): 707-746.

Simmel, Arnold. 1971. 'Privacy Is Not an Isolated Freedom', in J. Roland Pennock and John W. Chapman (eds.), *Privacy*, Nomos Series 13, Yearbook of the American Society for Political and Legal Philosophy. New York: Atherton Press.

Thomson, Judith Jarvis. 1975. 'The Right to Privacy', *Philosophy and Public Affairs* 4(4): 295.

Warren, Samuel D. and Louis D. Brandeis. 1890. 'The Right to Privacy', *Harvard Law Review* 4(December): 195.

Westin, Alan. 1967. *Privacy and Freedom*. New York: Atheneum.

2. Privacy from a Legal Perspective

Bart van der Sloot[1]

2.1 Introduction

This chapter adopts a 'Western' perspective, focusing on the United States of America, and in particular Europe. It will focus primarily, but not exclusively, on the informational aspect of privacy.[2] Section 2 will discuss the role and function of privacy in the legal realm; it will engage with the origins of privacy in the legal realm, the way it is protected in both national and international legal orders and set out some general characteristics of the right to privacy. Section 3 will provide an overview of the most important legal principles; it will look specifically at the basis which underpins privacy in Europe and the USA. Section 4 will recount some of the traditional debates in legal research; such as whether people have the right to control or even sell their personal data. Section 5, discussing new challenges, will engage with the tensions between privacy protection and developments known as Big Data. Finally, section 6 concludes and provides some suggestions for further reading.

Before discussing the role of privacy within the legal realm, it is important to discuss five general characteristics of the legal realm itself. This section will discuss the notion of regulation, the regulator, norms, laws and fields of law.

2.1.1 What is regulation?

Without regulation, there would be anarchy. Most societies do not want anarchy, so they regulate. Regulation is based on norms. Law is one way to regulate. Laws are always the mitigating factor between fact and fiction, between practice and norm, between the situation that is (for example, a society in which there is violence and murder) and the desired situation (for example, a society in which no violence exists). Obviously, law is never fully successful in this endeavour. Although the legal regime provides that murder

1 Thanks for Huw Roberts, Michael Collyer and Aviva de Groot for commenting on earlier drafts of this chapter.

2 A difference is often made between different types of privacy, such as bodily privacy, locational privacy (including the protection of the home), relational privacy (including the protection of family life) and informational privacy (including the protection of personal data and the secrecy of correspondence). Roessler 2005. Koops 2017

is prohibited (the norm), people still are being murdered (fact). To ensure compliance with the law, various tools of en*force*ment exist – these mostly depend on force (by the state). Consequently, the legal domain is always a combination of two elements:[3] norm and force. Law enforcement can be achieved through various means, such as imprisonment, fines, naming-and-shaming, and capital punishment. Law has traditionally focused on retroactive forms of enforcement, that is, when a person violates the law she is sanctioned. There is a trend, however, to enforce the law proactively, that is before the law is violated. Methods employed to enforce the law proactively include imposing sanctions on people who are believed to pose a high risk to society (such as suspects of terrorism), by proactively steering behaviour of citizens (for example nudging in smart cities), by laying down codes of conduct, or by embedding law in technological code[4] (for example, when online platforms simply block curse words, i.e. make it impossible to violate the norm). Law regulates citizens (natural persons), but also companies and other organizations (legal persons), including the state itself.

2.1.1.1 Who regulates?

Individuals as well as groups of people (family, friends) set norms. Organizations such as book clubs and companies have rules which may, for example, specify that an employee cannot arrive to work drunk. These forms of regulation and norm-setting are not, however, traditionally understood to fall under the legal regime. A law is seen as an instrument of the state or ruler (such as a dictator). It supposes a centralized form of order and authority. Within the state, the classical Western ideal is that there should be separation of powers.[5] Before, in medieval Europe, the monarch commonly embodied every aspect of state power – he could make rules and laws, he acted as the head of the police and military and operated as the ultimate judge. Because this led to abuse of power, most states currently separate three powers in three different bodies: the law-making power (traditionally granted to the parliament, which ideally should have democratic legitimation), the executive power (the government), and the judicial power (the judges and courts).[6]

3 Derrida 1989.

4 Lessig 1999.

5 Montesquieu 1989.

6 Obviously, there are exceptions and mixed forms. Referenda may take up part of the legislative process. Also, in many countries, the executive power has a big influence on the legislative process. The judiciary is often dependent in the sense that the members of the highest court are selected by parliament and/or the executive branch. And courts, and judges often engage in law-making.

2.1.1.2 *Who decides on norms?*

One of the classical legal debates regards the question of whether all laws are man-made. So-called legal positivists stress that indeed they are, while proponents of natural law theories suggest that there are laws that are not man-made. The former stress that laws are the rules which are enforced by the executive power and that are generally followed by the population – they adopt a primarily descriptive stance.[7] Natural law theories stress that there are laws that precede and supersede man-made law; these might either be the laws of God,[8] or the laws derived from human nature.[9] There is no uniform answer to the question of which natural laws or norms precede and supersede man-made laws, but reference is often made to legal principles such as human dignity, individual autonomy and personal freedom. Natural law theories provide the theoretical underpinning of human rights in the legal realm. Because natural rights are said to exist in the so-called state of nature (when there was no government and there were no man-made laws), they are believed to be intrinsic to being human.

The question inspired by the horrors of the Second World War is as follows: suppose a regime came to legitimate power and adopted laws, which on the one hand followed the correct constitutional procedures and had democratic legitimation, but on the other hand stated that all people of a certain religious denomination or with a certain ethnic background should be exterminated. Are those laws legal? Should citizens obey those laws? No, natural law theories would say, because there are higher laws than the man-made laws; if man-made laws contradict those, for example because they trample upon basic human dignity, they are simply null and void. In any democracy, a majority may rule over minorities; but there should be limits to the law-making capacities of the democratic majority.

The valid critique of the positivists is: who decides what these mystical, 'higher' norms are? Should judges decide on what higher norms exist and if so, what is their methodology for selecting these norms? If, on the other hand, these norms are selected through democratic means, how exactly do they differ from normal laws adopted by man? How is it that if these norms are supposedly innate to man (the claim of human or natural rights), that every region in the world has its own selection and interpretation human rights? In addition, they point to the fact that in the history of mankind, human rights have been violated more often than not. Are they really inalienable?

7 Bentham 1970; Austin,1995; Hart 1994.

8 Aquinas 1914-1942.

9 Locke 1988.

2.1.1.3 *General characteristics of the law*

Concerning man-made laws generally, there is no single doctrine on how laws should be adopted. Typically, democracies require a majority in parliament for adopting laws and qualified majorities (for example two-thirds of parliament) for adopting or amending constitutions.

There are certain general characteristics that have been ascribed to laws:[10]

- Laws should be relatively stable, so that people know the rules and can take them into account (which becomes impossible if the norms change by the hour).
- Laws should be proactive and not applied retroactively (a law adopted in May 2019, for example prohibiting wearing headscarves in public buildings, cannot be used to sanction a person that wore a headscarf in a public building in January 2019).
- Laws should not ask the impossible of people (for example, a law simply stating 'citizens are prohibited from drinking water or other fluids').
- A law should be general ('Jack Black cannot enter this building' is generally not considered to be a law; a rule saying 'People cannot enter this building' can be).
- Laws should be publicized and generally accessible to the people.
- The rules in the law should also be understandable (they need not be written in layman's terms, but generally understandable for people who want to).
- Laws should not contradict each other.
- Laws should generally be enforced (if laws are not enforced, they are symbolic only).

2.1.1.4 *Fields of law*

There are four different fields law on a national level:

- Civil law: regulates the dealings between citizens/companies among themselves. Examples are tort law, contract law, marital law, and consumer law.
- Criminal law: also regulates the dealings between citizens and companies among themselves. Unlike civil law, which is seen as protecting the private interests of citizens and companies, criminal law is enforced by the state because the rules protect public interests. Public order provides the clearest example of this, with murder, rape, theft, and hate speech all prohibited.

10 Fuller 1969.

- Administrative law: procedural principles that regulate the bodies of the state and their dealings.
- Constitutional law: the constitution is seen as the highest 'law' in a country (though not all countries have a constitution, for example the United Kingdom). It usually contains constitutional rights, such as freedom of speech and the right to privacy, and regulates the relationship between and dealings of the three branches of government (the legislative power, the executive power and the judicial power).

Typically, there are three types of courts in a country:
- Lower Court: deals with a claim or a complaint in first instance. (In civil law cases, two private parties – citizens and/or private organisations – stand against each other. In criminal law cases, a private party – a citizen or an organisation – is prosecuted by the state. In administrative or constitutional law cases, a private party – a citizen or a private organisation – complains about the behaviour or a decision of the state. Civil law cases are called horizontal; criminal, administrative and constitutional cases are called vertical. Criminal, administrative and constitutional law is part of what is sometimes called 'public law', contrasting with 'civil law', which regulates horizontal relationships).
- Court of Appeal: deals with appeals (either party may object to the decision of the lower court).
- Constitutional Court/High Court/Supreme Court: deals with cases in final instance and can be the court of first instance for specific cases, such as those revolving around the constitutionality of laws (not all countries allow the high court to receive such cases). There is usually only one such court in a country; its decisions set precedents that should be followed by the lower courts and the courts of appeal.

Then there are so-called human rights documents. These documents are perceived as higher than national laws and even constitutions. Some international courts overseeing those documents can invalidate national laws; citizens can appeal to these international courts even when their national supreme court has denied their request or delivered an unfavourable decision. Four prominent examples are:
- Universal Declaration on Human Rights (UDHR) (1948) by the United Nations (UN) – no court oversees this document.
- European Convention on Human Rights (ECHR) (1950) by the Council of Europe (CoE) – the European Court of Human Rights (ECtHR) oversees this document.

- International Covenant on Civil and Political Rights (ICCPR) (1966) by the United Nations – is monitored by the United Nations Human Rights Committee.
- Charter of Fundamental Rights (CFR) by the European Union (EU) (2000)[11] – is monitored by the European Court of Justice (ECJ), like all regulation by the EU. The Charter can be compared to the constitution of the EU; the EU has competence to adopts laws on almost every aspect of society.

2.2 Meaning and function of privacy

This section will give a brief introduction into the role and function of privacy in the legal domain. Section 2.1 will recount the origins of privacy as a juridical concept; section 2.2 will introduce the forms through which privacy is protected in the national legal orders of a number of 'Western' countries; section 2.3 will give an overview of the most important privacy doctrines in human rights documents; and section 2.4 will discuss some of the general characteristics of the right to privacy and the right to data protection.

2.2.1 Origins of privacy in the legal realm

Privacy is perhaps the oldest legal principle. It pertains to the separation of the public and private domain. Where that boundary lies exactly differs from culture to culture, epoch to epoch, and country to country, but there always is one. In ancient times, the ruler or king had authority over the public domain, while the household fell under the rule of the *pater familias*, the male breadwinner of the family, who reigned over his family members like a king.[12] The separation of the public domain from the private domain, meant that public laws, in principle, held no sway

11 The CoE and the EU are different organizations. While 47 countries have ratified the ECHR (including countries such as the UK, Russia and Turkey), the European Union only has 28 members (27 when the UK leaves the EU). Traditionally, the difference between the two institutions was simple. The CoE regulated the field of human rights and the EU adopted legislation in the socio-economic area. However, the EU has entered the human rights realm as well, among others by adopting the Charter of Fundamental Rights. In principle, EU law and the decisions by the ECJ should take into account the standards contained in the ECHR and the jurisprudence of the ECtHR. Europe, as a continent, consists of about 53 countries.

12 Kantorowicz 2016.

Fig. 2.1: Countries of the EU

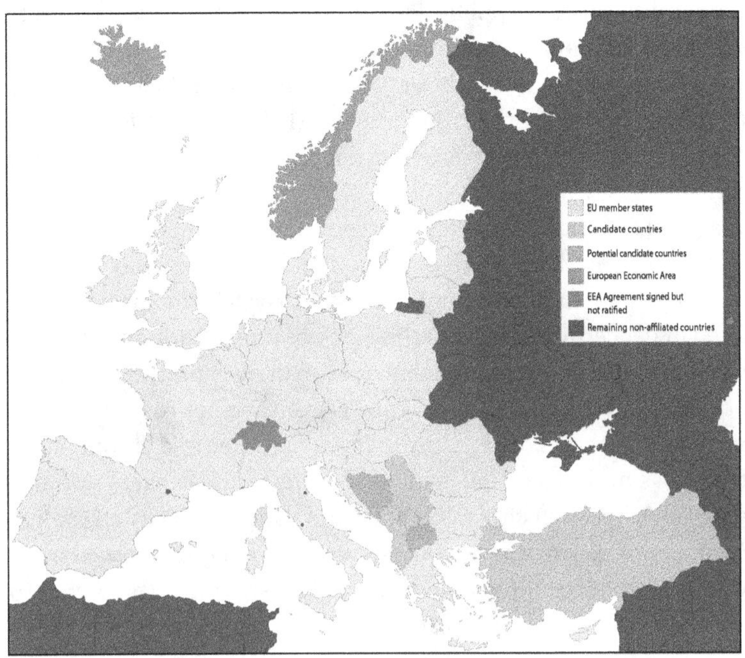

Fig. 2.2: Countries of the Council of Europe

over the household. Privacy derives from private and the Latin *privare*, taking something out of the public domain, and is thus the exact opposite of *publicare*, taking something from the private to the public domain.[13] A problematic consequence of the separation between the two spheres was that abuse of power by the father was mostly left unsanctioned – still until recently, rape within marriage was not a formal offence in a variety of countries.

The classical function of privacy was consequently to protect citizens from states entering the private domain. States held no sway over the household, or, in later time, could only enter the private domain for specific reasons and under certain conditions. Privacy was thus seen as an obligation of states not to abuse or overstretch their power; privacy protected citizens from totalitarian regimes.[14] One of the classic theories to explain this principle is that in the state of nature, people were free and autonomous, but as there was no state, no law and no law enforcement, there was also notable violence between citizens (sometimes called a 'war of all against all').[15] People then, so goes the hypothesis, decided to lay down their arms and give the state a monopoly of violence. The state had the power to adopt laws and enforce them; citizens could not use violence against each other. This 'social contract', however, only regarded the public domain, the domain where citizens interacted with each other, and not with respect to the private domain. Thus, the state had no or limited power to enter the latter domain.

2.2.2 National protection of privacy

Besides the protection of the home and private land ('my home is my castle'), the right to privacy traditionally included bodily integrity, private communication (secrecy of letters), and the family life. To some degree, the protection of one's reputation and good name is also encompassed. Such types of protection have been incorporated in national constitutional orders ever since the 13th century. It is impossible to clearly demarcate the

13 Aries & Duby 1988.

14 Totalitarian, in this sense, refers to states that regulate society in its totality, including both the private and the public domain.

15 Hobbes 2006.

right to privacy from a legal perspective – in some countries, it includes the right to found a family, while in others this is not regarded a legal right. In some constitutions, it also includes bodily integrity, while in others, the inviolability of the human body is a separate doctrine. The same applies to the protection of reputation and other aspects of private life.

Besides constitutional rights, countries can protect privacy through various fields of law. For example, in civil law, businesses that gather personal information about citizens while misleading or mistreating them can be brought to justice through tort or consumer law. Privacy can also be regulated through criminal law: rape is an offence, so is entering a person's home without permission. Stalking is increasingly penalized, and in some countries, violating a person's reputation is sanctioned by criminal law.

Some selected examples of how privacy is protected in the constitutions of states are provided below. The Dutch and Italian constitution have different articles on different aspects of privacy, the German constitution contains a personality right, Spain has one longer article with paragraphs that protect several aspects of privacy, and the USA does not really have one specific article that is referred to for privacy protection (see in more detail section 3).

Dutch Constitution	German Constitution	Italian Constitution	Spanish Constitution	Amendments to the constitution of the USA
Article 10 1. Everyone shall have the right to respect for his privacy, without prejudice to restrictions laid down by or pursuant to Act of Parliament. 2. Rules to protect privacy shall be laid down by Act of Parliament in connection with the recording and dissemination of personal data. 3. Rules concerning the rights of persons to be informed of data recorded concerning them and of the use that is made thereof, and to have such data corrected shall be laid down by Act of Parliament	Article 2 [Personal freedoms] 1. Every person shall have the right to free development of his personality insofar as he does not violate the rights of others or offend against the constitutional order or the moral law. 2. Every person shall have the right to life and physical integrity. Freedom of the person shall be inviolable. These rights may be interfered with only pursuant to a law.	Article 13 Personal liberty is inviolable. No one may be detained, inspected, or searched nor otherwise subjected to any restriction of personal liberty except by order of the Judiciary stating a reason and only in such cases and in such manner as provided by the law. In exceptional circumstances and under such conditions of necessity and urgency as shall conclusively be defined by the law, the police may take provisional measures that shall be referred within 48 hours to the Judiciary for validation and which, in default of such validation in the following 48 hours, shall be revoked and considered null and void. Any act of physical and moral violence against a person subjected to restriction of personal liberty shall be punished. The law shall establish the maximum duration of preventive detention.	Section 18 1. The right to honour, to personal and family privacy and to the own image is guaranteed. 2. The home is inviolable. No entry or search may be made without the consent of the householder or a legal warrant, except in cases of flagrante delicto. 3. Secrecy of communications is guaranteed, particularly regarding postal, telegraphic and telephonic communications, except in the event of a court order. 4. The law shall restrict the use of data processing in order to guarantee the honour and personal and family privacy of citizens and the full exercise of their rights.	First Amendment Congress shall make no law respecting an establishment of religion, or prohibiting the free exercise thereof; or abridging the freedom of speech, or of the press; or the right of the people peaceably to assemble, and to petition the Government for a redress of grievances.

Dutch Constitution	German Constitution	Italian Constitution	Spanish Constitution	Amendments to the constitution of the USA
Article 11 Everyone shall have the right to inviolability of his person, without prejudice to restrictions laid down by or pursuant to Act of Parliament.	Article 6 [Marriage – Family – Children] 1. Marriage and the family shall enjoy the special protection of the state. 2. The care and upbringing of children is the natural right of parents and a duty primarily incumbent upon them. The state shall watch over them in the performance of this duty. 3. Children may be separated from their families against the will of their parents or guardians only pursuant to a law, and only if the parents or guardians fail in their duties or the children are otherwise in danger of serious neglect. 4. Every mother shall be entitled to the protection and care of the community. 5. Children born outside of marriage shall be provided by legislation with the same opportunities for physical and mental development and for their position in society as are enjoyed by those born within marriage.	Article 14 The home is inviolable. Personal domicile shall be inviolable. Home inspections, searches, or seizures shall not be admissible save in the cases and manners complying with measures to safeguard personal liberty. Controls and inspections for reason of public health and safety, or for economic and fiscal purposes, shall be regulated by appropriate laws.		**Third Amendment** No Soldier shall, in time of peace be quartered in any house, without the consent of the Owner, nor in time of war, but in a manner to be prescribed by law.

Dutch Constitution	German Constitution	Italian Constitution	Spanish Constitution	Amendments to the constitution of the USA
Article 12 1. Entry into a home against the will of the occupant shall be permitted only in the cases laid down by or pursuant to Act of Parliament, by those designated for the purpose by or pursuant to Act of Parliament. 2. Prior identification and notice of purpose shall be required in order to enter a home under the preceding paragraph, subject to the exceptions prescribed by Act of Parliament. 3. A written report of the entry shall be issued to the occupant as soon as possible. If the entry was made in the interests of state security or criminal proceedings, the issue of the report may be postponed under rules to be laid down by Act of Parliament. A report need not be issued in cases, to be determined by Act of Parliament, where such issue would never be in the interests of state security.	Article 10 [Privacy of correspondence, posts and telecommunications] 1. The privacy of correspondence, posts and telecommunications shall be inviolable. 2. Restrictions may be ordered only pursuant to a law. If the restriction serves to protect the free democratic basic order or the existence or security of the Federation or of a Land, the law may provide that the person affected shall not be informed of the restriction and that recourse to the courts shall be replaced by a review of the case by agencies and auxiliary agencies appointed by the legislature	Article 15 Freedom and confidentiality of correspondence and of every other form of communication is inviolable. Limitations may only be imposed by judicial decision stating the reasons and in accordance with the guarantees provided by the law		**Fourth Amendment** The right of the people to be secure in their persons, houses, papers, and effects, against unreasonable searches and seizures, shall not be violated, and no Warrants shall issue, but upon probable cause, supported by Oath or affirmation, and particularly describing the place to be searched, and the persons or things to be seized.

Dutch Constitution	German Constitution	Italian Constitution	Spanish Constitution	Amendments to the constitution of the USA
Article 13 1. The privacy of correspondence shall not be violated except in the cases laid down by Act of Parliament, by order of the courts. 2. The privacy of the telephone and telegraph shall not be violated except, in the cases laid down by Act of Parliament, by or with the authorization of those designated for the purpose by Act of Parliament.				**Fifth Amendment** No person shall be held to answer for a capital, or otherwise infamous crime, unless on a presentment or indictment of a Grand Jury, except in cases arising in the land or naval forces, or in the Militia, when in actual service in time of War or public danger; nor shall any person be subject for the same offence to be twice put in jeopardy of life or limb; nor shall be compelled in any criminal case to be a witness against himself, nor be deprived of life, liberty, or property, without due process of law; nor shall private property be taken for public use, without just compensation. **Eleventh Amendment** The Judicial power of the United States shall not be construed to extend to any suit in law or equity, commenced or prosecuted against one of the United States by Citizens of another State, or by Citizens or Subjects of any Foreign State.

Dutch Constitution	German Constitution	Italian Constitution	Spanish Constitution	Amendments to the constitution of the USA
				Fourteenth Amendment Section. 1. All persons born or naturalized in the United States and subject to the jurisdiction thereof, are citizens of the United States and of the State wherein they reside. No State shall make or enforce any law which shall abridge the privileges or immunities of citizens of the United States; nor shall any State deprive any person of life, liberty, or property, without due process of law; nor deny to any person within its jurisdiction the equal protection of the laws.

2.2.3 Privacy in human rights documents

Human rights documents also contain the right to privacy. A distinction is sometimes made between the first wave of human rights documents, such as the Magna Carta from 1215, the second wave of human rights documents, such as the United States Bill of Rights and the French *Declaration* of the *Rights* of the Man and of the Citizen from the 18th century, the third wave of human rights documents, including the UDHR, the ECHR, and the ICCPR from the 20th century and the post 20th century documents, such as the Charter of Fundamental Rights, forming the fourth wave. Only in the third wave of human rights documents is the right to privacy explicitly mentioned; the older documents did contain prohibitions for states in relation to the abuse of power and conditions for entering the private domain, but this was not coined in terms of privacy. The first document that did was the UDHR, but even in there, the original title (in a draft of the document) of the privacy provision was simply 'Freedom from wrongful interference'. Provided below are some of the most important examples of Human Rights documents that contain a right to privacy:

UDHR	ECHR	ICCPR	American Convention on Human Rights (1969)
Article 12 No one shall be subjected to arbitrary interference with his privacy, family, home or correspondence, nor to attacks upon his honour and reputation. Everyone has the right to the protection of the law against such interference or attacks.	Article 8 Right to respect for private and family life 1. Everyone has the right to respect for his private and family life, his home and his correspondence. 2. There shall be no interference by a public authority with the exercise of this right except such as is in accordance with the law and is necessary in a democratic society in the interests of national security, public safety or the economic well-being of the country, for the prevention of disorder or crime, for the protection of health or morals, or for the protection of the rights and freedoms of others.	Article 17 1. No one shall be subjected to arbitrary or unlawful interference with his privacy, family, home or correspondence, nor to unlawful attacks on his honour and reputation. 2. Everyone has the right to the protection of the law against such interference or attacks.	Article 11 Right to Privacy 1. Everyone has the right to have his honor respected and his dignity recognized. 2. No one may be the object of arbitrary or abusive interference with his private life, his family, his home, or his correspondence, or of unlawful attacks on his honor or reputation. 3. Everyone has the right to the protection of the law against such interference or attacks.

UDHR	ECHR	ICCPR	American Convention on Human Rights (1969)
Article 16 1. Men and women of full age, without any limitation due to race, nationality or religion, have the right to marry and to found a family. They are entitled to equal rights as to marriage, during marriage and at its dissolution. 2. Marriage shall be entered into only with the free and full consent of the intending spouses. 3. The family is the natural and fundamental group unit of society and is entitled to protection by society and the State.	Article 12 Right to marry Men and women of marriageable age have the right to marry and to found a family, according to the national laws governing the exercise of this right.	Article 23 1. The family is the natural and fundamental group unit of society and is entitled to protection by society and the State. 2. The right of men and women of marriageable age to marry and to found a family shall be recognized. 3. No marriage shall be entered into without the free and full consent of the intending spouses. 4. States Parties to the present Covenant shall take appropriate steps to ensure equality of rights and responsibilities of spouses as to marriage, during marriage and at its dissolution. In the case of dissolution, provision shall be made for the necessary protection of any children.	Article 17 Rights of the Family 1. The family is the natural and fundamental group unit of society and is entitled to protection by society and the state. 2. The right of men and women of marriageable age to marry and to raise a family shall be recognized, if they meet the conditions required by domestic laws, insofar as such conditions do not affect the principle of nondiscrimination established in this Convention. 3. No marriage shall be entered into without the free and full consent of the intending spouses. 4. The States Parties shall take appropriate steps to ensure the equality of rights and the adequate balancing of responsibilities of the spouses as to marriage, during marriage, and in the event of its dissolution. In case of dissolution, provision shall be made for the necessary protection of any children solely on the basis of their own best interests. 5. The law shall recognize equal rights for children born out of wedlock and those born in wedlock.

UDHR	ECHR	ICCPR	American Convention on Human Rights (1969)
			Article 18 Right to a Name Every person has the right to a given name and to the surnames of his parents or that of one of them. The law shall regulate the manner in which this right shall be ensured for all, by the use of assumed names if necessary. Article 19 Rights of the Child Every minor child has the right to the measures of protection required by his condition as a minor on the part of his family, society, and the state.

African Charter on Human and Peoples' Rights (1981)	EU Charter of Fundamental Rights	Association of Southeast Asian Nations (ASEAN) Human Rights Declaration (2012)
Article 18 1. The family shall be the natural unit and basis of society. It shall be protected by the State which shall take care of its physical health and moral. 2. The State shall have the duty to assist the family which is the custodian of morals and traditional values recognized by the community. 3. The State shall ensure the elimination of every discrimination against women and also censure the protection of the rights of the woman and the child as stipulated in international declarations and conventions. 4. The aged and the disabled shall also have the right to special measures of protection in keeping with their physical or moral needs.	Article 3 Right to the integrity of the person 1. Everyone has the right to respect for his or her physical and mental integrity. 2. In the fields of medicine and biology, the following must be respected in particular: – the free and informed consent of the person concerned, according to the procedures laid down by law, – the prohibition of eugenic practices, in particular those aiming at the selection of persons, – the prohibition on making the human body and its parts as such a source of financial gain, – the prohibition of the reproductive cloning of human beings. Article 7 Respect for private and family life Everyone has the right to respect for his or her private and family life, home and communications. Article 8 Protection of personal data 1. Everyone has the right to the protection of personal data concerning him or her. 2. Such data must be processed fairly for specified purposes and on the basis of the consent of the person concerned or some other legitimate basis laid down by law. Everyone has the right of access to data which has been collected concerning him or her, and the right to have it rectified. 3. Compliance with these rules shall be subject to control by an independent authority. Article 9 Right to marry and right to found a family The right to marry and the right to found a family shall be guaranteed in accordance with the national laws governing the exercise of these rights.	19. The family as the natural and fundamental unit of society is entitled to protection by society and each ASEAN Member State. Men and Women of full age have the right to marry on the basis of their free and full consent, to found a family and to dissolve a marriage, as prescribed by law.

2.2.4 Some general characteristics of the right to privacy and the right to data protection

This section provides some of the more general characteristic of the right to privacy and data protection. Especially, it will briefly reflect upon how these doctrines have changed over the last decennia.

It is important to underline three transitions with respect to the right to privacy:

- Horizontalization: Privacy as a human and constitutional right was originally coined as a vertical right, which means that it regulates the relationship between the citizen and the state. However, there has been a trend of so-called 'horizontalization' of human rights (horizontal cases are between citizens and/or private organizations): in civil law cases, for example tort law or conflicts arising from consumer law, constitutional and human rights are taken into account.
- Positive freedom: Privacy as a human and constitutional right was originally seen as a negative right, as freedom from interference (for example, protection against a government entering one's home), and not as a positive freedom, one that gives a right to engage in certain activities. Currently, however, many of the privacy provisions are interpreted as also including positive rights, meaning a freedom to do something, such as the right to develop social relationships, the right to actively communicate with others and the right to develop one's personality to the fullest.
- Positive obligation: Correspondingly, privacy as a human and constitutional right was originally seen as laying down a negative obligation for the state. The state had to abstain from abusing its power, while positive obligations require states to use their power in the best interests of the people. Currently, many privacy provisions in constitutions and human rights documents are interpreted in a way that states should also actively use their power to protect privacy (for example in horizontal relations) or to facilitate the personal development of its citizens.

The origins of the right to data protection lie partially in the data protection rules of northern European countries,[16] which arose in several nations in the 1970s on the one hand, and the Council of Europe's Resolutions on data processing on the other.[17] In parallel with this, data protection was emerging

16 Below is based on Van der Sloot 2014.

17 Secretary's Advisory Committee on Automated Personal Data Systems, Records, Computers and the Rights of Citizens (1973) <https://www.hsdl.org/?view&did=479784>.

in the USA with the realization of the so called Fair Information Practices (FIPs), which were developed because the right to privacy was thought to be unfit for the 'modern' challenges of large automated data processing. The increased use of large databases (primarily by governmental organizations) raised a number of problems for the traditional concept of the right to privacy. First, data processing often does not handle private or sensitive data, but public and non-sensitive data such as car ownership, postal codes, number of children, etc. Secondly, and related to that, privacy doctrines at that time emphasized the right of the data subject as having an important role in deciding the nature and extent of her self-disclosure (which will be discussed in more detail in the next section).

However, because data processing often does not deal with private and sensitive data, the right to control by the data subject was felt undesirable. This is because governments need general data to develop, among other things, adequate social and economic policies. In addition, it was felt unreasonable, because in contrast to private and sensitive data, data subjects have no or substantially less direct and personal interest in controlling (partially) public and general information. Consequently, the term 'personal data' also included public and non-sensitive data, but instead of granting a right to control, the focus of data protection principles was on the fairness and reasonableness of the data processing.

Although data protection instruments were introduced to complement the right to privacy, early data protection instruments were explicitly linked to the right to privacy; the right to data protection was seen either as a sub-set of privacy interests or as a twin-right. As an example, the first frameworks for data protection on a European level were issued by the Council of Europe in 1973 and 1974. They regarded the data processing taking place in the private and in the public sector: the Resolution 'on the protection of the privacy of individuals vis-à-vis electronic data banks in the private sector'[18] and the Resolution 'on the protection of the privacy of individuals vis-à-vis electronic data banks in the public sector'.[19] Here, data processing issues are still explicitly seen as a part of and related to the right to privacy. The Resolution on the public sector also stated explicitly 'that the use of electronic data banks by public authorities has given rise to increasing concern about the protection of the privacy of individuals'.

18 <https://wcd.coe.int/com.instranet.InstraServlet?command=com.instranet.CmdBlobGet&InstranetImage=58940 2&SecMode=1&DocId=646994&Usage=2>.

19 <https://wcd.coe.int/com.instranet.InstraServlet?command=com.instranet.CmdBlobGet&InstranetImage=59051 2&SecMode=1&DocId=649498&Usage=2>.

The Convention for the Protection of Individuals with regard to Automatic Processing of Personal Data of 1981 by the Council of Europe did not contain the word privacy in its title but specified in its preamble:

> Considering that it is desirable to extend the safeguards for everyone's rights and fundamental freedoms, and in particular the right to the respect for privacy, taking account of the increasing flow across frontiers of personal data undergoing automatic processing; Reaffirming at the same time their commitment to freedom of information regardless of frontiers; Recognising that it is necessary to reconcile the fundamental values of the respect for privacy and the free flow of information between peoples.

Also, Article 1 of the Convention, laying down the object and purpose of the instrument, made explicit reference to the right to privacy: 'The purpose of this Convention is to secure in the territory of each Party [each member state to the Council of Europe] for every individual, whatever his nationality or residence, respect for his rights and fundamental freedoms, and in particular his right to privacy, with regard to automatic processing of personal data relating to him ("data protection").' In addition, the explanatory memorandum to the Convention mentioned the right to privacy a dozen times.[20] Thus, although the reference to privacy in the title was omitted, it is still obvious that the rules on data protection as laid down in the Convention must be seen in light of the right to privacy.

Gradually, however, the EU started to engage in the field of data protection and the European Union has traditionally adopted a different take on data protection. In the EU, data processing was partially treated as an economic matter, with the EU being the traditional guardian of the internal economic market, while the main focus of the Council of Europe has been to protect human rights on the European continent. The original mandate to regulate data protection by the EU was also found in market regulation. Still, however, in the rhetoric of the EU, the right to data protection was initially strongly connected to the right to privacy. This was also reflected in the Data Protection Directive from 1995, which makes reference to the right to privacy 13 times and in Article 1, concerning the objective of the Directive, holds: 'In accordance with this Directive, Member States shall protect the fundamental rights and freedoms of natural persons, and in particular their right to privacy with respect to the processing of personal

20 <https://rm.coe.int/CoERMPublicCommonSearchServices/DisplayDCTMContent?documentId=09000016800ca 434>.

data. Member States shall neither restrict nor prohibit the free flow of personal data between Member States for reasons connected with the protection afforded under paragraph 1.'

However, in the General Data Protection Regulation (GDPR) from the EU, which has replaced the Data Protection Directive as per May 2018, a radical choice was made. All references to the right to privacy have been deleted. Common terms such as 'privacy by design' have been renamed to 'data protection by design' and 'privacy impact assessments' have become 'data protection impact assessments'. This is reflected on a higher regulatory level as well. In 2000, the European Union adopted a Charter of Fundamental Rights, which came into force in 2009. In it, the right to privacy and the right to data protection are separated and treated as two independent fundamental rights.

Besides the disconnection between the right to privacy and the right to data protection, it is important to underline three general transitions with respect to data protection:

- Increased scope: Data protection rules apply when 'personal data' are processed. More and more data is considered 'personal'. The sentence 'that person next to the garbage bin, with the black hat' can be considered personal data, even if the name or exact identity of a person is unknown. All data that relates to a person, or can be used to affect her, can be considered personal data. In addition, data which is currently not identifying anyone, but is likely to do so in the future can still be considered personal data.
- Fundamentalisation: The two Resolutions of the Council of Europe merely recommended member states of the CoE to adopt rules to protect the principles contained in the Resolutions. They had a code of conduct or soft law like status. It was at the Member States' liberty to implement sanctions or rules regarding liability. Only in the Convention of 1981 was it explicitly provided that: 'Each Party undertakes to establish appropriate sanctions and remedies for violations of provisions of domestic law giving effect to the basic principles for data protection set out in this chapter.' Moreover, the Convention explicitly provided a number of rules regarding the application and enforcement of the rule on cross-border data flows, the cooperation between states and the national Data Protection Authorities. Adopting an EU-wide Directive in 1995 aimed at bringing uniformity in the national legislations of the different countries, which was promoted, among others, by providing further and more detailed rules for cross-border data processing. The member states of the EU were obligated to adopt the rules from the

Directive in their legal order. As of May 2018, the GDPR has replaced the Directive. The fact that data protection rules are now contained in a 'Regulation' instead of a 'Directive' has important legal implications. A Regulation needs not be implemented by the member states of the EU – it has 'direct effect', which means that people and organisations can rely on the GDPR as such, while previously, they had to refer to the national implementation of the Directive. Finally, as has been stressed, in the Charter of Fundamental Rights, the EU has decided to make data protection a fundamental right of its own, next to such rights as privacy, the freedom of expression and the freedom from discrimination. The GDPR is seen as an implementation of article 8 of the Charter, which contains the fundamental right to data protection.

– Juridification: Not only the material scope, but also the provisions in the instruments, providing the rights and obligations for the different parties involved with data processing activities, have extended quite significantly. The two Resolutions from 1973 and 1974 contained 8 and 10 articles respectively. The Convention from 1981 contained 27 provisions, the Directive 34 and the GDPR 99. While the two Resolutions were literally one-pagers, the Regulation consists of 88 pages.

2.3 Classic texts and authors

Discussing classic authors is a bit different for law than for most other academic disciplines. Law is made by legislators and partially by judges, not by scholars.[21] Scholars reflect on legal texts and jurisprudence by courts. That is why this section will primarily refer to the legal texts and jurisprudence (which are called primary sources) and only marginally to texts by scholars (which are called secondary sources). This section briefly discusses the approach to privacy protection in the USA (section 3.1) and more thoroughly engages with the approach to privacy protection within the CoE (section 3.2), and the approach to data protection within the EU (section 3.3).

21 Although this is a bit different for so called common law countries (such as the US, the UK, Canada, and Australia), which rely on judge-made law to a significant extent, than for civil law countries (such as most countries in Europe and Latin America) that rely predominantly on laws by parliament. In common law countries, there is more room for authors to develop new interpretations of rights and doctrines. The difference between common law and civil law countries is unrelated to the distinction between 'civil law', 'criminal law', 'constitutional law', etc.

2.3.1 The protection of privacy in the USA

This section introduces three classic American authors (section 3.3.1), the most important privacy laws and rules (section 3.3.2), and five landmark cases of the American Supreme Court (section 3.3.3).

2.3.1.1 Classical authors

There has been a number of authors that had an effect on the development of privacy doctrines in the US. Three of the most important are:

1. Warren and Brandeis: arguably introduced the right to informational privacy in the US. They did so by distilling from existing doctrines and case law a new principle, namely the right to be 'let alone'.
2. Prosser: distinguished between four types of tort that may be used for the protection of privacy, which were derived from the existing case law of various American courts. These are:
   a. Intrusion upon the plaintiff's seclusion or solitude, or into his private affairs.
   b. Public disclosure of embarrassing private facts about the plaintiff.
   c. Publicity which places the plaintiff in a false light in the public eye.
   d. Appropriation of the plaintiff's name or likeness.
3. Westin: wrote one of the first comprehensive books about informational privacy. He defined privacy as the claim of individuals, groups, or institutions to determine for themselves when, how, and to what extent information about them is communicated to others.

2.3.1.2 Privacy laws

It is difficult to discuss the legislation of privacy in the USA because it is rather scattered. While in the EU, there is one general framework for data protection, and in the ECHR, there is one specific article on the right to privacy, this does not hold true for American Privacy Law.

2.3.1.2.1 The American Constitution

1. The First Amendment, providing the freedom to assembly and speech is sometimes invoked in privacy cases, when claims relate to positive privacy rights and freedoms.
2. The Fourth Amendment provides protection against arbitrary searches and seizures. It is seen as, inter alia, protecting the home of citizens against unlawful intrusion by the governement.
3. The Fifth Amendment provides procedural protection in criminal law cases, which may have an effect on the privacy rights of citizens.

4. The Ninth Amendment provides that enumeration in the constitution, of certain rights, shall not be construed to deny or disparage others retained by the people, such as possibly the right to privacy.
5. The Fourteenth Amendment provides protection to privacy rights to the extent they are related to due process.

2.3.1.2.2 Federal law

There have been several attempts to draw up omnibus privacy legislation in the USA. So far, however, these endeavours have been unsuccessful. That is why a patchwork framework exists of sectoral laws and privacy provisions for specific circumstances, five of which are:

1. The Federal Trade Commission Act: provides privacy protection in consumer relations and grants the Federal Trade Commission (FTC), the governmental body overseeing the sector, significant powers to enforce these provisions. The FTC is seen as the main governmental organization enforcing privacy in the USA.[22]
2. The Children's Online Privacy Protection Act (COPPA): regulates the online collection of information concerning children and is enforced by the FTC.
3. The Health Insurance Portability and Accountability Act (HIPAA): specifies rules for gathering and processing medical information.
4. The Fair Credit Reporting Act: regulates consumer-reporting agencies that use consumer reports or provide consumer-reporting information.
5. The Electronic Communications Privacy Act: contains rules for the interception of, inter alia, electronic communications.

2.3.1.2.3 Constitutions of States

Some constitutions of states contain a right to privacy, such as:

- Article 1 of Alaska's constitution: 'The right of the people to privacy is recognized and shall not be infringed. The legislature shall implement this section'.
- Article 1 of the Californian constitution: 'All people are by nature free and independent and have inalienable rights. Among these are enjoying and defending life and liberty, acquiring, possessing, and protecting property, and pursuing and obtaining safety, happiness, and privacy'.
- Article 1 of Florida's constitution: 'Every natural person has the right to be let alone and free from governmental intrusion into the person's private life except as otherwise provided herein. This section shall not

22 Hoofnagle 2016.

be construed to limit the public's right of access to public records and meetings as provided by law'.
- Article 2 of the constitution of Montana: 'The right of individual privacy is essential to the well-being of a free society and shall not be infringed without the showing of a compelling state interest'.

2.3.1.2.4 State law

Finally, there are privacy laws by states, which only apply on the territory of the state. Most important in this respect is the state of California, as most tech-companies are based there. An example is the California Electronic Communications Privacy Act.

2.3.1.3 Landmark cases

It is impossible to give a full overview of landmark cases by the Supreme Court on the right to privacy. Five influential cases are:
1. Olmstead v. United States (1928): concerned the use of wiretapped telephone conversations by the police without judicial approval. The use of the information obtained as evidence in a court case was declared not to be in violation of the Fourth and Fifth Amendment.
2. Griswold v. Connecticut (1965): concerned a Connecticut law that prohibited the use of, inter alia, contraception. The law was invalidated by the Supreme Court with a reference to 'marital privacy'.
3. Katz v. United States (1967): overturned the Olmstead case. Extended the notion of 'search' to include technological means of gathering evidence and underlined the doctrine of the 'reasonable expectation of privacy'.
4. Roe v. Wade (1973): on the basis of the 14th amendment, the Supreme Court accepted the so-called 'decisional privacy' doctrine, which grants women the right to decide over their own body, including the right to abortion.
5. Riley v. California (2014): concerned the warrantless search and seizure of a phone's contents during an arrest. This was declared unconstitutional by the Supreme Court.

2.3.2 The European Protection of Privacy

Below is a discussion of the jurisprudence of the European Court of Human Rights (ECtHR) on Article 8 of the European Convention on Human Rights (ECHR). It is important to stress that although privacy is a human right, and even although human rights are the highest legal norms there are, legal rights are never absolute. Rights are subject to a double conditionality: first the conditions for the applicability of a right and second the conditions under which the right can be curtailed. This section will discuss in further detail two conditions for applicability, the concepts of *ratione personae* (section 3.2.1) and *ratione materiae* (section 3.2.2). *Ratione personae* refers to the question of personal scope – can the claimant indeed invoke the right she is relying on; for example, in most jurisdictions, a person cannot complain about the police entering the house of her friends uncle. *Ratione materiae* refers to the question of material scope – does the matter complained of fall under the protective sphere of the article relied on. For example, the fact that a person's car is stolen will normally not be considered a privacy violation. This section will also describe the conditions for curtailing the right to privacy (section 3.2.3). Finally, it will touch upon some of the landmark cases by the ECtHR (section 3.2.4).

2.3.2.1 Ratione personae

Three phases can be distinguished with respect to the doctrine of *ratione personae* under the European Convention on Human Rights: the original text of the ECHR, the interpretation of the Court roughly between 1960 and 2000, and the interpretation of the European Court of Human Rights after 2000. The doctrine of *ratione personae* sets limits to who can invoke a right to privacy.

2.3.2.1.1 Original text of the ECHR

The text of the Convention contains two modes of complaints: (1) inter-state complaints (for example Norway submits a case against Sweden for violating human rights) and (2) individual complaints (for example, Mr Brown or Brown Bread Company submits a claim against Spain). The right to individual petition is open to three types of complainants: (2a) individuals, (2b) non-governmental organizations, and (2c) groups of individuals. Claims can only be brought against states. The focus originally was on inter-state complaints, as the ECHR was drafted against the backdrop of the Second World War. The core focus of the Convention was not to protect particular interest of particular individual claimants, but to prevent large and systematic abuse of power by states.

The Convention supervision consisted of a two-tiered system. First, the European Commission on Human Rights (ECmHR), which no longer exists today, would decide on the admissibility of cases and function as a filtering system. It was only with the Commission that the mechanism of individual complaints existed. Even if a case was brought before the Commission by an individual complainant, and even if the Commission declared the application admissible, the applicant (natural person, legal person or group) had no right to submit it for review to the Court. This could only be achieved on initiative of the Commission or a Member State of the Council of Europe. The idea was that only those cases that transcended the mere individual complaint of an applicant, i.e. cases that concerned a large issue or a public interests, would be sent to the ECtHR. The ECtHR is the second tier; it deals with the cases in substance, and decides on the question of whether the Convention has been violated or not.

2.3.2.1.2 ECtHR's approach between 1960-2000

Over time, however, the Convention has been revised on a number of points so that, inter alia, individual complainants (individuals, groups, and legal persons) have direct access to the Court to complain about a violation of their privacy (the task of the Commission being reassigned to a separate chamber of the Court – the two-tiered system still exists). Moreover, over time, the Court has strongly emphasized individual interests and personal harm when it assesses a case regarding a potential violation of Article 8 ECHR, therewith transforming the ECHR from a document that was focussed on preventing large scale abuse of power by governments and protecting general and societal interests, into an instrument that mainly provided protection to the specific interests of an individual claimant. To give a few examples:

- So-called *in abstracto* claims will in principle be declared inadmissible by the ECtHR. These are claims that regard the mere existence of a law or a policy, without them having any concrete or practical effect on the claimant.
- *A priori* claims are rejected as well, as the Court will usually only receive complaints about injury which has already materialized. Claims about future damage will in principle not be considered.
- Hypothetical claims regard damage which might have materialized, but about which the claimant is unsure. The Court usually rejects such claims because it is unwilling to provide a ruling on the basis of presumed facts.
- The ECtHR will in principle also not receive an *actio popularis*, a case brought up by a claimant or a group of claimants, not to protect their own interests, but that of others or society as a whole.
- According to the European Court of Human Rights, what distinguishes the right to privacy, under the interpretation of the ECtHR, from other rights under the Convention, such as the freedom of expression, is that it in principle only provides protection to individual interests. Cases that do not regard such matters, but mainly concern societal issues

or public interests, are rejected by the Court when it regards Article 8 ECHR.

- This focus on individual interests has also had an important effect on the types of applicants that are able to submit a complaint about the right to privacy. Although the Court has accepted that churches may invoke the freedom of religion (Article 9 ECHR) and that press organizations may rely on the freedom of expression (Article 10 ECHR), because Article 8 ECHR only protects individual interests, the Court has said that in principle, only natural persons can invoke a right to privacy.
- The Court has rejected the capacity of groups to complain about a violation of human rights. Contrary to the intention of the authors of the Convention, it has stressed that only individuals who have been harmed personally and significantly by a specific violation or infringement can bundle their claims.
- Finally, the last non-individual mode of complaint under the Convention, the possibility of inter- state complaints, has had almost no significance under the Convention's supervisory mechanism. Although there are more than 20,000 judgements by the ECtHR on claims submitted under the Convention, less than 50 are the result of interstate complaints.

2.3.2.1.3 ECtHR's approach from 2000 onwards

Recently, however, the Court has been willing to relax its focus on the individual and individual interests somewhat and has allowed for occasional exceptions, for example:

- The Court has been willing to allow for some twenty complaints by legal persons under Article 8 ECHR, inter alia when their business premises was searched by police officials without a warrant.
- The European Court of Human Rights has been willing to provide protection to minority rights under the right to privacy; though not granting a right of a group to submit a claim, there are steps towards that direction.
- In exceptional cases, the ECtHR has been willing to allow for *in abstracto* claims, in particular when there is a law that provides uncontrolled power to intelligence services to execute blanket mass surveillance programmes.
- Such *in abstracto* claims can be seen as *a priori* claims, because no damage has yet materialized. The mere existence of a law or policy is addressed.

- They may also be seen as shifting the focus from individual interest, towards general interests related to the abuse of power.
- And they may be seen as a form an *actio popularis*, as these cases aim to protect the population at large.

2.3.2.2 *Ratione materiae*

The right to privacy under the ECHR has witnessed an significant extension in terms of its material scope. While the right to privacy was originally conceived as a quite narrow and limited doctrine, the ECtHR has extended its scope and meaning considerably. Article 8 ECHR is no longer interpreted as laying down negative rights for citizens only, it also includes many positive rights; it not only requires states to abstain from abusing their powers, but also to use them to certain positive ends. In general, Article 8 ECHR has provided protection to almost anything that is remotely related to the personal interest of the individual. Article 8 ECHR contains four elements of privacy, namely 'private life', 'family life', 'home', and 'correspondence'. Each of those terms has been interpreted in a very broad and all-inclusive manner by the ECtHR. In addition, the right to privacy has tended to overshadow some of the other provisions contained in the ECHR, such as the right to fair trial and the right to marry and found a family. Article 8 ECHR has been interpreted by the Court to include certain elements that were explicitly rejected by the authors of the Convention. And the ECtHR has brought new rights and freedoms under the scope of the right to privacy that were not envisaged when the ECHR was drafted.[23]

1. Broadening of the terms in Article 8 ECHR:
   a. Private life: Private live is perhaps the broadest notion under the European Convention on Human Rights. Although it was originally interpreted in narrow terms, relating to personal affairs in the private domain, it currently provides protection to almost every aspect of a person's life. The ECtHR has interpreted Article 8 ECHR as a very broad provision, that provides protection to a variety of matters, such as personal development, education, engaging in social relationships, and even the protection, at least under certain circumstances, from being fired at work (because the ECtHR holds that work is important for a person's development).
   b. Family life: Again, although the notion of family life was originally only applied to the traditional family unit, over time, the ECtHR

23 See for a full overview: Van der Sloot 2015.

has extended this notion quite considerably. According to it, family life is a broad concept that may incorporate relations with aunts, nephews, grandparents, siblings, family in law, stepfamily, and may even relate to the relationship between a child and her legal representative or custodian. It not only provides the freedom from interference with those relationships, but also the positive freedom to develop such relationships.

c. Home: Although in its early case law, the European Court of Human Rights took a very traditional view on what falls under the concept of home, it now holds that a home is not only a house. The term may refer to any object in which a person lives. For example, under circumstances, a car may function as a person's home, if she sleeps in it. Interestingly, the Court has stressed that business premises may also fall under the concept of home, protecting companies against police searches.
d. Correspondence: Again, a similar transition can be witnessed with respect to the term correspondence. According to the ECtHR, the term correspondence not only includes communication through traditional means, but also when use is made of modern technological devices or services, such as the internet. Consequently, Article 8 ECHR also provides protection to meta-data about communication over the internet.

2. Article 8 ECHR overshadows some of the other provisions in the ECHR, such as:
   a. Right to marry and found a family: Article 12 ECHR provides: 'Men and women of marriageable age have the right to marry and to found a family, according to the national laws governing the exercise of this right'. This provision has been interpreted very restrictively by the Court, while Article 8 ECHR has been granted a very wide scope. Consequently, most issues relating to gay marriage, artificial insemination, adoption, and other non-traditional forms of marriage and procreation are dealt with under the scope of the right to privacy instead of Article 12 ECHR.
   b. Right to a fair trial: The right to a fair trial is protected under Article 5 and especially Article 6 ECHR. Although these provisions are still highly influential and most cases under the ECHR relate to Article 6 ECHR, when issues of due process, procedural safeguards, and fair trial are related to privacy matters, the ECtHR is willing to discuss such elements under the right to privacy itself.

Inter alia, it has stressed that it 'is true that Article 8 contains no explicit procedural requirements, but this is not conclusive of the matter. The local authority's decision-making process clearly cannot be devoid of influence on the substance of the decision, notably by ensuring that it is based on the relevant considerations and is not one-sided and, hence, neither is nor appears to be arbitrary. Accordingly, the Court is entitled to have regard to that process to determine whether it has been conducted in a manner that, in all the circumstances, is fair and affords due respect to the interests protected by Article 8'.[24]

c. The protection of reputation: Article 8 ECHR is based on Article 12 UDHR, which provides protection to one's reputation, besides the protection of private and family life, home and communication. This element was excluded from the scope of Article 8 ECHR by the authors of the Convention and moved to the second paragraph of Article 10 ECHR. Paragraph 1 of Article 10 ECHR grants the right to freedom of expression and paragraph 2, like paragraph 2 of Article 8 ECHR, provides for the conditions for limiting this right. Consequently, the protection of reputation was not intended as a subjective right of citizens, but as a ground on the basis of which governments can (and not must) limit the freedom of expression. Although the ECtHR has respected this principled choice for a long time, from 2009 onwards, the right to the protection of one's reputation, honour, and good name is said to fall under the scope of Article 8 ECHR, making it a subjective privacy right of citizens.[25]

a. Bodily integrity: A final example is the right to bodily integrity, which is not explicitly mentioned in Article 8 ECHR, although Article 2 (the right to life) and Article 3 (the prohibition of torture) do protect elements of one's bodily integrity. Still, the court usually turns to Article 8 ECHR when discussing issues relating to the body, such as medical procedures, mandatory vaccination schemes, and euthanasia.

24 ECtHR, B. v. the United Kingdom, Application no. 9840/82, 8 July 1987, § 63-64.

25 A subjective right (*droit subjectif*) means that a person can invoke it. An objective right (*droit objectif*) is a legal principle that has general effect, but cannot be invoked by an individual claimant.

3. Article 8 ECHR provides protection to freedoms explicitly left outside the scope of the ECHR, such as:
   a. Right to property: The right to property was explicitly rejected from the ECHR.[26] In addition, proposals to include under Article 8 ECHR the protection of private property were rejected during the drafting process of the Convention. Still, the European Court of Human Rights has overturned that decision from the start and has consistently included the protection of private property under the scope of Article 8 ECHR, such as with respect to inheritance, destruction of private property, and even, as indicated above, the right to work.
   b. Right to education: As with the right to property, the right to education was not included in the European Convention on Human Rights, but moved to an additional protocol, the signing of which was optional. Still, the ECtHR has included under the right to privacy, inter alia, the right of families to decide on the education of their children, for example in terms of language.
   c. Personality rights: Although the UDHR contains several provisions that protect one's personality, these were left outside the scope of the ECHR because they were believed to be too vague and unspecific. Currently, however, Article 8 ECHR functions as a personality

26 The ECHR only contains so called first generation human rights (not to be confused with the different waves of human rights). While first generation or civil and political rights require states not to interfere with certain rights and freedoms of their citizens in an arbitrary way (right to privacy, freedom of speech, freedom from discrimination), socioeconomic or second generation rights such as the right to education, to property and to a standard of living require states not to abstain from action, but to actively pursue and impose such freedoms by adopting legal measures or by taking active steps. The second generation rights were transferred to the 1th Protocol of the Convention, signing of which was non-mandatory. When the ECHR was drafted, the so called third generation rights, which revolve around intercultural and intergenerational solidarity, such as group rights, cultural rights and the right to a healthy living environment, did not yet exist. However, as will be explained below, the ECtHR has regarded the ECHR to be a so called 'living instrument', which means that the Convention should be interpreted in present daylight. The Court has provided protection to such third generation human rights by referring to existing provisions in the ECHR, in particular Article 8 ECHR. Reference can also be made to those tentatively describing the development of 'fourth generation human rights'. It does not matter whether reference is made to a right to general 'information management', the 'rights of indigenous peoples', the 'right to sustainable development of the future generation', 'women's rights, the rights of future generations, rights of access to information, and the right to communicate' or rights needed due to 'phenomena like the great developments in the area of biotechnology or the Internet'. Most, if not all, of these 'new' fourth generation human rights, suggested by different authors and commentators, would presumably, if accepted, be approached by the Court from the angle of Article 8 ECHR. Vasek 1977.

right – it provides protection to almost every aspect of a person's life, development, and flourishing.

d. Right to nationality: Although some of the other human rights documents do contain a right to residence, a right to nationality or a similar provision, such was excluded from the ECHR. The ECtHR has, however, included a right to residence in a certain country, or the prohibition to be expulsed, inter alia when such would have consequences for the family life of an immigrant (for example, a Tunisian immigrant has married an Italian woman, with whom he has children, but is threatened with extradition by the Italian government).

4. Article 8 ECHR is used by the ECtHR to include new rights and freedoms, that were not considered when drafting the ECHR, such as:
   a. The right to data protection: Although the ECHR does not contain reference to a right to data protection, the ECtHR often refers to CoE's Convention from 1981, the EU Charter and other EU documents in this field. Although it does not provide a similar level of data protection to the EU, the Court has incorporated a number of elements traditionally part of the data protection regimes under the scope of the right to privacy.
   b. The right to a clean and healthy living environment: Although the European Court of Human Rights does not accept a fully-fledged right to live in a clean and healthy living environment, it is prepared to deal with cases under Article 8 ECHR. This is true if the cases revolve around noise pollution, air pollution, scent pollution, and other forms of environmental damage, so long as the pollution affects the 'quality of life' of the applicant (which the Court agrees is a very vague and broad term).
   c. Minority rights: states may be under the positive obligation to take active measures to respect and facilitate the development of minority identities. Like environmental rights, minority rights are not included in the European Convention on Human Rights. The Court, however, provides protection to both, with reference to the right to privacy.
   d. Right to a name: a final example may be the right to a name and the right to change one's name. This right too is provided protection by the ECtHR with reference to Article 8 ECHR. It includes not only the right to alter one's first name or family name, but also to change one's identity, for example with respect to being transgender.

2.3.2.3 *Conditions for curtailing the right to privacy*

The previous two sections discussed two conditions for the applicability of the right to privacy: *ratione personae* and *ratione materiae*. There are a number of other conditions under the ECHR for the right to privacy to apply, but these are the most important ones. When the right to privacy applies, that is when it can be invoked by a claimant, the second question is whether there was a violation of this right in a particular circumstance. The right to privacy under the European Convention on Human Rights is a so-called qualified right.[27] This means that Article 8 ECHR specifies under which conditions the right can be legitimately curtailed by the government; these conditions are listed in paragraph 2 of Article 8 ECHR, which specifies: 'There shall be no interference by a public authority with the exercise of this right except such as is in accordance with the law and is necessary in a democratic society in the interests of national security, public safety or the economic wellbeing of the country, for the prevention of disorder or crime, for the protection of health or morals, or for the protection of the rights and freedoms of others.'

Consequently, if the government limits a person's privacy, for example by entering her home, this need not be illegitimate or a violation of her privacy. The infringement can be deemed in harmony with the European Convention on Human rights when it abides by three cumulative requirements: (1) the infringement must have a legal basis; (2) must serve one of the legitimate goals as listed in the second paragraph of Article 8 ECHR; and (3) must be necessary in a democratic society. The ECtHR may find that although there has been an infringement of the right to privacy (as provided in paragraph 1 of Article 8 ECHR), this was a legitimate one and thus not in violation of Article 8 ECHR. The ECtHR only reaches this conclusion if all three requirements (legal basis, legitimate aim, necessary in a democratic society) have been fulfilled; if the government fails to fulfil either one of these requirements, a violation of the right to privacy will be found.

The Court may find that an infringement was not prescribed for by law for a number of reasons – the 'law', in this sense, is always the national law of a country. The ECtHR uses a quite wide definition of law, it includes not only legislation, but also judge-made law typical of common law jurisdictions and secondary sources, such as royal decrees and internal regulations. First, a violation of the Convention will be found on this point if the actions of governmental officials are not based on a legal provision granting them the authority to act in the way they did. Second, a violation will be established

27 This sub-section is based on: Van der Sloot 2017B.

if the conditions as specified in the law for using certain authority have not been complied with, for example, if police officials have no warrant for entering the home of a citizen. Third, the actions of the governmental officials may be prescribed for by law, but the law itself may not be sufficiently accessible to the public. Fourth, the law may be so vague that the consequences of it may not be sufficiently foreseeable for ordinary citizens. Fifth and finally, the ECtHR has in recent years developed an additional ground, namely that the law on which actions are based will be deemed invalid if it does not contain sufficient safeguards against the abuse of power by the government. This typically applies to laws authorizing mass surveillance activities by intelligence agencies that set virtually no limits on their capacities, specify no possibilities for oversight by (quasi-) judicial bodies, and grant no or very limited rights to individuals, with respect to redress.

The Court may also find a violation of Article 8 ECHR if the infringement serves no legitimate aim. The second paragraph specifies a number of legitimate aims, primarily having to do with security related aspects, such as national security, public safety, and the prevention of crime and disorder. These terms are sometimes used interchangeably by the Court, but in general 'national security' is applied in more weighty cases than 'public safety', and 'public safety' in more weighty cases than the 'prevention of crime and disorder'. The right of privacy may also be legitimately curtailed to protect the rights and freedoms of third parties; for example, a child may be placed out of home (an infringement on the right to family life of the parents), because the parents molest the child. The protection of health and morals may be invoked to limit the right to privacy, though this category is applied hesitantly by the ECtHR, because of the fear that the majority imposes its moral believes on the minority. Still with respect to controversial medical or sexual issues, such as euthanasia or BDSM, the ECtHR sometimes allows a country to rely on this ground to curtail the right to privacy. Finally, a country can rely on the 'economic wellbeing of the country'; this ground can only be found in Article 8 ECHR and in no other provision under the Convention. It is invoked by countries in a number of cases; for example, if an applicant complains about the fact that a factory or airport in the vicinity of her home violates her right to private life, the country can suggest that running a national airport is in fact necessary for the economic wellbeing of a country.

Much more can be said about the use, extent and interpretation of these aims, but this is unnecessary, because this requirement plays no role of significance. This is due to two factors. First, the ECtHR is often very unspecific about which term exactly applies, stressing that an infringement 'clearly had a legitimate aim', or that 'it is undisputed that the infringement served one of the aims as contained in Article 8 ECHR'. It often combines categories, underlining that the infringement served a legitimate aim, such as 'the prevention of crime', 'the economic well-being of the country' or 'the rights of others' or it merely lists all different aims and holds that one of these grounds applies in the case at hand. Furthermore, it introduces new aims, not contained in Article 8 ECHR, especially in cases revolving around positive obligations for states. Second, the Court almost never finds a violation of Article 8 ECHR on this point. It usually allows the government a very wide margin of appreciation with respect to the question of whether and which of the aims apply in a specific case and whether the infringement did actually serve that aim. In many cases, it simply ignores this requirement when analysing a potential violation of the right to privacy or incorporates it in the question of whether the infringement was necessary in a democratic society. Thus, only in 20 cases was a violation of Article 8 ECHR found on this point.

Finally, the third requirement that must be fulfilled by a government wanting to curtail the right to privacy is that the infringement must be necessary in a democratic society. This question is approached by the Court primarily as a question of balancing the different interests at stake. 'This test requires the Court to balance the severity of the restriction placed on the individual against the importance of the public interest.'[28] Consequently, to determine the outcome of a case, the Court balances the damage a specific privacy infringement has done to the individual interest of a complainant against its instrumentality towards safeguarding a societal interest, such as national security.

2.3.2.4 Landmark cases by the ECtHR

This chapter can not provide a full overview of the cases of the ECtHR on the right to privacy, as there are some 2,000 cases (second tier, meaning those cases that have been declared admissible) and more than 4,000 applications (first tier). Some of the most memorable cases include:[29]

- Klass and others v. Germany (1978): The case concerned German legislation that allowed for the monitoring of citizen's correspondence and telephone communications without an obligation to inform them subsequently of the measures taken against them. Although the Court

28 Ovey & White 2002, p. 209.
29 www.echr.coe.int/Documents/FS_Data_ENG.pdf.

did not find a violation of Article 8 ECHR (the infringement was considered legitimate because the three conditions for limiting the right were met), it did stress that powers of secret surveillance of citizens are only allowed in so far as strictly necessary for safeguarding the democratic institutions.

- P.G. and J.H. v. the United Kingdom (2001): The case concerned the recording of the applicants' voices at a police station. The Court stressed that the gathering of personal data fell under the scope of the right to privacy and found a violation of Article 8 ECHR because there was no legal basis for such data gathering.
- S. and Marper v. the United Kingdom (2008): The case regarded the indefinite retention in a database of fingerprints, cell samples, DNA profiles, and similar data after criminal proceedings, even when suspects were acquitted. The ECtHR stressed that such a regime was disproportionate and consequently, could not be regarded as 'necessary in a democratic society'.
- Delfi v. Estonia (2015): Central to this case was an Internet service provider that was held liable for user comments on its news website, because those violated the right to reputation of a person that was in the news. The ECtHR stressed that such a limitation on the freedom of speech was legitimate in light of the protection of the right to privacy (reputational harm).
- Zakharov v. Russia (2015): The case concerned secret surveillance powers in Russia. There was no or limited judicial control on the use of power nor parliamentary control. The ECtHR allowed for an *in abstracto* claim and held Russia in violation of the Convention.
- Szabó and Vissy v. Hungary (2016): This case regarded Hungarian legislation on secret antiterrorist surveillance. Like with Zakharov, the complaint was directed at the lack of control and checks and balances against the potential abuse of power. Again, the Court found a violation of Article 8 ECHR.

2.3.3 The European Data Protection Framework

This section will discuss the data protection principles by introducing the so-called Fair Information Principles (section 3.3.1), the rules contained in the EU General Data Protection Regulation (section 3.3.2) and some of the landmark cases by the EU Court of Justice (section 3.3.3). The focus is on the EU because it has the most elaborate and influential rules on data protection in the world

2.3.3.1 *Fair Information Principles (FIPs)*

The two classic texts on informational privacy are probably the Guidelines on the Protection of Privacy and Transborder Flows of Personal Data from 1980 by the Organization for Economic Co-operation and Development's (OECD), an intergovernmental economic organization with 35 mostly 'Western' member states, and the previously mentioned CoE's Convention for the Protection of Individuals with regard to Automatic Processing of Personal Data, from 1981. Those contain the so-called Fair Information Practices. The OECD guidelines mention eight:

1. Collection Limitation Principle: There should be limits to the collection of personal data and any such data should be obtained by lawful and fair means and, where appropriate, with the knowledge or consent of the data subject.
2. Data Quality Principle: Personal data should be relevant to the purposes for which they are to be used, and, to the extent necessary for those purposes, should be accurate, complete, and kept up to date.
3. Purpose Specification Principle: The purposes for which personal data is collected should be specified not later than at the time of data collection and the subsequent use limited to the fulfilment of those purposes or such others as are not incompatible with those purposes and as are specified on each occasion of change of purpose.
4. Use Limitation Principle: Personal data should not be disclosed, made available or otherwise used for purposes other than those specified in accordance with the Purpose Specification Principle, except:
    a. with the consent of the data subject; or
    b. by the authority of law.
5. Security Safeguards Principle: Personal data should be protected by reasonable security safeguards against such risks as loss or unauthorized access, destruction, use, modification, or disclosure of data.
6. Openness Principle: There should be a general policy of openness about developments, practices, and policies with respect to personal data. Means should be readily available of establishing the existence and nature of personal data, and the main purposes of their use, as well as the identity and usual residence of the data controller.
7. Individual Participation Principle: An individual should have the right:
    a. to obtain from a data controller, or otherwise, confirmation of whether or not the data controller has data relating to him;
    b. to have communicated to him, data relating to him within a reasonable time; at a charge, if any, that is not excessive; in a reasonable manner; and in a form that is readily intelligible to him;

   c. to be given reasons if a request made under subparagraphs (a) and (b) is denied, and to be able to challenge such denial; and
   d. to challenge data relating to him and, if the challenge is successful to have the data erased, rectified, completed, or amended.
8. Accountability Principle: A data controller should be accountable for complying with measures which give effect to the principles stated above.

2.3.3.2 *Rules contained in the GDPR*

In the EU, the general data protection framework is provided by the General Data Protection Regulation. The GDPR replaces the Data Protection Directive from 1995. The GDPR will most likely have a worldwide effect (also called the Brussels effect), because of its large scope and broad requirements.

2.3.3.2.1 *When does the GDPR apply?*

There are five general conditions for the applicability of the GDPR.[30]

1. The activity must involve 'personal data', which is defined as: 'any information relating to an identified or identifiable natural person ('data subject'); an identifiable natural person is one who can be identified, directly or indirectly, in particular by reference to an identifier such as a name, an identification number, location data, an online identifier or to one or more factors specific to the physical, physiological, genetic, mental, economic, cultural, or social identity of that natural person'. As stressed, almost all data is or can be personal data.
2. These data must be 'processed', which is defined as 'any operation or set of operations which are performed on personal data or on sets of personal data, whether or not by automated means, such as collection, recording, organisation, structuring, storage, adaptation or alteration, retrieval, consultation, use, disclosure by transmission, dissemination or otherwise making available, alignment or combination, restriction, erasure or destruction'. Consequently, almost everything that can be done with personal data, such as storing, analysing, selling, and even correcting or deleting personal data, falls within the definition of 'processing'.
3. The rules in the GDPR apply primarily to the 'data controller' and partially on the 'data processor'. The data subject is the person who can be identified through the personal data. There is always a data controller and always a data subject; there may or may not be a data

30 Articles 1-4, 23, and 85-91 GDPR.

processor. The data controller is the natural or legal person who, alone or jointly with others, determines on the one hand the purposes and on the other hand the means of the processing of personal data. Simply put, the data controller is the person or organisation that decides that data should be processed and how. The controller is primarily responsible for upholding the data protection principles. The processor is the party that processes data on behalf of the data controller, for example a cloud provider that stores data on behalf of the data controller. The processor has to abide by a number of obligations of its own, but in principle, the data controller is responsible for the data processing by the data processor. If the latter makes a mistake, the former is responsible.

4. Obviously, the EU must have territorial competence for the GDPR to apply. There are four instances in which the EU claims competence:
   a. When personal data are being processed in the context of the activities of an establishment of a controller or a processor in the Union, regardless of whether the processing takes place in the Union or not.
   b. When data controllers or data processors are not established in the EU, but offer goods or services, irrespective of whether a payment of the data subject is required, to such data subjects in the Union.
   c. When data controllers or data processors are not established in the EU, but use personal data for monitoring the behaviour of EU citizens (for example by using cookies), as far as their behaviour takes place within the Union.
   d. When an EU Member State has an embassy or similar organization outside the EU, that organization falls under the scope of the GDPR.
5. There are exceptions and limitations to the applicability of the GDPR, examples of which are:
   a. When processing personal data is for a purely personal or household activity, such as keeping a list of telephone numbers and addresses of acquaintances.
   b. Processing activities concerning national security (such as by secret services or intelligence agencies), over which the EU has no competence.
   c. Processing takes place by EU institutions. The GDPR does not apply, but another Regulation does, which incorporates the same basic principles.
   d. When processing activities take place by law enforcement authorities (such as the police). The GDPR does not apply, but a separate Directive (called the Police Directive) does, adopted at the same

time as the GDPR. This Directive contains the same basic principles as the GDPR, but allows for more limitations and exceptions when this is necessary in terms of protecting public order and combating crime.

e. Then there are several fields in which the GDPR does apply, but for which Member States to the EU may make special arrangements, such as:
   i. Freedom of expression;
   ii. Archiving purposes;
   iii. Scientific research;
   iv. Governmental transparency; and
   v. Re-use of public sector information

If personal data are processed by a data controller and the EU has territorial competence and no exception applies, the GDPR will be applicable.

2.3.3.2.2 *When is processing of personal data legitimate?*

The GDPR contains its own version of the FIPs, specifying that personal data must be:[31]

1. processed lawfully, fairly, and in a transparent manner in relation to the data subject ('lawfulness, fairness, and transparency');
2. collected for specified, explicit, and legitimate purposes ('purpose specification') and not further processed in a manner that is incompatible with those purposes ('purpose limitation');
3. adequate, relevant, and limited to what is necessary in relation to the purposes for which they are processed ('data minimisation');
4. accurate and, where necessary, kept up to date ('accuracy');
5. kept in a form which permits identification of data subjects for no longer than is necessary for the purposes for which the personal data are processed ('storage limitation'); and
6. processed in a manner that ensures appropriate security of the personal data ('integrity and confidentiality').

Each of these six basic principles must be respected; otherwise, the data processing will not be deemed legitimate. How these principles must be interpreted depends partially on the circumstances of the case.

The purpose specification principle requires a specific purpose to be designated before processing personal data. A specific purpose may exist,

31 Article 5 GDPR.

for example, when a pizza delivery service asks a customer for her address. An unspecific (and hence illegitimate) purpose are vague terms such as 'improving customer experience' or 'innovation and product development'. Data may subsequently only be used for purposes directly related to this specific purpose. The pizza delivery service may also use the data to deliver hamburgers and perhaps, depending on the circumstances, send advertisements about new pizza deals. Nonetheless, it may not sell these data, for example, to a hotel, who then offers cheap vacations to the customer.

Only the data that is needed in relation to the specific purpose can be processed by the data controller. The pizza delivery service can ask for the address and a person's name, but not her gender, political believes or sports interests. These are simply unrelated to and unnecessary for the purpose of delivering a pizza.

Data should be accurate and kept up to date. This is the responsibility of the data controller. Thus, if the pizza delivery service retains the address and name of a person, the next time the person orders a pizza, it is up to the pizza delivery service to ask whether the address has remained the same.

In principle, when the pizza is delivered, the pizza delivery service should delete the name and address of the customer. If it decides to store the name and address of a regular customer, it may only do so for a reasonable period of time. For example, if that person has not ordered a pizza for a consecutive six months, it might be reasonable to delete the data.

Finally, if personal data is stored by the data controller, it must ensure that these data are maintained safely and confidentially. This means that it must take measures to protect the databased from being hacked; in addition, data may be encrypted or pseudonymised, so that if data fall into the wrong hands, they are of no or little value to the that party. Also, the data controller must ensure that only those people within the organisation that need to access the personal data can do so and that others do not have permission or authorisation to enter the database (i.e. a need to know basis).

The GDPR gives further guidance on when data processing can be legitimate for three situations: (1) when personal data are being processed, (2) when so called 'sensitive personal data' are being processed and (3) when personal data (sensitive or not) are transferred from the EU to countries outside the EU.

1. The GDPR exhaustively lists six grounds for processing personal data, one of which must apply for a processing initiative to be legitimate.
2. For sensitive personal data, the GDPR specifies: 'Processing of personal data revealing racial or ethnic origin, political opinions, religious or philosophical beliefs, or trade union membership, and the processing

of genetic data, biometric data for the purpose of uniquely identifying a natural person, data concerning health or data concerning a natural person's sex life or sexual orientation shall be prohibited.' The general thought behind this provision is that because this data is so sensitive, they simply should not be processed. Still, 10 grounds are contained in the GDPR that provide an exception to this prohibition.

3. Finally, with respect to the transfer of personal data, the basic principle is that personal data should not leave the territory of the EU. This is because with the GDPR, the EU has laid down the highest standards for data protection in the world. Transferring the data to other areas would mean that the strict rules could be circumvented. That is why the GDPR holds that the data can only be transferred to a country outside the EU when more or less the same principles as contained in the GDPR are upheld. The GDPR provides three grounds on the basis of which there can be legitimate transfer:
   a. When there is a so-called adequacy decision by the European Commission (which can be compared to the government of the EU), in which the Commission determines that a certain non-EU country, for example Switzerland, has an adequate level of data protection and data may be legitimately transferred to that country.
   b. When there are appropriate safeguards. This means that not the country to which the data are transferred has an adequate level of data protection, but a specific organisation within that country has. This commitment is laid down, for example, in a contract between the EU-based organisation and the organisation based outside the EU, the latter receiving the personal data from the former.
   c. For specific cases (for example when one file of one person is transferred to a country outside the EU), derogations may apply.

Six grounds for processing personal data*	**Ten exceptions to the prohibition to process sensitive personal data****	**Three grounds on the basis of which personal data (including sensitive data) may be transferred to countries outside the EU*****
(1) the data subject has given **consent** to the processing of his or her personal data for one or more specific purposes;	(1) the data subject has given **explicit consent** to the processing of those personal data for one or more specified purposes	(1) Adequacy decision 'A transfer of personal data to a third country or an international organisation may take place where the Commission has decided that the third country, a territory or one or more specified sectors within that third country, or the international organisation in question ensures an adequate level of protection. Such a transfer shall not require any specific authorisation.'
(2) processing is necessary for the performance of a **contract** to which the data subject is party or in order to take steps at the request of the data subject prior to entering into a contract;	(2) processing is necessary for the purposes of carrying out the obligations and exercising specific rights of the controller or of the data subject in the field of **employment and social security** and social protection law	(2) Transfers subject to appropriate safeguards Appropriate safeguards can be achieved through either of the following means: 1. A legally binding and enforceable instrument between public authorities or bodies; 2. Binding corporate rules; 3. Standard data protection clauses adopted by the Commission 4. Standard data protection clauses adopted by a supervisory authority and approved by the Commission 5. An approved code of conduct 6. An approved certification mechanism 7. Subject to the authorization from the competent supervisory authority, contractual clauses between the controller or processor and the controller, processor, or the recipient of the personal data in the third country 8. Subject to the authorization from the competent supervisory authority, provisions to be inserted into administrative arrangements between public authorities or bodies which include enforceable and effective data subject rights.

Six grounds for processing personal data*	Ten exceptions to the prohibition to process sensitive personal data**	Three grounds on the basis of which personal data (including sensitive data) may be transferred to countries outside the EU***
(3) processing is necessary for compliance with a **legal obligation** to which the controller is subject;	(3) processing is necessary to protect the **vital interests of the data subject** or of another natural person where the data subject is physically or legally incapable of giving consent;	(3) Derogations for specific situations 1. the data subject has **explicitly consented** to the proposed transfer, after having been informed of the possible risks of such transfers for the data subject due to the absence of an adequacy decision and appropriate safeguards 2. the transfer is necessary for the **performance of a contract between the data subject and the controller** or the implementation of pre-contractual measures taken at the data subject's request; 3. the transfer is necessary for the **conclusion or performance of a contract concluded in the interest of the data subject** between the controller and another natural or legal person; 4. the transfer is necessary for **important reasons of public interest;** 5. the transfer is necessary for the **establishment, exercise, or defence of legal claims;** 6. the transfer is necessary in order to protect the **vital interests of the data subject** or of other persons, where the data subject is physically or legally incapable of giving consent; 7. the transfer is made from a register which according to Union or Member State law is intended to **provide information to the public** and which is open to consultation either by the public in general or by any person who can demonstrate a legitimate interest, but only to the extent that the conditions laid down by Union or Member State law for consultation are fulfilled in the particular case. 8. When the transfer of personal data cannot be based on either of the exceptions above, the GDPR specifies: 'a transfer to a third country or an international organization may take place only if the transfer is not repetitive, concerns only a limited number of data subjects, is necessary for the purposes of **compelling legitimate interests pursued by the controller** which are not overridden by the interests or rights and freedoms of the data subject, and the controller has assessed all the circumstances surrounding the data transfer and has on the basis of that assessment provided suitable safeguards with regard to the protection of personal data'.

Six grounds for processing personal data*	Ten exceptions to the prohibition to process sensitive personal data**	Three grounds on the basis of which personal data (including sensitive data) may be transferred to countries outside the EU***
(4) processing is necessary in order to protect the **vital interests of the data subject** or of another natural person;	(4) processing is carried out in the course of its legitimate activities with appropriate safeguards by a foundation, association or any other **not-for-profit body with a political, philosophical, religious, or trade union aim** and on condition that the processing relates solely to the members or to former members of the body or to persons who have regular contact with it in connection with its purposes and that the personal data are not disclosed outside that body without the consent of the data subjects;	
(5) processing is necessary for the performance of a task carried out in the **public interest** or in the exercise of official authority vested in the controller;	(5) processing relates to personal data which are **manifestly made public by the data subject**;	
(6) processing is necessary for the purposes of the **legitimate interests pursued by the controller** or by a third party, except where such interests are overridden by the interests or fundamental rights and freedoms of the data subject which require protection of personal data, in particular where the data subject is a child. This ground does not apply to processing carried out by public authorities in the performance of their tasks.	(6) processing is necessary for the **establishment, exercise or defence of legal claims** or whenever courts are acting in their judicial capacity;	

Six grounds for processing personal data*	Ten exceptions to the prohibition to process sensitive personal data**	Three grounds on the basis of which personal data (including sensitive data) may be transferred to countries outside the EU***
	(7) processing is necessary for reasons of **substantial public interest**, on the basis of Union or Member State law which shall be proportionate to the aim pursued, respect the essence of the right to data protection, and provide for suitable and specific measures to safeguard the fundamental rights and the interests of the data subject; (8) processing is necessary for the purposes of **preventive or occupational medicine**, for the assessment of the working capacity of the employee, medical diagnosis, the provision of health or social care, or treatment or the management of health or social care systems and services (9) processing is necessary for reasons of public interest in the area of **public health**, such as protecting against serious cross-border threats to health or ensuring high standards of quality and safety of healthcare and of medicinal products or medical devices (10) processing is necessary for archiving purposes in the public interest, **scientific or historical research purposes, or statistical purposes**	

* Article 6 GDPR.

** Article 9 GDPR.

*** Articles 44-50 GDPR.

When processing personal data, the data controller must ensure that one of the grounds provided in the left column applies. If not, data processing will be considered illegitimate. When processing sensitive personal data, the same counts for the exceptions in the middle column. When transferring data from the EU to countries outside the EU, one of the grounds mentioned in the right column must apply for the transfer to be legitimate. If personal data are transferred, both a ground in the left and in the right column must apply. If sensitive data are transferred to countries outside the EU, both a ground in the middle and in the right column must apply. It is important to stress that these requirements come on top of the Fair Information Principles. Both the FIPs and the rules on legitimacy must be respected to be GDPR compliant.

It is often stressed that informational privacy or data protection is about informed consent or control over data by data subjects. This is untrue for the European legislation. A data controller can be fully GDPR-compliant without asking for consent a single time. Consent is one of the six grounds on which the processing of data can be based and only one of the 10 exceptions to the prohibition to process sensitive data. In addition, under the GDPR, the requirements for consent are tight to such an extent that it will be difficult to obtain legitimate consent from a data subject. Consent must be informed, specific, unambiguous and freely given. The burden of proof lies on the data controller to demonstrate that all these conditions have been met. If privacy policies or terms and conditions are written in juridical jargon or are overly long, data subjects that consent will not be deemed to have been properly informed. Consent is thus invalid. If consent is given for broad and vague processing activities, such as 'we process personal data for a variety of activities related to our services and in order to optimize customer experience', consent will not be deemed to be specific. If consent is given as part of a larger contract, in which the data subject gives consent to a variety of matters, consent will not be deemed to be given unambiguously. When consent is mandatory for a data subject to enter a site or service, consent may not be deemed to be given freely. And even if all these conditions are met, the data subject may always revoke its consent. Finally, it is important to note that consent cannot be used to curtail the FIPS. If the data subject consents, for example, to the processing of more data than the data controller strictly needs to fulfil its goal, it still conflicts with the data minimisation principle and hence is a violation of the GDPR.

Consent	Conditions for consent*	Conditions applicable to child's consent in relation to information society services**
'Consent' of the data subject means any freely given, specific, informed and unambiguous indication of the data subject's wishes by which he or she, by a statement or by a clear affirmative action, signifies agreement to the processing of personal data relating to him or her;***	Where processing is based on consent, the controller shall be able to demonstrate that the data subject has consented to processing of his or her personal data.	Where point (a) of Article 6(1) applies, in relation to the offer of information society services directly to a child, the processing of the personal data of a child shall be lawful where the child is at least 16 years old. Where the child is below the age of 16 years, such processing shall be lawful only if and to the extent that consent is given or authorised by the holder of parental responsibility over the child. Member States may provide by law for a lower age for those purposes provided that such lower age is not below 13 years.
Consent should be given by a clear affirmative act establishing a freely given, specific, informed and unambiguous indication of the data subject's agreement to the processing of personal data relating to him or her, such as by a written statement, including by electronic means, or an oral statement. This could include ticking a box when visiting an internet website, choosing technical settings for information society services or another statement or conduct which clearly indicates in this context the data subject's acceptance of the proposed processing of his or her personal data. Silence, pre-ticked boxes or inactivity should not therefore constitute consent. Consent should cover all processing activities carried out for the same purpose or purposes. When the processing has multiple purposes, consent should be given for all of them. If the data subject's consent is to be given following a request by electronic means, the request must be clear, concise and not unnecessarily disruptive to the use of the service for which it is provided.****	If the data subject's consent is given in the context of a written declaration which also concerns other matters, the request for consent shall be presented in a manner which is clearly distinguishable from the other matters, in an intelligible and easily accessible form, using clear and plain language. Any part of such a declaration which constitutes an infringement of this Regulation shall not be binding.	The controller shall make reasonable efforts to verify in such cases that consent is given or authorised by the holder of parental responsibility over the child, taking into consideration available technology.

Consent	Conditions for consent*	Conditions applicable to child's consent in relation to information society services**
It is often not possible to fully identify the purpose of personal data processing for scientific research purposes at the time of data collection. Therefore, data subjects should be allowed to give their consent to certain areas of scientific research when in keeping with recognised ethical standards for scientific research. Data subjects should have the opportunity to give their consent only to certain areas of research or parts of research projects to the extent allowed by the intended purpose.*****	The data subject shall have the right to withdraw his or her consent at any time. The withdrawal of consent shall not affect the lawfulness of processing based on consent before its withdrawal. Prior to giving consent, the data subject shall be informed thereof. It shall be as easy to withdraw as to give consent.	Paragraph 1 shall not affect the general contract law of Member States such as the rules on the validity, formation or effect of a contract in relation to a child.
Children merit specific protection with regard to their personal data, as they may be less aware of the risks, consequences and safeguards concerned and their rights in relation to the processing of personal data. Such specific protection should, in particular, apply to the use of personal data of children for the purposes of marketing or creating personality or user profiles and the collection of personal data with regard to children when using services offered directly to a child. The consent of the holder of parental responsibility should not be necessary in the context of preventive or counselling services offered directly to a child.******	When assessing whether consent is freely given, utmost account shall be taken of whether, *inter alia*, the performance of a contract, including the provision of a service, is conditional on consent to the processing of personal data that is not necessary for the performance of that contract.	

Consent	Conditions for consent*	Conditions applicable to child's consent in relation to information society services**
Where processing is based on the data subject's consent, the controller should be able to demonstrate that the data subject has given consent to the processing operation. In particular in the context of a written declaration on another matter, safeguards should ensure that the data subject is aware of the fact that and the extent to which consent is given. In accordance with Council Directive 93/13/EEC (1) a declaration of consent pre-formulated by the controller should be provided in an intelligible and easily accessible form, using clear and plain language and it should not contain unfair terms. For consent to be informed, the data subject should be aware at least of the identity of the controller and the purposes of the processing for which the personal data are intended. Consent should not be regarded as freely given if the data subject has no genuine or free choice or is unable to refuse or withdraw consent without detriment.*******		

Consent	Conditions for consent*	Conditions applicable to child's consent in relation to information society services**
In order to ensure that consent is freely given, consent should not provide a valid legal ground for the processing of personal data in a specific case where there is a clear imbalance between the data subject and the controller, in particular where the controller is a public authority and it is therefore unlikely that consent was freely given in all the circumstances of that specific situation. Consent is presumed not to be freely given if it does not allow separate consent to be given to different personal data processing operations despite it being appropriate in the individual case, or if the performance of a contract, including the provision of a service, is dependent on the consent despite such consent not being necessary for such performance.********		

* Article 7 GDPR.
** Article 8 GDPR.
*** Article 4 GDPR.
**** Recital 32 GDPR.
***** Recital 33 GDPR.
****** Recital 38 GDPR.
******* Recital 42 GDPR.
******** Recital 43 GDPR.

2.3.3.2.3 What additional obligations do data controllers have?

Data controllers have to respect the FIPS, have to obtain a legitimate ground for processing personal data, sensitive data or transferring them and have to abide by a number of more specific requirements provided below. There are conditions for and exceptions to each of those obligations; these are too detailed to describe here. Instead, the basic requirements are provided. There are six mandatory requirements:[32]

1. The GDPR introduces a general obligation for data controllers to keep records of their processing activities, in which they describe meticulously what data they have, about whom, for what reasons they are processed, with whom they are shared, etc.
2. The GDPR requires data controllers to demonstrate transparency regarding their processing activities. They should provide data subjects (on their own initiative) with the information about the data processing activity, e.g. what data is processed about the data subject, why, by whom, how long they will be processed, which technical and organisational safety measures have been adopted, etc.
3. There must be appropriate technical and organisational safeguards applicable to the processing of personal data. Such security measures can include pseudonymization, encryption and protecting databases against hackers. Such organisational measures may include introducing authentication systems for entering databases, limiting access-rights to a small number of people within the organisation and logging which employees have entered databases and when.
4. The GDPR requires a data controller to notify the relevant Data Protection Authority (DPA – its role is explained in more detail below) when there has been a data breach (data have fallen into the hands of third parties, for example hackers, has been accidentally lost, or someone within the organization has had unauthorized access) and the data subject has to be informed when the data breach is likely to affect her.
5. A Data Protection Officer (DPO) must be appointed by public authorities processing data and by private organizations when they are processing sensitive data, systematically monitoring citizens on a large scale or perform other risk-prone processing operations. A data protection officer has the responsibility to ensure that the data protection principles are respected within an organization. The officer has an independent position and should be fully equipped by the organisation to allow her

32 Articles 24-43 GDPR.

to assess to what extent the organisation is GDPR-compliant and what measures should be adopted to ensure compliance.

6. When there are risk-prone processing operations, an organization has to perform a so-called Data Protection Impact Assessment (DPIA), in which it assesses the impact of its intended data-processing operation. It has to adopt precautionary measures to mitigate risks when they follow from such an Impact Assessment. When the risks cannot be mitigated, the data controller should abstain from its intended data-processing operation or ask the DPA for permission.

In addition, there are two optional clauses in the GDPR:

7. There is no obligation, but a possibility for data controllers to draw up a code of conduct. A code of conduct is a primarily sectorial instrument, which specifies in further detail how the principles in the GDPR should be interpreted in specific contexts/sectors. If an association (e.g. the association for European universities) has adopted such a code of conduct (which is in itself optional), all members of that association (e.g. the specific universities being member of the association for European universities) are obliged to abide by the rules in the code of conduct.
8. The GDPR promotes, but does not oblige, self- and co-regulation through self-certification. A certificate can only be given to an organisation by an officially authorised certification body. A certificate may, for example, state: 'This organisation has adopted sufficient organisational and technical security measures to be, on this point, GDPR-complaint'.

2.3.3.2.4 What are the rights of data subjects?

Data subjects have rights, which the data controller (and the data processor to some extent), needs to respect.[33] Most of these rights correlate with the obligations of the data controllers. Thus, only if the data controller ignores its duties, which is a violation of the GDPR in itself, will the data subject have a legitimate reason to invoke its rights. The right to information of the data subject correlates with the obligation of the data controller to provide data subjects with information on its own behalf. The right to rectify personal data of the data subject correlates with the obligation of the data controller to keep data correct and up to date. The right to erasure (sometimes called the right to be forgotten) by the data subject can only be invoked when the data controller is processing data illegitimately. The right to object to the processing of data only applies when the data controller has no legitimate

33 Articles 12-22 GDPR.

ground for processing the data. And finally, the right of the data subject not to be subjected to autonomic decision making, including profiling, is in fact an obligation of the data controller not to make decisions without human assessment, at least when the decision affects the data subject significantly.

Consequently, if the data controller follows the rules of the GDPR, data subjects will not have a legitimate claim to any of their rights. There are two exceptions: rights that do not correlate with independent duties of the data controller, which can be invoked by the data subject even if the data controller has not violated any obligations under the GDPR. (1) The right to copy gives the data subject the right to not only request information about the data that is being processed about her, but also a right to obtain a copy of that information. This is especially important in the medical sector. (2) The right to data portability, which only applies when data subjects have given personal data to a data controller (e.g. Facebook) themselves and when the ground for processing this data is the consent of or a contract with the data subject (e.g. 'I agree to be on Facebook under the following conditions'). When a person decides to leave the data controller (e.g. leaves Facebook in order to join another social network), the data subject can take the data that she has provided with her or ask the data controller to send the data to the new data controller she is going to (right to data portability).

2.3.3.2.5 How are the rules in the GDPR enforced?

If the data protection rules are not followed by the data controller, and the data protection officer has been unable to correct the situation, the data subject may submit a complaint to either a judge or to the DPA. The DPA is a governmental agency that has a variety of tasks; it can be compared to a market regulator, such as exist in inter alia the telecommunications sector. The DPA can also take measures on its own initiative, that is without the complaint of a data subject. The DPA will in principle only take action when the data controller has neglected its obligations as specified in the GDPR.

A general problem with data protection provisions before the introduction of the GDPR has been that they have lacked adequate enforcement. This is tackled by the GDPR, in particular in five ways:[34]

1. A general problem was that the EU Data Protection Directive 1995 needed to be implemented by each Member State. This meant that there existed differentiation in the rules among countries. Data controllers were often established in countries where the rules applicable to its business endeavours would be least strict. This is addressed by the

34 Articles 51-84 and 92-93 GDPR.

GDPR because a Regulation, as opposed to a Directive, has direct effect throughout the EU. This means that data subjects can rely directly on the GDPR, without having to refer to the national implementation of the EU rules (as was the case with the Data Protection Directive).

2. A general problem was that the enforcement of the data protection rules was mostly in the hands of national governments and the Data Protection Authority, which each country needed to install. However, countries differed in their approach to enforcement, some being more lenient than others. Again, data controllers were often established in countries were the level of enforcement was low. Under the GDPR, there is enhanced cooperation between the different DPAs and one DPA can be assigned authority over a company with respect to its establishments and activities throughout the whole EU.
3. In addition, there are several ways for the European Commission and other EU institutions, such as the European Data Protection Board, in which all national DPAs have a seat, to engage in monitoring and norm-setting, to further harmonize regulation and provide more specific provisions on data-processing activities.
4. Not all DPAs were well equipped prior to the GDPR; some of them were also lacking independence from the government. The GDPR guaranties the independence of DPAs and gives them wide authority on a number of accounts.
5. Finally, a general problem has been that the fines that could be imposed on companies that violated the data protection principles were considered low, especially when considering the high profits made by tech-companies. The GDPR addresses this problem and allows for sanctions that may run up to 20 million euro or, in the case of an undertaking, up to 4% of the total worldwide annual turnover of the preceding financial year, whichever is higher.

2.3.3.3 Landmark cases by the ECJ

It is impossible to give a full overview of the case law of the ECJ on the right to data protection. Instead, four recent and influential cases will be briefly touched upon:

1. Digital Rights Ireland (2014): Concerned an EU Directive which required states to retain data for a period of time on, inter alia, citizen's Internet use. The ECJ rendered this Directive invalid, because it was considered an illegitimate infringement on the rights to privacy and data protection.
2. Google Spain (2014): Concerned the request of a citizen about whom compromising information could be found by using Google's search

engine. The Court ruled that there may be an obligation of an operator of a search engine to remove from the list of results links to web pages, published by third parties, also in a case where that name or information is not erased beforehand or simultaneously from those web pages, and even, when its publication in itself on those pages is lawful.
3. Schrems (2015): Concerned an adequacy decision (known as 'Safe Harbour') of the European Commission in which the United States of America was considered, with respect to some data-processing operations, to provide an adequate level of data protection. The ECJ declared that decision invalid, because it was not convinced that the US did have an adequate level of protection.
4. Tele 2 (2016): Concerned the EU e-Privacy Directive and the obligation to retain data about, inter alia, Internet traffic. The ECJ stressed that the rights to privacy and data protection preclude national legislation which provides for general and indiscriminate retention of all traffic and location data of all subscribers and registered users relating to all means of electronic communication.

2.4 Traditional debates and dominant schools

As stressed before, scholarly debates are less important in law than in most other fields of research. There is some discussion, but these mainly stem from the differences in legal regulation of privacy in various countries and jurisdictions. Below a brief introduction to five of those discussions.

2.4.1 Privacy as control

Some authors feel that privacy and data protection are about control of the individual, either over her data, or, for example, control over who has access to the house. This school mainly focuses on individual rights and individual interests mainly. It presupposes that the individual can practically take control over her privacy and personal data. In part, this school is inspired by the so-called census decision by the German Constitutional Court, who has introduced the notion of 'informational self-determination'.[35]

Others stress that privacy and data protection are in essence not individual rights that protect individual interests, but obligations of states and data controllers not to abuse their powers and/or to use their powers in a

35 Bundesverfasungsgericht 15 December 1983.

good and careful manner. Privacy and data protection, in this school, are seen as only partially protecting individual interests, and mainly focussed on the public interest. Furthermore, scholars have pointed to the fact that individuals are simply unable to control their data, because there are simply too many data-processing initiatives that contain one's personal data.

2.4.2 Privacy as property

In addition to seeing privacy and data protection in terms of individual control, a few scholars have argued that people should have a property right over their personal data. Seeing that large companies make high profits by gathering, processing, and selling personal data or profiles distilled from those data, scholars have argued that property rights over personal data may be introduced, so that individuals could at least have a share in the profits that are being made by using their data.

Others stress that it is impossible to give property rights over personal data to individuals. Personal data are all data, also data that can be gathered by walking in the street – 'that man with the black shawl' may be considered 'personal data'. It is simply undoable to restrict the use of these types of data by subjecting them to property rights. In addition, why would anyone be legitimized to claim a property right over information like 'man' or 'black shawl'? On the other hand, some scholars stress that if there are personal data that can be seen as so intrinsic to a person's identity or personality that she should have a right of control, then it would be simply unethical to treat these as an economic and 'tradable' good. A person just cannot sell herself into slavery, because the body is not a transferable good that can be owned by another – personal data shouldn't be traded either.

2.4.3 Privacy as a personality right

Some stress that privacy should be seen as a personality right. They point to the German constitution, in which a personality right is firmly engrained, and to the trends in the various jurisdictions, such as the case law of the European Court of Human Rights, under which the right to privacy has been transformed into such a personality right. They point to the fact that personality rights have a bigger material scope and thus provide for more protection and grant more freedom to citizens.

Others point to the fact that the bigger a right or doctrine is, in general, the weaker it becomes. By including all types of remotely related interests under the same doctrine, more exceptions and limitations will be necessary.

In addition, some scholars stress that privacy and personality rights are simply two different doctrines that should not be mixed up. Privacy rights are about 'freedom from', while personality rights concern the 'freedom to'.

2.4.4 Privacy and data protection

There is discussion about whether there is a difference between privacy and data protection or not. For many American scholars, the protection of personal data falls under the scope of informational privacy. Some feel that the European scope of the notion of 'personal data' is too broad, others feel that the obligations on data controllers are too strict and place too many hurdles for innovative companies and start-ups that base their business models on the processing of personal data.

For many European scholars, however, there is a clear distinction between privacy rights and data protection principle – although within the Council of Europe laws and jurisprudence, this distinction is less strict than in the European Union. Many scholars around the world have praised the General Data Protection Regulation as an attempt not so much to protect the privacy of citizens, but to curtail the gathering and processing of data by companies and other organizations, and the growing power and information imbalances that this entails. The GDPR is seen as a highly 'proceduralistic' instrument, to the dismay of some, while being lauded by others. In any case, to many Europeans, data protection legislation is of a different nature than privacy laws: they have different scopes, different obligations, rights and different approaches (as explained in section 3.2 and 3.3 of this chapter).

2.4.5 Balancing

As has been stressed, one of the most common methodologies used by courts, but also politicians and researchers, to determine the outcome of a case or a conflict between doctrines and principles is 'balancing'. Through this methodology, one right or principle is balanced against the other, for example the right to privacy against the right to freedom of speech or individual autonomy against national security. The outcome is determined on a case-by-case basis, by weighing one interest against the other, taking account of the circumstances of the case.

Others have argued that balancing is a nonsensical metaphor in the legal realm. Privacy has no weight, nor does security. There is no objective methodology of weighing and there is no base unit (such as a kilogram) to express weights of legal principles. Still others have underlined that when applied in privacy cases, it normally means that privacy is outweighed by security, because privacy is limited to an individual interest, while security, so it is said, relates to the interests of the entire population.

2.5 New challenges and topical discussions

There are many challenges and topical discussions concerning the rights to privacy and data protection in the legal realm. Mostly, they relate to new data surveillance techniques, smart applications, and the internet of Things (IoT). Big Data is the overarching term that is used to describe many of the societal, economic, and technical changes, such as the technical capacity to gather data in all types of structures, the reduced costs of storing and analysing data, and the interest of many companies and governments to apply data-driven innovation. It is impossible to give an exact definition of Big Data, but in general it is described as an asset with the following affordances (in how far these are real is a matter of debate): large quantities of data that can be gathered without a concrete or specified reason. These data will subsequently be analysed to see which data is valuable, and computer algorithms can find patterns and distil correlations that go beyond human hypotheses. Data can be reused for new purposes and combined with existing databases, offline or online, or complemented with data from open sources, for example by scraping the Internet. By analysing large quantities of data, statistical correlations may be found and group profiles can be developed. It is obvious that this trend will conflict with a number of principles of the current privacy and data protection regime. Three examples will be provided. Section 5.1 will discuss data protection principles in light of Big Data developments, section 5.2 will analyze the focus on the individual in the current legal framework and section 5.3 will discuss legal regulation as such in light of recent technological developments.[36] Section 5.4 will provide a brief discussion.

2.5.1 Big Data challenges to Data Protection principles

- Personal data must be collected for specified, explicit, and legitimate purposes, while Big Data and new data technologies enable the indiscriminate gathering of personal data.

36 These sections are partially based on: Van der Sloot and Van Schendel 2016.

- Personal data may not be further processed in a way that is incompatible with the original purpose, while the key adage of Big Data is that data can always have a second life and be reused for purposes previously unforeseen.
- The current data protection regime is based on the principle of data minimization, while the trend with Big Data technologies is rather to collect as much data as possible and store them for as long as possible.
- Under the legal framework, data should be treated confidentially and should be stored in a secure manner, while this principle is challenged because data is increasingly shared between different organizations and/or made available online (open data).
- The current framework also specifies that the data should be accurate and kept up to date. It is, however, becoming less and less important for data analytics to work with correct and accurate data about specific individuals, because the correlations found and group profiles made transcend the individual. A general correlation or group profile can be distilled from messy data sets. 'Quantity over quality of data', so the saying goes.
- Data subjects have the right to request information about whether data relating to them is processed, how, and by whom. This principle is also at odds with the rise of Big Data, partly because data subjects often simply do not know that their data are being collected and are therefore not likely to invoke their right to information. This applies equally to the other side of the coin: the transparency obligation for data controllers. For them, it is often unclear to whom the information relates, where the information came from, and how they could contact the data subjects, especially when the processes entail merging different databases and the reuse of information.

Consequently, Big Data challenges many of the classic Fair Information Principles and Data Protection principles.

2.5.2 Big Data and the individual

The current privacy and data protection paradigm focuses to a large extent on the individual, on subjective rights, and personal interests. This is put under pressure by new data technologies.

- The principle of *ratione personae* seems hard to maintain in Big Data processes, because these processes do not focus on specific individuals, but on large groups of people or potentially everyone. Briefly put, many Big Data processes and applications based thereon are general, large-scale projects that have an impact on big groups or on society as a whole, while the link to individuals and individual interests is increasingly vague and abstract. The problem with large scale data processing activities, such as data gathering by intelligence agencies, is not so much that a specific individual is affected, but that communication data are intercepted about thousands or even millions of people. It regards a structural and societal problem.

- The principle of *ratione materiae* is also challenged in Big Data processes because it is increasingly unclear whether a particular right is at all involved with a certain practice. To give an example, the application of data protection instruments depends on whether personal data are processed. However, increasingly, data is no longer stored and processed on the individual level; rather, the trend is to work with aggregated data and to generate general patterns and group profiles. These statistical correlations or group profiles cannot be qualified as personal data, but can be used to change the environment in which people live significantly. An individual as part of a group or as assigned to a particular category may not be identifiable directly herself, but can nonetheless be affected by the data processing.
- The current legal system places much emphasis on subjective individual rights. The question is whether this focus can be maintained in the age of Big Data. It is often difficult for individuals to demonstrate personal injury or an individual interest in a particular case; individuals are often unaware that their rights are being violated or even that their data has been gathered. In the Big Data era, data collection will presumably be so widespread that it is impossible for individuals on a practical level to assess each data process to determine whether it includes their personal data, if so whether the processing is lawful, and if that is not the case, to go to court or file a complaint.

Consequently, the focus on privacy as an individual right providing protection to individual interest is put under pressure by Big Data innovations.

2.5.3 Big Data and legal regulation

Finally, Big Data and other modern data technologies challenge the legal regulation of privacy as such. This is because law is always dependent of legally defined categories and concepts, which are becoming increasingly blurry and vague in the age of Big Data. Examples are:

- Data processing is becoming increasingly transnational. This implies that more and more agreements must be made between jurisdictions and states. Making these agreements legally binding is often difficult due to the different traditions and legal systems. Rapidly changing technology means that specific legal provisions can easily be circumvented and that unforeseen problems and challenges arise. The legal reality is often overtaken by events and technical developments.
- The fact that many of the problems resulting from Big Data processes predominantly revolve around more general social and societal issues makes it difficult to address the Big Data issues within specific legal doctrines, which are often aimed at protecting the interests of individuals, of legal subjects. That is why more and more national governments are looking for alternatives or additions to traditional black letter law when regulating Big Data – for example self-regulation, codes of conduct, and ethical guidelines.
- The legal framework often depends on static concepts and divisions. These are put under pressure by Big Data processes. For example, the current legal regime is based on different levels of protection for different types of data. Article 8 ECHR protects private data (which do not necessarily have to be sensitive) and sensitive data (which do not have to be private) and provides limited protection only to other personal data and metadata. The GDPR distinguishes between ordinary personal data, sensitive personal data, anonymous data (which fall outside the scope of the GDPR), and pseudonymous data. However, it is increasingly questionable whether these distinctions are still tenable in the age of Big Data. Increasingly, these categories are merely temporary stages, because data can almost always be linked back to an individual or can be de-anonymized or re-identified. Overall, while the current legal system is focused on relatively static stages of data and links to these stages a specific level of protection, in practice, data processing is becoming a circular process: data are linked, aggregated, and anonymized and then again de-anonymized enriched with other data in order to create sensitive profiles, etc.

In conclusion, the possibility of protecting privacy through legal means is put under pressure by the developments known as Big Data.

2.5.4 Discussion

There is discussion about what these challenges should mean for the legal regulation of privacy and data protection. In general, several positions can be distinguished, five of the most influential ones being:

1. Big Data and similar technologies should simply be prohibited, as they are contrary to the rights to privacy and data protection.
2. The regulation of privacy and data protection is outdated and only hampers innovation. Consequently, the laws should be changed or left unenforced.
3. Big Data is only a hype – so far, there is little evidence that Big Data technologies actually are effective. Thus, no changes to the legal regime are necessary.
4. Middle ground needs to be found to allow for new data technologies, while still respecting most of the privacy and data protection principles.
5. The current privacy and data protection regime should remain intact, but there should be a special and separate privacy and data protection regime for Big Data and similar technologies.

2.6 Conclusion

In conclusion, privacy, from a legal perspective, is a concept that originally demarcated the private and the public domain. The king or ruler held sway over the public domain, the *pater familias* ruled as king over the household. Privacy has been protected in the legal domain throughout the ages, for example by granting a special legal status to the home of an individual, private correspondence and bodily integrity. Privacy is protected though civil law, such as tort and consumer law, through criminal law, and more recently, through constitutional law. How privacy is protected and what falls under its scope differs from jurisdiction to jurisdiction.

More recently, privacy has been incorporated in human rights instruments such as the Universal Declaration on Human Rights and the European Convention on Human Rights. The ECtHR has granted the right to privacy, provided under Article 8 ECHR, a very broad scope. The EU Charter of Fundamental Rights contains a right to data protection, in addition to a right to privacy. Data protection is regulated in more detail in the EU by the General Data Protection Regulation. The GDPR provides detailed rules on how and when data controllers may legitimately process personal data of citizens. There is considerable discussion among scholars about how privacy could and should be approached, such as seeing it as a personality right, a right that grants control over data, or even as a property right. The legal approach to privacy protection is challenged by new data technologies such as Big Data.

Further reading

General literature

Aquinas, T. (1914-1942). *The 'Summa theologica' of St Thomas Aquinas.* City: Burns Oates and Washbourne.

Aries, P. & Duby, G. (1988). *A History of Private Life: Revelations of the Medieval World* (Harvard: Belknap).

Austin, J. (1995). *The Province of Jurisprudence Determined.* Cambridge: Cambridge University Press.

Bentham, J. (1970). *Of Laws in General.* City: Athlone Press.

Derrida, J. (1989-1990). 'Force of *Law:* The *Mystical Foundation* of Authority', translated by Mary Quaintance, *Cardozo Law Review* 11, pages.

Fuller, L.L. (1969). *The Morality of Law.* City: Yale University Press.

Hart, H.L.A. (1994). *The Concept of Law.* City: Clarendon Press.

Hobbes, T (2006). *Leviathan.* Cambridge: Cambridge University Press.

Kantorowicz, E. (2016). *The King's Two Bodies: A Study in Medieval Political Theology.* Princeton: Princeton University Press 2016.

Koops et. al (2017). A Typology of Privacy. *University of Pennsylvania Journal of International Law,* 38(2).

Lessig, L. (1999). *Code and Other Laws of Cyberspace.* City: Basic Books.

Locke, J. (1988 [1689]). *Two Treatises of Government.* Cambridge: Cambridge University Press.

Montesquieu. (1989). *The Spirit of the Laws.* Cambridge: Cambridge University Press.

Rössler, B. (2005). *The Value of Privacy.* Cambridge: Polity Press.

Vasak, K. (1977). 'Human Rights: A Thirty-Year Struggle: the Sustained Efforts to give Force of law to the Universal Declaration of Human Rights', *UNESCO Courier* 30(11).

Privacy in the United States of America

Agre, P.E. and M. Rotenberg. (2001). *Technology and Privacy: The New Landscape.* City: MIT Press.

Allen, A.L. (2011). *Unpopular Privacy: What Must we Hide?* Oxford: Oxford University Press.

Benn, S.I. (1984). 'Privacy, Freedom, and Respect for Persons' in F. Schoeman (ed.), *Philosophical Dimensions of Privacy: An Anthology.* Cambrdige: Cambridge University Press.

Cohen, J.E. (2012). *Configuring the Networked Self: Law, Code, and the Play of Everyday Practice.* New Haven: Yale University Press.

Cohen, J.L. (2002). *Regulating Intimacy: A New Legal Paradigm.* Princeton: Princeton University Press.

DeCew, J.W. (1997). *In Pursuit of Privacy: Law, Ethics, and the Rise of Technology.* City: Cornell University Press.

Gavison, R. (1980). 'Privacy and the Limits of Law', *Yale Law Journal* 89, p. 455.

Gray, D. and D. Citron. (2013). 'The Right to Quantitative Privacy', *Minnesota Law Review* 101.

Hoofnagle, C.J. (2016). *Federal Trade Commission Privacy Law and Policy.* Cambridge: Cambridge University Press.

Nissenbaum, H. (2010). *Privacy in Context: Technology, Policy, and the Integrity of Social Life.* City: Stanford University Press.

Prosser, W.L. (1960). 'Privacy', *Californian Law Review* 48(383).

Regan, P.M. (1995). *Legislating Privacy: Technology, Social Values, and Public Policy.* City: University of North Carolina Press.

Solove, D.J. (2008). *Understanding Privacy.* City: Harvard University Press.

Warren, S.D. and L.D. Brandeis. (1890). 'The Right to Privacy', *Harvard Law Review* 4(5).

Westin, A.F. (1967). *Privacy and Freedom.* City:Atheneum.

Privacy under the Universal Declaration on Human Rights

Johnson, M.G. and J. Symonides. (1998). *The Universal Declaration of Human Rights: A History of its Creation and Implementation.* City: Unesco.

Robinson, N. (1958). *The Universal Declaration of Human Rights: Its Origin, Significance, Application, and Interpretation.* City: World Jewish Congress.

Schabas, W.A. (ed.). (2013). *The Universal Declaration of Human Rights: the travaux préparatoires.* (Cambridge: ambridge University Press.

Verdoodt, A. (1964). *Naissance et signification de la Déclaration universelle des droits de l'homme.* City: Warny.

Privacy under the European Convention on Human Rights

Arai-Takahashi, Y. *The Margin of Appreciation Doctrine and the Principle of Proportionality in the Jurisprudence of the ECHR.* City: Intersentia.

Dijk, P. van, F. van Hoof, A. van Rijk, and L. Zwaak (eds.). (2006). *Theory and Practice of the European Convention on Human Rights.* City: Intersentia.

Greer, S. (1997). *The Exceptions to Articles 8 to 11 of the European Convention on Human Rights.* City: Council of Europe.

Vande Lanotte, J. and Y. Haeck. (2004). *Handboek EVRM. Dl.2 Artikelsgewijze commentaar, Vol. 1, Antwerpen.* City: Intersentia.

Vande Lanotte, J. and Y. Haeck. (2004). *Handboek EVRM. Dl.2 Artikelsgewijze commentaar, Vol.* 2 (Intersentia 2004).

Ovey, C. and R.C.A. White. (2002). *European Convention on Human Rights.* Oxford: Oxford University Press.

Sloot, B. van der. (2015). 'Privacy as Personality Right: Why the ECtHR's Focus on Ulterior Interests Might Prove Indispensable in the Age of Big Data', *Utrecht Journal of International and European Law* 31(80).

Sloot, B. van der. (2017B). 'Where is the harm in a privacy violation? Calculating the damages afforded in privacy cases by the European Court of Human Rights', *JIPITEC* (4).

Data Protection, especially in the European Union

Burkert, H. (1983). *Freedom of Information and Data Protection.* City: Gesellschaft für Mathematik und Datenverarbeitung.

Bygrave, L.A. (2002). *Data Protection Law: Approaching Its Rationale, Logic and Limits.* City: Kluwer Law International.

Bygrave, L.A. (2014). *Data Privacy Law: an International Perspective.* Oxford: Oxford University Press.

Dammann, U., O. Mallmann, and S. Simitis (eds.). (1977). *Data Protection Legislation: An International Documentation: Engl.-German: eine internationale Dokumentation: Die Gesetzgebung zum Datenschutz.* City: Metzner.

Fuster, G.G. (2014). *The Emergence of Personal Data Protection as a Fundamental Right of the EU.* City: Springer.

Hondius, F.W. (1975). *Emerging Data Protection in Europe.* City: Elsevier.

Hijmans, H. (2016). *The European Union as a Constitutional Guardian of Internet Privacy and Data Protection: the Story of Article 16 TFEU.* University of Amsterdam Dissertation.

Sloot, B. van der. (2014). 'Do Data Protection Rules Protect the Individual and Should They? An Assessment of the Proposed General Data Protection Regulation', *International Data Privacy Law* 4.

Literature on debates and challenges

Privacy as control

Deci, E.L. and R.M. Ryan. (1985). 'The General Causality Orientations Scale: Self-determination in Personality', *Journal of Research in Personality* Volume(issue), pages.

Hornung, G. and C. Schnabel. (2009). 'Data Protection in Germany I: the Population Census Decision and the Right to Informational Self-determination', *Computer Law & Security Review* volume(issue), pages.

Rossnagel, A. and P. Richter. (2016). 'Big Data and Informational Self-Determination. Regulative approaches in Germany: The Case of Police and Intelligence Agencies' in van der Sloot, B., Broeders, D. and Schrijvers, E. (eds.), *Exploring the boundaries of Big Data.* Amsterdam: Amsterdam University Press.

Rouvroy, A. and Y. Poullet,. (2009).'The Right to Informational Self-determination and the Value of Self-development: Reassessing the Importance of Privacy for Democracy' in S. *Gutwirth*, Y. Poullet, P. de Hert, C. de Terwangne, and S. Nouwt, *Reinventing Data Protection?* City: Springer.

Privacy as property

Epstein, R.A. (1977). 'Privacy, Property Rights, and Misrepresentations', *Georgia Law Review* 12(456), pages.

Hermalin, B.E. and M.L. Katz. (2006). 'Privacy, Property Rights and Efficiency: the Economics of Privacy as Secrecy', *Quantitative Marketing and Economics* volume(issue), pages.

Murphy, R.S. (1995). 'Property Rights in Personal Information: an Economic Defense of Privacy', *The Georgetown Law Journal* volume(issue), pages.

Posner, R.A. (1977). 'The Right of Privacy', *Georgia Law Review* volume(issue), pages.

Post, R.C. (1990). 'Rereading Warren and Brandeis: Privacy, Property, and Appropriation', *The Case Western Reserve Law Review* volume(issue), pages.

Purtova, N. (2012). *Property Rights in Personal Data: a European Perspective.* City: Kluwer Law International.

Privacy as personality right

Bloustein, E.J. (1964). 'Privacy as an Aspect of Human Dignity: An Answer to Dean Prosser', *New York University Law Review* 39(962), pages.

Eberle, E.J. (1997). 'Human Dignity, Privacy, and Personality in German and American Constitutional Law', *Utah Law Review* 963, pages.

Pound, R. (1915). 'Interests of Personality', *Harvard Law Review* 28(4), 1pages.
Schwartz, P.M. and K.-N. Peifer. (2010). 'Prosser's Privacy and the German Right of Personality: Are Four Privacy Torts Better than One Unitary Concept?', *California Law Review* 98(1925), pages.
Strömholm, S. (1967). 'Right of Privacy and Rights of the Personality: a Comparative Survey', Nordic Conference on Privacy organized by the International Commission of Jurists, Stockholm.
Whitman, J.Q. (2004). 'The Two Western Cultures of Privacy: Dignity versus Liberty', *Yale Law Journal* 113(1151), pages.

Privacy and data protection

Fuster, G.G. and R. Gellert. (2012). 'The Fundamental Right of Data Protection in the European Union: in Search of an Uncharted Right', *International Review of Law, Computers & Technology* 26, pages.
Gellert, R. and S. Gutwirth. (2013). 'The Legal Construction of Privacy and Data Protection', *Computer Law & Security Review* 29, pages.
Hert, P. de. (1998).'Human Rights and Data Protection. European Case-Law 1995-1997', [Mensenrechten en bescherming van persoonsgegevens. Overzicht en synthese van de Europese rechtspraak 1955-1997] in *Jaarboek ICM 1997*. City: Maklu.
Kokott, J. and C. Sobotta. (2013). 'The Distinction between Privacy and Data Protection in the Jurisprudence of the CJEU and the ECtHR', *International Data Privacy Law* 3, pages.
Lynskey, O. (2014). 'Deconstructing Data Protection: the "Added-value" of a right to data protection in the EU legal order', *International and Comparative Law Quarterly* 3, pages.

Balancing

Aleinikoff, T.A. (1987). 'Constitutional Law in the Age of Balancing', *Yale Law Journal* 97(5), pages.
Alexy, A. (2003). 'Constitutional Rights, Balancing, and Rationality', *Ratio Juris* 16, pages.
Cali, B. (2007). 'Balancing Human Rights? Methodological Problems with Weights, Scales and Proportions', *Human Rights Quarterly* 29(1), pages.
Greer, S. (2004). '"Balancing" and the European Court of Human Rights: a Contribution to the Habermas-Alexy Debate', *The Cambridge Law Journal* 63, pages.
Habermas, J. (1996). *Between Facts and Norms.* City: Polity.
Sloot, B. van der. (2016). 'The Practical and Theoretical Problems with "Balancing": Delfi, Coty and the Redundancy of the Human Rights Framework', *Maastricht Journal of European and Comparative Law* 3, pages.
Sloot, B. van der. (2017A). 'Ten Questions about Balancing', *European Data Protection Law Review* 2, pages.
Tsakyrakis, S. (2008). 'Proportionality: an Assault on Human Rights?', *Jean Monnet Working Paper* 09/08.

Big Data and privacy regulation

Ambrose, J. and M. Leta. (2015). 'Lessons from the Avalanche of Numbers: Big Data in Historical Context' *I/S: A Journal of Law and Policy for the Information Society*, http://ssrn.com/abstract=2486981.
Anderson, C. (2008). 'The End of Theory: The Data Deluge Makes the Scientific Method Obsolete', *Wired Magazine* 16 July, pages.
Andrejevic, M. (2014). 'The Big Data Divide', *International Journal of Communication* 8, pages.
Bollier, D. (2010). 'The Promise and Peril of Big Data',

www.emc.com/collateral/analyst-reports/10334-ar-promise-peril-of-big-data.pdf.
Boyd, D. and K. Crawford. (2011). 'Six Provocations for Big Data', http://papers.ssrn.com/sol3/papers.cfm?abstract_id=1926431.
Boyd, D. and K. Crawford. (2012). 'Critical Questions for Big Data: Provocations for a Cultural, Technological, and Scholarly Phenomenon', *Information, Communication & Society* 5, pages.
Crawford, C. and J. Schultz. (2014). 'Big Data and Due Process: Toward a Framework to Redress Predictive Privacy Harms', *Boston College Law Review* 55, pages.
Davis, K. and D. Patterson. (2012). *Ethics of Big Data: Balancing Risk and Innovation.* City: O' Reilly Media.
Sloot, B. van der and S. van Schendel. (2016). 'International and Comparative Legal Study on Big Data', WRR-rapport, working paper 20, www.wrr.nl/publicaties/publicatie/article/international-and-comparative-legal-study-on-big-data/.

Legal documents

United States of America

Legal documents

US constitution and amendments https://www.usconstitution.net/const.html.
Federal Trade Commission Act https://www.law.cornell.edu/uscode/text/15/41.
The Children's Online Privacy Protection Act (COPPA) https://www.law.cornell.edu/uscode/text/15/6501.
The Health Insurance Portability and Accountability Act (HIPAA) https://www.law.cornell.edu/uscode/text/42/1301b.
The Fair Credit Reporting Act https://www.law.cornell.edu/uscode/text/15/1681.
The Electronic Communications Privacy Act https://www.law.cornell.edu/uscode/text/18/2510.
Constitution of Alaska https://www.commerce.alaska.gov/web/Portals/4/pub/AK%20CONSTITUTION-Citizens%27%20Guide.pdf.
Constitution of California http://leginfo.legislature.ca.gov/faces/codes_displayText.xhtml?lawCode=CONS&division=&title=&part=&chapter=&article=I.
Constitution of Florida http://dos.myflorida.com/media/693801/florida-constitution.pdf.
Constitution of Montana https://courts.mt.gov/portals/113/library/docs/72constit.pdf.
California Electronic Communications Privacy Act https://leginfo.legislature.ca.gov/faces/billNavClient.xhtml?bill_id=201520160SB178.

Supreme Court cases

Supreme Court, Olmstead v. United States, 277 U.S. 438 1928.
Supreme Court, Griswold v. Connecticut 381 U.S. 479 1965.
Supreme Court, Katz v. United States 389 U.S. 347 1967.
Supreme Court, Roe v. Wade 410 U.S. 113 1973.
Supreme Court, Riley v. California 13-132, 573 U.S. 2014.

United Nations

Universal Declaration on Human Rights www.un.org/en/universal-declaration-human-rights/.
International Covenant on Civil and Political Rights www.ohchr.org/EN/ProfessionalInterest/Pages/CCPR.aspx.

Council of Europe

Legal documents

Council of Europe, Original European Convention on Human Rights www.echr.coe.int/library/annexes/CEDH1950ENG.pdf.

Council of Europe, Current European Convention on Human Rights www.echr.coe.int/Documents/Convention_ENG.pdf

Council of Europe, Committee of Ministers, Resolution (73) 22 On the Protection of the privacy of individuals vis-à-vis electronic data banks in the private sector. (Adopted by the Committee of Ministers on 26 September 1973 at the 224th meeting of the Ministers' Deputies).

Council of Europe, Committee of Ministers, Resolution (74) 29 On the Protection of the privacy of individuals vis-à-vis electronic data banks in the public sector. (Adopted by the Committee of Ministers on 20 September 1974 at the 236th meeting of the Ministers' Deputies).

Council of Europe, Convention for the Protection of Individuals with regard to Automatic Processing of Personal Data Strasbourg, 28 January 1981. http://conventions.coe.int/Treaty/EN/Reports/HTML/108.htm.

Council of Europe, Explanatory Report to the Convention for the Protection of Individuals with regard to Automatic Processing of Personal Data, Strasbourg, 28.I.1981 https://rm.coe.int/16800ca434.

Council of Europe, Protocol to the Convention for the Protection of Human Rights and Fundamental Freedoms Paris, 20.III.1952.

Council of Europe, Protocol No. 8 to the Convention for the Protection of Human Rights and Fundamental Freedoms, Vienna, 1985.

Council of Europe, Protocol No. 9 to the Convention for the Protection of Human Rights and Fundamental Freedoms Rome, 6 November 1990. This Protocol has been repealed as from the date of entry into force of Protocol No. 11 (ETS No. 155) on 1 November 1998.

Council of Europe, Protocol No. 11 to the Convention for the Protection of Human Rights and Fundamental Freedoms, restructuring the control machinery established thereby. Strasbourg, 11 June 1994. Since its entry into force on 1 November 1998, this Protocol forms an integral part of the Convention (ETS No. 5).

ECtHR jurisprudence

ECtHR, Klass and others v. Germany, application no. 5029/71, 6 September 1978.

ECtHR, B. v. the United Kingdom, Application no. 9840/82, 8 July 1987.

ECtHR, P.G. and J.H. v. the United Kingdom, application no. 44787/98, 25 September 2001.

ECtHR, S. and Marper v. the United Kingdom, application no. 30562/04 30566/04, 04 December 2008.

ECtHR, Delfi v. Estonia (Grand Chamber), application no. 64569/09, 16 June 2015.

ECtHR, Roman Zakharov v. Russia, application no. 47143/06, 04 December 2015.

ECtHR, Szabó and Vissy v. Hungary, application no. 37138/14, 12 January 2016.

European Union

Legal texts

Directive 95/46/EC of the European Parliament and of the Council of 24 October 1995 on the protection of individuals with regard to the processing of personal data and on the free movement of such data.

Charter of Fundamental Rights of the European Union (2000/C 364/01). www.europarl.europa.eu/charter/pdf/text_en.pdf.

2000/520/EC: Commission Decision of 26 July 2000 pursuant to Directive 95/46/EC of the European Parliament and of the Council on the adequacy of the protection provided by the safe harbour privacy principles and related frequently asked questions issued by the US Department of Commerce

Directive 2002/58/EC of the European Parliament and of the Council of 12 July 2002 concerning the processing of personal data and the protection of privacy in the electronic communications sector (Directive on privacy and electronic communications, also known as the e-Privacy Directive), *Official Journal L 201, 31/07/2002 P. 0037–0047.*

Directive 2006/24/EC of the European Parliament and of the Council of 15 March 2006 on the retention of data generated or processed in connection with the provision of publicly available electronic communications services or of public communications networks and amending Directive 2002/58/EC

Council Framework Decision 2008/977/JHA of 27 November 2008 on the protection of personal data processed in the framework of police and judicial cooperation in criminal matters, OJ L 350, 30 December 2008.

Regulation (EU) 2016/679 of the European Parliament and of the Council of 27 April 2016 on the protection of natural persons with regard to the processing of personal data and on the free movement of such data, and repealing Directive 95/46/EC (General Data Protection Regulation).

Directive (EU) 2016/680 of the European Parliament and of the Council of 27 April 2016 on the protection of natural persons with regard to the processing of personal data by competent authorities for the purposes of the prevention, investigation, detection or prosecution of criminal offences or the execution of criminal penalties, and on the free movement of such data, and repealing Council Framework Decision 2008/977/JHA.

ECJ case law

European Court of Justice, Digital Rights Ireland and Seitlinger and Others, Joined Cases C-293/12 and C-594/12, 8 April 2014.

European Court of Justice, Google Spain SL, Google Inc. v. Agencia Española de Protección de Datos (AEPD), Mario Costeja González, case C131/12, 13 May 2014.

European Court of Justice, Maximillian Schrems v. Data Protection Commissioner, joined party: Digital Rights Ireland Ltd, Case C362/14, 6 October 2015.

European Court of Justice, Tele2 Sverige AB (C203/15) v. Post-och telestyrelsen, and Secretary of State for the Home Department (C698/15) v. Tom Watson, Peter Brice, Geoffrey Lewis, interveners: Open Rights Group, Privacy International, The Law Society of England and Wales, Joined Cases C203/15 and C698/15, 21 December 2016.

Organization for Economic Co-operation and Development

Guidelines on the Protection of Privacy and Transborder Flows of Personal Data from 1980 www.oecd.org/sti/ieconomy/oecdguidelinesontheprotectionofprivacyandtransborder-flowsofpersonaldata.htm.

Regional human rights documents

American Convention on Human Rights adopted in 1969 www.oas.org/dil/treaties_B-32_American_Convention_on_Human_Rights.htm.

African Charter on Human and Peoples' Rights from 1981 www.humanrights.se/wp-content/uploads/2012/01/African-Charter-on-Human-and-Peoples-Rights.pdf.

Association of Southeast Asian Nations (ASEAN) Human Rights Declaration from 2012 http?aichr.org/?dl_name=ASEAN-Human-Rights-Declaration.pdf.

European constitutions

Dutch constitution https://www.government.nl/documents/regulations/2012/10/18/the-constitution-of-the-kingdom-of-the-netherlands-2008.

German Constitution https://www.btg-bestellservice.de/pdf/80201000.pdf.

Italian Constitution https://www.senato.it/documenti/repository/istituzione/costituzione_inglese.pdf.

Spanish Constitution www.congreso.es/portal/page/portal/Congreso/Congreso/Hist_Normas/Norm/const_espa_texto_ingles_0.pdf.

Three Dimensions of Privacy

Beate Roessler

In what follows, I want to look at a systematic account of privacy in which I differentiate, explain, and discuss three dimensions of privacy. Conceptions of privacy based upon a concept of autonomy or individual freedom provide the most interesting and forward-looking possibilities for a conceptualization of the term. Three such dimensions of privacy should be distinguished. These dimensions – not realms, not spaces – of privacy serve to protect, facilitate, and effectuate individual liberties in a variety of respects. Freedom-oriented theories of privacy are to be found within the whole range of theories of privacy, from those that deal with the privacy of (intimate) actions to those concerned with informational privacy or the privacy of the household. It makes sense, therefore, to discuss these different aspects of freedom and privacy individually.

1. Decisional privacy

It is only in recent years that decisional privacy, or the privacy of actions, has been a specialist term in the literature. A decisive factor here was the ruling of the US Supreme Court in the *Roe v. Wade* case, where for the first time in US legal history women were granted a right to physical, sexual self-determination and to terminate a pregnancy, this being grounded upon an appeal to a *right to privacy*. As the explanation formulated by Justice Blackmun famously put it, 'this right to privacy (…) is broad enough to encompass a woman's decision whether or not to terminate her pregnancy' (*Roe v. Wade*, 410 US 113 [1973] 153). This verdict and the discussions that preceded and followed it were hugely influential upon the conceptualization of privacy not only in the United States. As a result, feminist theory has treated sexual freedom of action, the privacy of intimate and sexual acts, and the woman's right of sexual self-determination as central elements in the theory of privacy. Decisive significance is given to the *privacy of the body* (Gatens 2004). This includes the woman's newly won right to conceive of her body as private to the extent that she can decide for herself whether or not to bear a child, and thus enjoys rights of reproductive freedom.

The idea of physical privacy in the sense of the privacy of actions that concern the intimate sphere of women and men lies at the heart of decisional privacy. We should here mention two further central aspects of this form of privacy, therefore, both of which also concern the link between sexuality, the body, and identity, and are decisive for the societal coding and meaning of decisional privacy: these relate to the issues of sexual harassment and sexual orientation.

Both protection from sexual harassment and the respect for diverse sexual orientations form aspects of decisional privacy because and to the extent that it is the privacy of the body that is here vulnerable to infringement (the most comprehensive discussion in Cohen 2002).

The reason that privacy with respect to intimate, sexual decisions is considered so vital is that these decisions form the core of very general decisions that may have far-reaching consequences in terms of who one wants to be and how one wants to live: the core, in other words, of one's freedom to form one's own authentic identity.

When decisional privacy is placed within such a context and understood as serving to secure the possibility of a self-determined, authentic life, of individual projects, individual ways of life, and an individual practical identity, it becomes clear that it is called upon to secure autonomy not only in one's most intimate sphere, but in private acts and behaviour in public contexts too. It emerges, that is, that the protection of decisional privacy is necessary so that freedom in social space and with respect to other individuals in society can be enjoyed in such a way that modes of action, ways of life, and projects can be pursued without undesired interference from others. Restraint, inattention, reserve, and indifference – as forms of respect for this decisional privacy – are expected from others when it comes to the private aspects of the life a person leads in public. One must here of course distinguish very different aspects of decisional privacy according to their social context, but the argument underlying the claim to protection of such privacy remains structurally the same. If one understands a person's self-determination and autonomy to consist in her right to be the (part-)author of her own biography, among other things this must mean that within very different social contexts the person can demand that her decisions and actions should be respected (in the sense that they are 'none of your business') both by social convention and state law. The limits to such privacy are regulated by convention and of course subject to constant renegotiation, yet this sort of respect for a person's 'privacy' – in public contexts as well – is especially relevant for women. The spectrum of decisional privacy thus extends from reproductive rights to freedom of conduct in public space (see Roessler 2017 on decisional privacy and religious freedom).

2. Informational privacy

Discussion about informational privacy also goes back to the interpretation of the US Constitution, in this case beginning with an essay written by Justices Warren and Brandeis after what they felt was an invasion of privacy by intrusive paparazzi at the wedding of Warren's daughter (in 1890). It was here for the first time that the *right to be left alone* was described as a constitutional right to privacy in the sense that information about a person is worthy of protection even when it involves something that occurs in public. This is grounded in an appeal to the protection of individual freedom and thus known as the *right to be left alone* (Warren and Brandeis 1890).

Of course, there have been enormous technological advances that have *radically transformed* not only the possibilities for keeping people under surveillance but also our concepts of privacy as well as freedom and autonomy, and that threaten to continue to do so. Especially in the age of Big Data, the surveillance of people as consumers by companies and social platforms, as well as the surveillance by the state is a permanent threat to or violation of informational privacy (Mayer-Schoenberger and Cukier 2013; Morozov 2013).

The idea of informational privacy, however, also incorporates a further framework. At issue here is not only the question of not wanting to have one's smartphone or other devices tapped, one's data kept or sold, or to be kept under surveillance, but also the more general point that people like to keep the knowledge that others have of them under control and within limits they can expect. This brings to light the deep-seated connection between informational privacy and autonomy: people like to have control of their own self-presentation and use the information others have about them to regulate their relationships and thus the roles they play in their various social spaces. If everyone knew everything about everyone else, differentiated relations and self-presentations would no longer be possible, and nor would autonomy and the freedom to determine one's own life. As the German Federal Constitutional Court argued as early as 1983:

> A person who cannot tell with sufficient certainty what information concerning him in certain areas is known to his social environment, or who is unable to assess in some measure the knowledge of his communication partners, may be substantially restricted in his freedom to make plans or take decisions in a self-determined way. (BVerfGE 65, 1 [43])

This form of privacy is relevant, first, in friendships and in love relationships, and serves both as protection *of* relationships and as protection *within* relationships. In some theories of privacy, this even constitutes the very heart of privacy, 'relational privacy' guaranteeing the opportunities for withdrawal that are constitutive for an authentic life (Fried 1968; Rachels 1975).

Of central importance nowadays, however, is the fact that we now live in the *digital society*. Overwhelming evidence suggests that new information and communication technologies (ICTs) are radically transforming our social and political relations: Twitter, Facebook and other social media are changing the ways the public sphere functions, 'Big Data' is accumulating ever more personal information, and whistle-blowers like Edward Snowden use privacy-enhancing technologies (PETs) to reveal 'deep' secrets states strive to shield from public scrutiny. Today, information is increasingly gathered by employers, security agencies, Internet service providers, online businesses, social networking sites (SNS), and citizens themselves on a tremendous scale. Mobile and wearable computing and reconfigurable sensor networks are used ubiquitously and are converging in an Internet of Things. This new confluence of socio-technical practices will lead to amassing ever-larger quantities of data, large parts of which relate to traceable persons, a development which has been described as a revolution in the history of information technologies (Floridi 2015). On the background of the essential link between individual autonomy and informational privacy, the protection of informational privacy in the digital society thus becomes ever more important as well as challenging.

3. Local privacy

With local privacy, we have now come to the classic, traditional place of privacy, its most genuine locus: one's own home, which for many people still intuitively represents the heart of privacy. It is within our own four walls that we can do just what we want, unobserved and uncontrolled (see Roessler 2017).

Yet it should be made clear from the outset that this form of local privacy is not derived from a 'natural' separation of spheres but from the idea that one of the vital conditions for protecting individual freedoms in modern liberal democracies is to be able to withdraw to within one's own four walls. This has nothing to do with 'nature', but a great deal to do with the notion that (culturally or conventionally constructed) opportunities for withdrawal are a constitutive element of a person's freedom.

Two different aspects of privacy are of relevance here: solitude and 'being-for-oneself' on the one hand, and the protection of family communities or relationships on the other.

People seek the solitude and isolation provided by the protection of their private dwelling in order to avoid confrontation with others. This brings us back to the privacy of the body and the desire to shield one's own body from the sight of other people, thus securing a realm of completely personal intimacy that may even be bound up with feelings of shame. Another aspect of such privacy comes to light, however, in the work of literary models such as Virginia Woolf or George Orwell, for both of whom the privacy of the room – the privacy to write or think – is a precondition for the possibility of self-discovery and an authentic life (Orwell 1954; Woolf 1977).

Secondly, and in a classic sense of the word, local privacy offers protection for family relationships: the privacy of the household provides the opportunity for people to deal with one another in a different manner, and to take a break from roles in a way that is not possible when dealing with one another in public. As is known, however, this dimension or sphere of privacy is especially prone to generate potentials for conflict. From the outset, this has been a particularly important starting-point for feminist criticism, which has associated this realm and the understanding of privacy that accompanies it with the oppression of women on account not only of the gender-specific division of labour, but also domestic violence, and in general the idea that one's home constitutes a pre-political space. What this means, however, is that in discussions about this local form of privacy it is especially important to recall the meaning and function of privacy, which is to protect and facilitate freedom and autonomy, and more specifically to protect and facilitate *equal* freedoms and *equal* opportunities to lead a rewarding life, for women and men alike. Conflicts can here arise with traditional conceptions of privacy as the loving family haven, which have nothing to do with demands for justice or equal rights (as Honneth 2004 argues; contrast Rawls 1999). Yet it should be clear by now that traditional conceptions of the gender-specific division of labour have nothing to do with a protection of privacy that is oriented towards the protection of individual freedom, and that such a reconceptualization thus has repercussions for the justice of the family (Okin 1989 and 1992).

To conclude, let me point out just one of the future problems: In recent debates on privacy, in addition to the focus on privacy as a social value (Solove 2011; Nissenbaum 2010; Roessler and Mokrosinska 2015), the Snowden leaks have highlighted the constitutive relevance of *privacy for democracy* and the ways in which violations of privacy undermine democratic citizenship. This debate around the democratic value of privacy already started in the 1970s, focusing on privacy as a 'constitutive element of a democratic society' (Simitis 1987, 732). The debate gathered momentum especially after the new surveillance laws and massive intrusions of privacy following 9/11. Different authors have analysed the ways in which these intrusions directly influence the democratic political process and change political relations within a society (Hughes 2015; Goold 2010; Solove 2008). It has also been pointed out that the presumption of innocence, one of the cornerstones of the democratic rule of law, is in danger with mass surveillance turning every democratic subject into a potentially guilty object (Roessler and Mokrosinska 2013). However, the precise nature of the relation between privacy and democracy, despite first attempts at conceptualization, remains in need of systematic conceptual and normative reconstruction. Democracy relies upon citizens who value their autonomy, both in public and in private. Threats to privacy, therefore, are always also threats to democracy.

3. Privacy from an Ethical Perspective

Marijn Sax

3.1 Introduction

Philosophy is a rich discipline consisting of many branches that focus on a wide range of questions. Epistemology, for instance, is the study of knowledge and focuses on questions like 'What conditions need to be fulfilled for something to count as knowledge?' Aesthetics is another example, which is the study of (the nature of) art and beauty. Ethics is the branch of philosophy that, in its most general sense, is concerned with the question of what we *ought* to do. More specifically, ethicists often focus on *normative* questions concerning (1) the value of certain goods, practices, or norms, and (2) how – given those values – we should act and relate to each other. The ethics of *privacy*, then, focuses on questions such as 'What is the value of privacy?' and 'What privacy norms should be respected by individuals (including ourselves), society, and the state?'

The formulation 'the ethics of privacy' might suggest that there is *one* ethics[1] of privacy. Nothing could be further from the truth. Precisely because ethics is concerned with normative questions, there are no fixed answers to any of these questions. The answer to a normative question admits to different degrees or plausibility, relative to the arguments provided. As a result, different ethicists develop and argue for different theories of the value of privacy, which, in turn, often implies that they also identify different norms that *should* regulate privacy-related behaviours and policies.

This chapter will focus on the most important and influential ethical theories of privacy. First, some of the important conceptual distinctions that figure prominently in the ethical literature on privacy will be discussed. Here, the *definition* and *function* of privacy are discussed. Second, the classical text that laid the foundation for all contemporary analyses of both the legal and moral right to privacy is discussed. Third, the most important and influential perspectives on privacy's *value*, and what that implies for the *norms* that should regulate our behaviour and policies,

1 Some philosophers insist that 'ethics' and 'morality' are two distinct fields of philosophy, while others use the terms interchangeably. In order not to introduce unnecessary complications, I will not make a distinction between ethics and morality.

are discussed. In this section, perspectives that are critical of (particular aspects of) privacy are discussed as well. Fourth, some of the important *contemporary ethical challenges* to privacy and how they are addressed in the literature are discussed. This section will mostly focus on technological developments and what they imply for privacy.

3.2 Privacy's meaning and function

3.2.1 Privacy's meaning: access and control-access

There is persistent disagreement in the literature on privacy's proper meaning and definition. It is, however, possible to identify two terms that figure prominently in discussions on privacy's meaning and definition: the terms of 'access' and 'control'.

Some authors define privacy solely in terms of *access*. Reiman, for instance, writes that 'privacy is the condition in which others are deprived of access to you' (Reiman 1995, 30). According to access definitions such as Reiman's, privacy is a function of the extent to which people can access you either physically, or can access information about you. In case people cannot access you in any way, you enjoy complete privacy. Most of the time, however, other people can either gain some access, or have to go through some trouble to gain (some) access to you. So formally speaking, people rarely enjoy complete privacy. This is not necessarily a problem. Seen from the perspective of ethics, we should not focus on access per se, but on the question of *how* access is gained, and to *what* one is gaining access. For example, every time you enter a public place others have 'access' to you and information about you; they can see what you are wearing, where you are going, how tall you are, and so on. This is usually not considered to be problematic.

Other authors point out that access definitions can lead to counterintuitive conclusions. Fried (1984, 209-210) argues that 'to refer [...] to the privacy of a lonely man on a desert island would be to engage in irony'. According to Fried, the judgment that a person stranded on a desert island enjoys complete privacy because no one can access her is a meaningless and absurd conclusion because 'the person who enjoys privacy is able to grant or deny access to others' (Fried 1984, 210). For Fried, privacy is an inherently interpersonal phenomenon, something the access definition does not properly capture. In order to remedy this shortcoming, a range of authors, including Fried, include *control* in their definition.[2] The resulting control-access definitions state that privacy is about the control one has over access to oneself. With control incorporated into the definition, it immediately becomes clear why the desert island example is, from this perspective, absurd. With no other people being present, there is no meaningful control to be exercised in the first place. But control is often precisely what we care about. Access to ourselves or our information is not undesirable per se; what matters is that we have control over this access. Consider two persons who are involved in a romantic relationship. As a constitutive part of their relationship, they share secrets. Under access definitions, we would have to conclude that they lack privacy due to this practice of sharing intimate secrets. Control-access theorists emphasize that the fact these two romantically involved persons have *chosen* to grant each other access is an ethically important feature of the situation. From a control-access perspective, then, a breach of privacy occurs when a person is not able to exercise control over access, or when the attempt to exercise control over access are ineffective or ignored.

While many authors employ a definition that incorporates the notions of access and/or control, there are also those who deny the possibility of defining privacy at all. Most prominent in this regard is Solove (2008, 2015), who calls privacy 'a concept in disarray' (Solove 2008, 1; Solove 2015, 73). According to Solove, we should stop pursuing a single definition of privacy and, instead, start 'understanding it with Ludwig Wittgenstein's notion of "family resemblances". Wittgenstein suggests certain concepts might not have a single common characteristic; rather, they draw from a common pool' (Solove 2008, 9).[3] Solove argues that privacy serves many different functions and has many different meanings in different contexts. These different functions and meanings are all related to each other, without necessarily sharing one common feature. As a result, Solove suggests that the pursuit of a single definition of privacy is misguided and unhelpful, since it will never be able to capture privacy's diverse nature.

2 Fried defines privacy as 'control over knowledge about oneself' (Fried 1984, 210). Roessler writes that 'Something counts as private if one can oneself control access to this "something"' (Roessler 2005, 8). Westin defines privacy as 'the claim of individuals, groups, or institutions to determine for themselves when, how and to what extent information about them is communicated to others' (Westin 1967, 7).

3 See Wittgenstein 1953: §§ 66-67 for his discussion of family resemblances.

3.2.2 Privacy's function: three dimensions

Solove's doubts concerning the possibility of defining privacy are understandable. There are so many things – spaces, bodies, information, behaviour, and so on – we call 'private'. It is indeed difficult to theorize about privacy in a structural and consistent manner. To help structure our reasoning, we can refer to different dimensions of privacy.

Roessler (2005) defines three dimensions of privacy which can be understood as 'possibilities for exercising control over "access"' and which describe 'three ways of describing the normativity of privacy' (Roessler 2005, 9). The three dimensions Roessler identifies are the local dimension,[4] the informational dimension, and the decisional dimension.[5]

The local dimension of privacy refers to our control over access to physical spaces or areas. Control over access to our own physical body can also be included in this dimension. It is easy to come up with examples of norms of local privacy. We have locks on our front doors. We put locks on bathrooms and, sometimes, bedrooms. We are not supposed to touch just any part of the body of the person sitting next to us on the bus. In all these examples, it is *not* the case that access to homes, bathrooms, bedrooms, and bodies should be strictly forbidden in all cases. Rather, we value our ability to determine who gets access under what conditions.

The informational dimension of privacy refers to 'control over what other people can *know* about oneself' (Roessler 2005, 111). With the fast developments in the domain of Information and Communication Technologies (ICTs), information is often understood as data. Although discussions concerning the collection, storage, analysis, and dissemination of (personal) data can indeed be understood from the perspective of privacy's informational dimension, it should be emphasized that not *all* information is necessarily data. Notice that the earlier mentioned example of looking at people in the streets is also about gaining access to information about other persons' appearances and behaviour.

The decisional dimension of privacy refers to our control over 'symbolic access' (Roessler 2005, 79) to our personal decisional sphere. Norms of decisional privacy are supposed to grant 'protection from unwanted access in the sense of unwanted interference or of heteronomy in our decisions and actions' (Roessler 2005, 9). This dimension of privacy gained prominence

4 Cohen (2008, 181) calls this dimension 'spatial privacy'.

5 Allen (2011, 5) identifies three additional dimensions of privacy which she calls 'proprietary privacy', 'associational privacy', and 'intellectual privacy'.

after the *Roe v. Wade* (410 U.S. 113) decision by the US Supreme Court, which ruled the decision to terminate a pregnancy a *private* decision protected by the right to privacy. In line with *Roe v. Wade*, the decisional dimension of privacy can also be said to include – but not be reducible to – *bodily privacy*, i.e. control one has over deciding who can (or cannot) do what to one's body. In essence, decisional privacy can be understood to be about those decisions for which we find it valuable that persons themselves are able to decide on the basis of which values, goals, and reasons they come to a decision. Decisions pertaining to our own bodies are one important example of such decisions. Other examples are decisions pertaining to whom we spend our life with, what type of ideological and political beliefs we adopt, and what type of lifestyle we adopt.

So far, we have seen that privacy is most often defined in terms of access and control. Moreover, we have seen that different dimensions of privacy are helpful conceptual tools when theorizing privacy. In discussing these conceptual issues, privacy's value has already been gestured at, without discussing it explicitly. For example, identifying control as an important component of privacy's definition seems to presuppose that it is, normatively speaking, desirable that people have control over access. In a similar vein, the different dimensions of privacy not only help *describe* privacy more precisely, they also help us to better understand privacy's value.

In the next section, the seminal text by Warren and Brandeis (1890) that started the discussion on the value of privacy is introduced. After the next section, we proceed to a more detailed discussion of the different theories on the value of privacy.

3.3 Classic Texts and Authors

This section focuses on Warren and Brandeis's seminal article 'the right to privacy' from 1890. Their contribution is the first to explicitly theorize a right to privacy and has been highly influential. Many of the other texts that can be considered 'classics' – and which will be briefly mentioned here before they are discussed in the next section – can be understood in relation to Warren and Brandeis' important contribution.

The story of origin of the article is a curious one, but also one that contains an important message. As Prosser (1960) explains, the article by Warren and Brandeis is likely the outcome of Warren's annoyance at the way in which 'the press had begun to resort to excesses in the way of prying that have become more or less commonplace today'[6] (Prosser 1960, 383). Prosser continues by explaining that 'the matter came to a head when the newspapers had a field day on the occasion of the wedding of a daughter' where many of the Boston elite of the time were present (Prosser 1960, 383).

Warren and Brandeis observe that the combination of 'instantaneous photographs' and an increasingly aggressive press constituted significant societal and technological changes as a result of which 'the sacred precincts of private and domestic life' came under such pressure that an intervention was needed (Warren and Brandeis 1890, 195). Suddenly, reporters with relatively small and easy-to-handle photo cameras could quickly capture images of everything they saw. Warren and Brandeis felt that this technological development, which allowed for a new level and type of privacy invasions, was serious enough to ask the question whether the legal protections of the time still offered enough protections to the individual. Their answer of this question was in the negative.

Law is, in their view, a system that needs 'from time and time to define anew the exact nature and extent of such protection [of the individual in person and property]' (Warren and Brandeis 1890, 194). In order to meet this new challenge of 'instantaneous photographs' and an aggressive press, they proposed it was high time to explicitly recognize a distinct right to privacy for individuals. It is interesting to emphasize at this point that their observations from 1890 feel surprising topical. More than 125 years later, it is still very much the case that technological developments challenge existing social norms, raising the question whether existing (legal) protections still suffice to protect individuals against (alleged) privacy intrusions.

How should this right to privacy as introduced by Warren and Brandeis be understood? They famously summarized this right to privacy as the right 'to be let alone' (Warren and Brandeis 1890, 195). Although this often-quoted formulation has almost become a slogan, it does not say much in itself. If we look behind the slogan, however, we encounter many observations concerning the role and value of privacy that are still relevant nowadays. They emphasize that:

> the intensity and complexity of life, attendant upon advancing civilization, have rendered necessary some retreat from the world, and man, under the refining influence of culture, has become more sensitive to publicity, so that solitude and privacy have become more essential to the individual; but modern enterprises and invention have, through invasions upon

6 Remember that Prosser wrote this in 1960.

> his privacy, subjected him to mental pain and distress, far greater than could be inflicted by mere bodily injury (Warren and Brandeis 1890, 196).

It is, first of all, interesting to focus on their formulation 'subjected him to mental pain and distress, far greater than could be inflicted by bodily injury'. This observation is in line with their repeated emphasis on the importance of not only the protection of the body and the property of an individual, but also its 'thoughts, sentiments, and emotions' (Warren and Brandeis 1890, 198). If we couple this observation to the emphasis on the necessity of a 'retreat from the world', it becomes clear that Warren and Brandeis think that individuals need – and also have the right to – a private sphere where they can think, feel, and be the way they want to, without having to worry about intrusions into this private sphere. They also stress that so long as the individual has not made any thought, emotions, or sentiments public, it is the individual who is 'entitled to decide whether that which is his shall be given to the public' (Warren and Brandeis 1890, 199). (Notice that this could be construed as a control-access definition of privacy).

At the foundation of their right to privacy, then, is the idea of the 'inviolate personality' (Warren and Brandeis 1890, 205), or, put differently, 'the more general right to the immunity of the person, – the right to one's personality' (Warren and Brandeis 1890, 207). It is ultimately up to the individual to decide how she wants to be, think, and act. For this to be possible, and in order to live a good life, the individual needs a private sphere free from intrusions and to which she herself can grant or refuse access. Without explicitly mentioning it, Warren and Brandeis essentially offer the contours of a theory of privacy that bases the value of privacy on its ability to enable the autonomy of the individual. As we will see in the next section, this intimate connection between privacy and personal autonomy has been further developed by a number of different privacy scholars.

It was Warren and Brandeis' article that started the still ongoing discussions on the right to, and the value of, privacy. Remarkably, many of the observations and arguments in their article are still as relevant today as they were at their time of publication in 1890. A further development of their ideas concerning the value of privacy to individuals can be found in a number of classical texts in the liberal traditions, such as Benn (1984), Fried (1984), Reiman (1995), and Roessler (2005). There is, however, also a range of classical texts that explores critiques of such theories of individual privacy, which have their origin in Warren and Brandeis. Here we have: the feminist critique with classical texts such as Allen (1988), MacKinnon (1989), and DeCew (1997); the reductionist critique as famously defended by Thomson

(1975); the communitarian critique of Etzioni (1999); classical texts on the social value of privacy such as Rachels (1975) and Regan (1995). Lastly, there is the 'modern classical text' of Nissenbaum (2010), who develops a theory of privacy that is very sensitive to changing technological circumstances. Although Nissenbaum's theory is in substance very different to Warren and Brandeis' theory, they share the fact that they are explicit answers to changing technologies. In the next section, all the classical texts that came after Warren and Brandeis will be discussed in greater detail.

3.4 Traditional debates and dominant schools

This section will provide an overview of the most important theories on privacy's value. By identifying different 'perspectives' on privacy's value, different authors who have developed theories that are in some important respect similar can be grouped together. First, theories that are predominantly liberal in nature and emphasize the value of privacy for individuals are discussed. Second, three critical perspectives that emerged in response to theories that emphasize privacy's value for individuals are discussed. Third, the literature on the social value of privacy – and which can be understood as a response to the various critiques – is discussed. The different perspectives discussed here are not necessarily mutually exclusive.

3.4.1 Privacy's value for individuals

A wide range of authors has focused on the value of privacy for individuals. Many of these authors understand privacy as being constitutive of, most importantly, personal liberty and autonomy (Benn 1984; Fried 1984; Schoeman 1984b; Allen 1988; Cohen 1992; Reiman 1995; Roessler 2005; Bennett and Raab 2006).

Fried (1984, 210) writes that 'privacy in its dimension of control over information is an aspect of personal liberty'. He provides an important illustration of this more general claim, by arguing that privacy is a necessary precondition for the possibility of friendship and love. The sharing of (very) private information that (nearly) no one else knows about is what makes friendships and intimate relationships special. However, for you to be able to share (very) intimate information it must, first, be the case that no one has access to the information in question, and, second, you yourself must be the one who can decide with whom to share it. This is exactly what privacy achieves – it makes it possible to give others the 'gift' of 'the intimacy of shared private information' (Fried 1984, 211).[7] The existence of privacy also provides 'means for modulating those degrees of friendship which fall short of love' (Fried 1984, 211). In short, because friendship and love are valuable aspects of our lives, privacy is valuable as well.

Benn (1984) emphasizes how respect for privacy expresses respect for persons and their personhood. Privacy protects you against unwanted observation and scrutiny. The respect of others for your attempts to enforce your right to privacy so as to ensure that you are not observed and scrutinized, is an expression of respect for your personhood. Why? Because, as Benn (1984, 242) explains, '[a] man's view of what he does may be radically altered by having to see it, as it were, through another man's eyes'. When you are observed – or suspect that you may be observed – in a place you deem private, you are forced to adopt an additional perspective (besides your own) on yourself. For Benn, this constitutes a lack of respect for the person in question, because for us to be able to act, think, and decide as we want, without having to always see ourselves through another person's eyes, is essential to our personhood. We need privacy precisely to afford us spaces free of observation and scrutiny in order to achieve various liberal personal ideals: the ideal of personal relations, the ideal of 'the politically free man', and the ideal of 'the morally autonomous man' (Benn 1984, 234). These three ideals will be used to structure the remainder of this section on privacy's value to individuals.

Where Fried focuses on the exclusivity of information (achieved by privacy) as a constitutive element of *personal relations*, Benn focuses primarily on the fact that '[p]ersonal relations are exploratory and creative' (Benn 1984, 236). He explains that all of our personal relations are largely regulated by role-expectancies. However, persons will also, first, 'fulfill them in different ways' (Benn 1984, 235), and, second, relations are not *completely* determined by role-expectancies. Privacy affords persons with a sphere in which to explore different ways of fulfilling roles, or to creatively shape relationships to the extent that they are not defined by role-expectancy. Without privacy, people would be less free to do so. Moreover, we need privacy to have a reasonable measure of control over how we present ourselves to others. Privacy, first of all, allows us to *separate* different roles to begin with (Cohen 2002). It is, next, important that we can have expectations of what others do and do not know about us, so we can determine how to present ourselves. The possibility to do so is important to us, because we need to play different roles in society (e.g. the roles of friend, co-worker, lover, stranger on the

7 See Inness 1992 and Cohen 2002 for other influential accounts of privacy and intimacy.

street, family member, and so on), and we would like to have meaningful control over *how* we choose to fulfil those roles (Roessler and Mokrosinska 2013; Marmor 2015).

The ideal of *political freedom* is explained by Benn by referring to 'the liberal ideal' that persons should enjoy 'an area of action in which he is not responsible to the state for what he does so long as he respects certain minimal rights of others' (Benn 1984, 240).[8] Privacy thus functions as a sort of 'shield' (Cohen 1992, 102), protecting a space where persons are not accountable to anyone but themselves. This so-called public/private distinction is central to liberalism, since it rules our private space (which can be defined somewhat differently by different authors) as off limits to the state.

Reiman (1995) provides a further elaboration of the relation between privacy and political freedom. Not respecting norms of privacy can lead to an 'extrinsic loss of freedom', by which Reiman means 'all those ways in which lack of privacy makes people vulnerable to having their behavior controlled by others' (Reiman 1995, 35). Much like Benn, Reiman argues that (the possibility of) observation and scrutiny of our behaviour can affect our actual behaviour. '[E]ven if they have reason to believe that their actions *may* be known to others and that those others *may* penalize them, this is likely to have a chilling effect on them that will constrain the range of their freedom to act' (Reiman 1995, 35). As a result of a lack of privacy, people may start to behave in ways they believe is in conformity with 'the lowest-common denominator of conventionality' (Reiman 1995, 41). If the lack of privacy is persistent enough, there is the risk of people becoming *different* – less willing and able to deviate from conventional norms, less willing and able to experiment, less willing and able to engage in political criticism. In a similar vein, Richards (2015, 95) argues that we need 'intellectual privacy' as 'protection from surveillance or unwanted interferences by others when we are engaged in the process of generating ideas and forming beliefs'. A severe lack of privacy would be inimical to political freedom as understood by liberals, since this freedom is premised on 'the autonomous individual, the one who acts on principles which she has accepted after critical review rather than simply absorbing them unquestioned from outside' (Reiman 1995, 42).

8 See Mill (1991 [1859]) for a classic elaboration of the liberal ideal, including the harm principle implicitly referred to here by Benn ('That the only purpose for which power can be rightfully exercised over any member of a civilized community, against his will, is to prevent harm to others' [Mill 1991 [1859], 14]).

The last remark provides a good transition to the third liberal ideal identified by Benn: *personal autonomy*. Roessler (2005) develops a systematic normative account of privacy to argue that privacy is constitutive of personal autonomy. The ideal of personal autonomy can be understood as providing a more concrete and more substantial interpretation of what normatively desirable freedom looks like. While we can ascribe freedom in a general sense to a person who is not obstructed in her acting and who can choose from a significant range of options, 'not every free action is an autonomous one' (Roessler 2005, 49). Personal autonomy is about one's practical relation to oneself – it is about 'the possibility of holding an attitude to oneself in general' (Roessler 2005, 51) by virtue of which one can critically reflect on one's reasons, goals, values, and projects. It thus becomes possible to ask 'oneself the "practical question" [...] how I want to live, what sort of person I want to be, and how I should strive for my own good in my own way' (Roessler 2005, 51). Freedom *as autonomy* should thus be understood as self-determination, which consists in developing – and at the same time is enabled by – the above-mentioned practical relation to oneself. Importantly, Roessler claims that living an autonomous life is more rewarding and desirable than living a non-autonomous life, 'for without this form of self-determination we would fail precisely to achieve our *own* good as our own' (Roessler 2005, 50).

As was described in the section on privacy's function, Roessler identifies three dimensions of privacy (local, informational, and decisional). The different dimensions help identify a range of different privacy norms that are supposed to protect and enable personal autonomy.

Norms of local privacy carve out spaces where one can go unobserved – or invite only those persons one wants present – in order to, among other things, engage in intimate relationships, experiment with new ways of doing, thinking, or living, and take a rest from the social demands of presenting oneself in certain ways in public.[9]

Norms of informational privacy allow one to control who knows what about oneself. It is important to have this kind of control, because the knowledge other people have about us shapes the ways in which we can present ourselves to others and act around others. Informational privacy thus affords space for autonomous freedom in choosing how to present ourselves to others and how to give shape to relationships.

9 See Goffman (1959) for a seminal analysis of self-presentation in social life. Similarly, Marmor (2015, 3-4) argues that the 'right to privacy is grounded in people's interest in having a reasonable measure of control over the ways in which they can present themselves (and what is theirs) to others'.

Norms of decisional privacy allow one to control access to one's decisional sphere. In practice, this means that for 'certain forms of behavior in public, as well as questions of lifestyles and more fundamental decisions and actions' we 'may with good reason tell other people that such-and-such a matter is none of their business' (Roessler 2005, 79). The relevance of decisional privacy for personal autonomy should be clear: it carves out a sphere where one can determine for oneself how to shape one's life and actions.

Thus far, theories that ground privacy's value in personal freedom and autonomy have been discussed. Moore (2010), however, adopts a different approach and starts from an account of human nature to explain privacy's value. In line with Aristotelean teleology, Moore explains that human nature is such that it allows humans to flourish in a particular human way. In order to flourish, humans need to develop those capacities and faculties that are unique to human nature, such as our rational faculties which allow us to, among other things, live an autonomous life. We also need favourable external conditions to flourish. In essence, Moore's argument is that rights to and norms of privacy are necessary to make human flourishing possible. For example, we need private places to relax, experiment, and think. Without the availability of private places, we would not be able to do these typically human activities that are conducive to human flourishing. Moore thus concludes that 'privacy is valuable *for beings like us*' (Moore 2010, 56, emphasis added).

A recurring idea – sometimes explicit, sometimes implicit – in many theories of privacy's value for individuals is that privacy is a special kind of value. Privacy is seen as an individual (and sometimes collective) right that expresses *respect* for persons and their personhood. One could even claim that respect for privacy is seen as acknowledging the *dignity* of persons. As early as 1890, Warren and Brandeis (1890, 214) referred to 'the dignity [...] of the individual' in discussing the right to privacy. Later, Bloustein (1964) criticized Prosser (1960), who suggested privacy should not be considered an independent value, but 'rather a composite of the interests in reputation, emotional tranquility and intangible property' (Bloustein 1964, 962). Bloustein's reply to Prosser is to suggest that Prosser's account of privacy's value is too superficial, because 'he neglects the real nature of the complaint; namely that the intrusion is demeaning to individuality, is an affront to personal dignity' (Bloustein 1964, 973). As we saw earlier, multiple authors in the liberal tradition emphasize the intrinsic connection between privacy and personal autonomy (Benn 1984; Reiman 1995; Roessler 2005). If personal autonomy is not possible without privacy, and if personal autonomy is the ground for the dignity of persons, then it follows that the right to privacy is a highly important right because it respects human dignity.

Because of privacy's supposed special importance, there is a reluctance to discuss privacy as 'just one value and right amongst many'. Seen from this perspective, privacy cannot simply be 'balanced' with other values and rights.[10] The image of 'balancing' suggests that one is balancing two things that are, in principle at least, equally important. Precisely this assumption is in many cases misleading, because some values and rights are more fundamental than others. Consider the following example. A person might claim that privacy can be violated as long as the violation is instrumental to generating enough monetary profits to 'tip the scales' in the right direction. Authors in the dignity tradition would see this judgment as fundamentally misguided. Why? Because the right to privacy protects and respects human dignity, whereas an increase of monetary profits is not (necessarily) constitutive of human dignity. To suggest that both values and rights are of the same kind and can therefore be 'traded' for each other given the right exchange rate, is to neglect the fact that some values – such as privacy – are (sometimes) categorically more important. Another example is the often-heard proposal to 'balance' privacy and security. Again, some would suggest that privacy is of special importance and that a simple balancing of privacy and security fails to acknowledge this.

3.4.2 Three critiques of privacy and its value

The writings on privacy's value to individuals have resulted in a number of different critiques, three of which will be discussed. First, the communitarian critique of privacy which questions the special importance that is ascribed to privacy. Second, the reductionist critique which suggests that the concept of privacy is redundant because it can be reduced to more basic values and rights. Third, the feminist critique which points out that privacy sometimes benefits particular groups more than others due to prevailing power structures.

3.4.2.1 The communitarian critique of privacy

In the introduction to his book with the telling title *The Limits of Privacy*, Etzioni (1999) announces that '[t]his is a book largely about the other side of the privacy equation' (Etzioni 1999, 2). So what are both sides of the equation that Etzioni is referring to here? One side of the equation – the one Etzioni criticizes – is the side that stresses privacy's unique value and, as a result,

10 For critical analyses of balancing as a method to answering normative questions, see Waldron 2003 and Van der Sloot 2017.

emphasizes the need for especially strong protections of the individual's moral and legal right to privacy (roughly the dignity position discussed above). The 'other side of the equation' he defends is 'about our investment in the common good, about our profound sense of social virtue, and most specifically about our concern for public safety and public health' (Etzioni 1999, 2). His book is an exploration into the question when 'serving the common good entails violating privacy' and when such violations of privacy are legitimate (Etzioni 1999, 2). It is important to emphasize that Etzioni does not wish to claim that privacy is *un*important. What he claims is that due to the strong focus on privacy as an (almost) inviolable individual right, we tend to forget about other values and rights that warrant our attention and protection as well. His critique is in line with the more general communitarian critique[11] on liberalism's strong focus on the individual and the individual rights that should protect her from unwanted interferences by society and the state. Communitarians seek to reclaim the value of living in a community that is not made up of atomistic liberal individuals pursuing maximum individual freedom. They emphasize the essential and valuable role our social surroundings play in forming and enabling our identity formation, a fact largely ignored by the liberal tradition. The reaffirmation of the function and value of community also comes with a stronger focus on 'the common good' and 'a sense of social virtue', as Etzioni puts it.

Etzioni's discussion of Megan's Laws[12] provides a good illustration. As he himself observes, arguments *against* such laws are often grounded in privacy considerations that are presented as knock-down arguments: 'They have paid their dues to society when they complete their jail sentence; [...] and they have the same inalienable rights to privacy and autonomy as the rest of us' (Etzioni 1999, 43-44). Given Etzioni's communitarian position, he does not take the individual's right to privacy as constituting reason enough to refute Megan's Laws. He discusses a great deal of empirical literature in an attempt to establish to what extent the violation of the sex offenders' privacy yields higher levels of security. If the increase of security *to the community at large* is substantial enough, he argues, it can justify violations of the privacy of individuals. In short: privacy is just one of the many values and

11 For important communitarian critiques of liberalism, see Sandel 1981; MacIntyre 1981; Walzer 1983; and Taylor 1989.

12 Megan's Laws refers to legislation that requires people who have been found guilty of sexual offences to register with local law enforcement, even if they have served their sentence. The resulting sex offender's registers are open to the public and some communities require people in the register to proactively inform others in their neighbourhood of their history as a sex offender (Etzioni 1999, 43-44).

rights that should be considered and privacy receives no special treatment vis-à-vis other values and rights.

3.4.2.2 The reductionist critique of privacy

As we have seen earlier, Solove (2008, 2015) suggests that there can be no single, unified definition of privacy because privacy protects too many different, diverse interests. Thomson (1975) defends the even stronger claim that the very idea of a *right* – both moral and legal – to privacy is conceptually superfluous. She offers a reductionist analysis of the right to privacy, arguing that it is made up of a cluster of other rights such as the right to property and the right over the person (which is similar to the right to bodily integrity and self-determination). Moreover, she argues that 'every right in the right to privacy cluster is also in some other right cluster' (Thomson 1975, 312). Every time we invoke our right to privacy, we can point to a different right that explains why the supposed right to privacy is important. As a result, 'the right to privacy is "derivative" in this sense: it is possible to explain in the case of each right in the cluster how come we have it without ever once mentioning the right to privacy" (Thomson 1975, 313). According to Thomson, introducing the term (and the right to) 'privacy' in discussions does not add any explanatory value, for everything we want to address when we discuss privacy can be addressed in terms of existing different rights.

Although Thomson's argument has been influential, it can be criticized.[13] In order to uphold her claim that *any* privacy right can be reduced to a different right, Thomson has to refer to an open-ended list which contains a large number of rights, some of which seem rather ad hoc and trivial. For example, we have a right not to be looked at, and a right not to be listened to. Thomson calls them 'un-grand' rights which, contrary to 'grand ones' like the right to life and the right to liberty, are not 'those that come to mind' when we speak of rights (Thomson 1975, 305). She maintains that they are relevant rights nonetheless and, moreover, that they help explain why privacy is a derivative right. However, by referring to an open-ended list of 'un-grand' rights, Thomson has introduced such a broad notion of rights that she can answer to any possible counterexample by introducing yet another highly specific, un-grand right. Do we need a right to privacy to explain that X is problematic? No, Thomson could reply, because we have a right not to be subjected to X.

Scanlon (1975) offers a direct reply to Thomson. He agrees that those violations we understand as privacy violations do not derive 'from any

13 See Scanlon 1975, Rachels 1975, and Reiman 1976 for critiques on Thomson's argument.

single overarching right to privacy' (Scanlon 1975, 315). Scanlon, however, argues that there is something *else* that unifies all the different privacy violations and the corresponding different rights: 'these rights have a common foundation in the special interests that we have in being able to be free from certain kinds of intrusions" (Scanlon 1975, 315). Put shortly, Scanlon argues that Thomson got the primary unit of analysis wrong. A satisfactory theory of privacy should start from the interests we have in privacy. These interests yield norms, conventions, and (legal and non-legal) rights supposed to protect them. Rights *can* indeed protect our interests, but there is not an intrinsic direct connection between the two. Sometimes our interests are (partly) harmed, without a right being violated. Scanlon thus concludes that Thomson's rights-based analysis cannot (always) adequately explain the interest we have in privacy. The concept of 'privacy' is thus still a valuable one to have in our vocabulary and does not need to be scrapped, as Thomson suggests. Reiman (1976) agrees with this conclusion when he writes that Thomson's argument is based on a 'large non sequitur': 'even if privacy rights were a grab-bag of property and personal rights, it might still be revealing, as well as helpful, in the resolution of difficult moral conflicts to determine whether there is anything unique that this grab-bag protects that makes it worthy of distinction from the full field of property and personal rights' (Reiman 1976, 28).

3.4.2.3 The feminist critique of privacy

A range of authors has formulated different feminist critiques of privacy (Allen 1988; MacKinnon 1989; Pateman 1989; Gavison 1992; DeCew 1997, 2015). Contrary to liberal scholars who praise privacy for its ability to provide us with a private sphere where the state cannot interfere with us, feminists argue that given unequal power relations shaped along gender lines, privacy can *dis*empower women, rather than empower them. In essence, all feminist critiques are founded on a similar observation: although privacy can indeed be considered valuable for many reasons, it can at the very same time shield off instances of violence, degradation, rape, and abuse, that take place in the private sphere from much-needed public scrutiny. The public/private distinction so essential to liberalism is therefore deeply problematic, for it perpetuates many gendered inequalities and injustices by allowing them to go unnoticed entirely or by labelling them as 'private issues' the state has no business in addressing. An additional problem addressed by feminist scholars is the 'naturalization' of the private and public sphere. Pateman (1989, 118-136) points out that liberal scholars often presume that there is a natural private sphere where family life takes place and a natural public sphere where social and political life takes place. Feminist critics emphasize that what counts as private and what counts as public is determined by *conventional* norms that can – and sometimes should – change. Insistence on the conventional nature of the public/private distinction affords feminist critics an important basis for critique, since conventional norms and boundaries can be (re)negotiated.

MacKinnon (1989) is most radical in her critique of the public/private distinction. After observing that the existence of a private sphere does not benefit both genders equally (to put it mildly), she concludes: 'This is why feminism has had to explode the private' (MacKinnon 1989, 191). She argues that the private has to be exploded, because 'women have no privacy to lose' (MacKinnon 1989, 191). If, as the liberal tradition teaches us, privacy is important because it allows for autonomous freedom, then privacy thus understood does not exist as long as women are subject to unequal power relations *within* the very private sphere that allows for the suppression of their autonomy. Accordingly, the private sphere should be exploded to allow for interventions aimed at gender equality.

Other feminist scholars have suggested that MacKinnon's dismissal of the private sphere in its entirety is implausible, because the dismissal is too rigorous. DeCew (1997, 86) agrees with MacKinnon that the public/private distinction *can*, and in fact often does, work to the detriment of women. However, proposing to completely collapse the private into the public is unattractive, since it implicitly assumes that privacy can *never* be attractive to women, not even under (more) ideal conditions. She refers to Gavison, who writes that 'it is rare to find feminists who argue consistently that everything should be regulated by the state, or that the family and all other forms of intimate relationships should disappear in favor of public communities [...] When pushed, feminists explicitly deny this ideal' (Gavison 1992, 28). So instead of arguing for exploding the private, DeCew, as well as Allen (1988), Gavison (1992), and Pateman (1989), argue for a more nuanced approach. Harmful practices that are allowed to go unnoticed because they take place in the private sphere should be remedied, for instance by allowing for more – but not complete – public scrutiny. But at the same time it should be observed that women *can* benefit from the existence of a private sphere, because women – just like men – have an interest in autonomous freedom enabled by a just private sphere. In sum, the very existence of a private sphere is not the problem, but unjust power structures that give rise to problematic gender norms structuring the private sphere are the problem.

3.4.3 The social value of privacy

The feminist critique of privacy has been an important source of inspiration for a branch of ethical literature that focuses on the social value of privacy. While few would disagree that privacy has value for individuals, authors in this tradition call attention to the fact that privacy is also valuable to social relations and society at large (Rachels 1975; Regan 1995; Solove 2008; Steeves 2009; Roessler and Mokrosinska 2013; Marmor 2015).

Rachels (1975) focuses on privacy's importance for social relations. The attentive reader might notice that in the previous section, different authors within the liberal tradition also emphasized privacy's importance for relations. These authors (Fried; Benn), however, focused on relations from the perspective of the individual. They argued that it is important to personal identity and personal autonomy to be able to shape relationships. Rachels' focus is somewhat different. He argues that privacy regulates all of our normal and ordinary social relationships: 'privacy is necessary if we are to maintain the variety of social relationships with other people that we want to have, and that is why it is important to us' (Rachels 1975, 326). To see why, consider your doctor, your close friend, and your co-workers. We behave differently with all of them and this is to a large extent regulated by the types of information we exchange with each of them. Social norms prescribe that it is certainly okay for you to reveal information about the private parts of your body to your doctor. It is, quite literally, her business to know these private facts (Rachels 1975, 331). At the same time, it would be weird – in the typical office space at least – to reveal the same private facts to your co-workers. Reversely, there are many things you could discuss with your co-workers that would be weird to share with your doctor. Rachels' argument is that privacy norms regulate the different types of appropriate information disclosures. Privacy thus allows us to maintain different relations with different persons and that, in turn, is what allows society to function in a way that is valuable to us all.

Roessler and Mokrosinska (2013) provide a further refinement of Rachels' argument. They focus specifically on different types of social interactions associated with different types of relationships (private relationships with friends, family, and intimates; professional relationships; and interactions between strangers in public). For each type of relationship, they show how norms of informational privacy regulate information exchanges and how this is a precondition for these different types of relationships to be able to exist alongside each other. Norms of informational privacy ensure that you can generally expect that people do not know something about you, unless you have chosen to disclose the information (or know that someone else has done so). Besides enabling different types of relationships, the resulting control you have over disclosures of information also enables you to autonomously decide how you want to give shape to the relationships you enter in to. So, 'by facilitating social interaction, norms of privacy contribute to creating social conditions that are required for the successful exercise of individual autonomy' (Roessler and Mokrosinksa 2013, 785).

Privacy's value to society can also be understood from the perspective of democracy. A range of authors has suggested that privacy is a necessary precondition for the proper functioning of democracy (Gavison 1980; Simitis 1987; Regan 1995; Reiman 1995; Lever 2006; Goold 2009, 2010; Hughes 2015; Lever 2015; Richards 2015). Gavison (1980), for instance, argues that '[p]rivacy is also essential to democratic governance because it fosters and encourages the moral autonomy of the citizen, a central requirement for democracy' (Gavison 1980, 455). A similar argument can be found in Reiman (1995), as was discussed earlier.

An important implication of the literature on the social value of privacy is that 'it is not always reasonable to assume a conflict between individual privacy on the one hand and society on the other' (Roessler and Mokrosinska 2013, 785). Privacy does not just place annoying *restrictions* on society's room for action, it is just as much an enabler of many valuable social practices.

3.5 New challenges and topical discussions

This section discusses a range of new challenges to privacy, most of which arise due to new technological developments that challenge existing norms, laws, and customs. To provide structure to the discussion of the wide range of technologies and challenges, the section is divided in three sections: challenges to local privacy, challenges to informational privacy, and challenges to decisional privacy.

3.5.1 Challenges to local privacy

One of the prominent contemporary challenges to local privacy is the rise of 'ambient technology' and 'smart devices' that (try to) find their way into our homes. Traditionally, access to the home is severely regulated and restricted by locks and social norms alike. Ambient technology and smart devices seem to be hardly bothered by tradition, as they gain access to a sphere that used to be impenetrable (e.g. Brey 2005; De Vries 2010; Van Dijk 2010; Roux and Falgoust 2013; Etzioni and Etzioni 2016).

By now, a smart thermostat is no longer a niche product that only a few enthusiasts have installed – it is starting to become the default. This seems to be only the beginning. At the latest (2018) Consumer Electronics Show, 'smart technology' took centre stage, with nearly every company present showing some kind of smart solution for the house. As Wired commentator David Pierce writes: 'Everything is a gadget now! A smart washing machine doesn't seem ridiculous anymore [...] All of it more powerful than last year's model, more connected, more deeply integrated into your everyday life' (Pierce 2018). The 'digital assistants' are another example. The 'big four' all try to push their digital assistant to become the standard: Apple with its Siri, Amazon with its Alexa, Facebook with its M, and Google with its Google Assistant. These digital assistants all aim to be present in your house and to become your go-to device for questions, suggestions, and for controlling other 'smart devices' in your home (Ezrachi and Stucke 2016; Stucke and Ezrachi 2017).

The very fact that an increasing number of devices – often connected to the Internet and thus to a fundamentally open, public sphere – occupy our private spaces, is not necessarily a reason for worries. What is, according to many, worrisome, is the fact these devices also challenge our informational and decisional privacy within our homes. Many of these devices are explicitly designed to collect, store, and analyse large amounts of data. Moreover, these devices often come with 'smart' functions aimed at making suggestions, or even at making choices for us. These worries pertaining to informational and decisional privacy will be discussed in the next sections.

In terms of local privacy, we should ask to what extent these devices threaten to destroy something of value in our private spaces. Recall that norms of local privacy are important because they allow persons to, among other things, take a break from performing different social roles; to experiment with different ideas, thoughts, and practices; to perform acts that would not be possible – or become less valuable – with spectators present. The presence of devices that constantly gather, store, and analyse data, often in order to make suggestions or make decisions for you, could potentially disturb these practices. As the presence of such devices in our private spaces grows, they might end up making us feel less free to experiment, and to engage in activities that require no uninvited spectators to be present. We might, moreover, end up feeling like we have to always incorporate the presence and abilities – that is to say, the perspective – of these devices into our view of and deliberations about ourselves; even within our homes, the one place where this should not be the case.

3.5.2 Challenges to informational privacy

Many of the contemporary challenges to informational privacy have their roots in technological developments as well. We have seen the rise of futuristically sounding phenomena such as big data, the Internet of things, social media, the quantified self, and smart cities. Although these phenomena are different in many respects, they share at least one thing: they all perpetuate the rapid 'datafication' of our life world (Van Dijck 2014). It has led many commentators to write things like 'the amount of data is growing fast, outstripping not just our machines but our imaginations' (Mayer-Schönberger and Cukier 2013, 8). Simply put, enormous amounts of data are collected, and those data can be put to work in increasingly smart ways.

To understand why this development raises privacy concerns, we should make at least three observations. First, it should be observed that not just more data are collected; it is equally important to observe that data about an increasing amount of different domains of life and activities are collected. We all know by now that our smartphones generate a great variety of data throughout the day. Some of us wear wearables that measure, for instance, heart rate and number of steps. The cities we live in are becoming smarter as well, datafying mundane activities. Consider Wi-Fi tracking of customers in stores (Gibbs 2016) and billboards that can film and generate data on people passing by (Ember 2016). In a world that is rapidly being filled with all kinds of sensors, one could ask whether persons can still keep track of – let alone exercise meaningful control over – all the different types of data that are generated.

Second, new techniques allow for the exploitation of all these data in increasingly sophisticated ways. Consider big data's promise to extract qualitatively *new* and *unexpected* insights from existing data; big data promises to let us see things we previously could not see (Sax 2016). Even if you have never disclosed a piece of information, big data analytics may allow others to still *infer* the information from existing data.

Third, the previous two developments are further exacerbated by the inherent properties (or: affordances) of bits, the 'material' that data are made of. Bits are persistent, replicable, scalable, and searchable (boyd 2010). As a result, once data is created, it can be easily shared (and exist in two or more places at once) and used for different purposes in different contexts.

Taken together, these developments lead to privacy concerns that can be understood from the perspective of privacy's informational dimension. People feel like they lose control over information that is about them. Why is this problematic? People are often quick to point out that there is a risk of unwanted access to *sensitive* information. This claim is then often followed by an especially embarrassing example involving love, intimacy, and/or sex. For example, a billboard at a train station might film a person in the company of her secret lover, a fact she does not want other people to know about. More formally put, the abundance of sensors collecting data and big data technology analysing data might erode a person's effective control over her sensitive information. It should be emphasized, however, that it is not just sensitive information that people (should) want control over. As was discussed earlier, people play different social roles in different contexts which are regulated by different social norms. This practice is enabled by the general expectation that people do *not* know certain information about us, unless we have shared the information deliberately. It is precisely this expectation that is threatened by contemporary technological developments, with the possible (partial) collapse of social boundaries between contexts as a result.

Besides people's ability to perform different roles, the current developments also put pressure on context-dependent interpretations of the meaning of anything that can be stored as data. As Miller (2016) describes, information that is produced in context A with intended meaning X, could, due to data's inherent properties, be reproduced in context B and interpreted to mean Y or Z. This can occur due to a lack of people's control over the original data, i.e. due to a lack of informational privacy. Sometimes, this will not result in serious harms. But lack of informational privacy can lead to serious harms, as is explained by Turow (2011) when he describes the inner workings of the online advertising industry. Due to people's limited informational privacy, an enormous amount of information about people is available to advertisers, which, in turn, allows them to build profiles of individuals. The built profiles can, next, be used to target particular persons with personalized offers. Zuboff (2015) even talks of 'surveillance capitalism' to indicate that the surveillance of consumers has become a dominant commercial strategy aimed at generating value. Turow also explains how these practices can lead to serious – often unintended – social discrimination. If people end up in the 'waste' category of advertisers, their opportunities will be narrowed as a result of receiving less interesting and useful offers.[14] A lack of informational privacy can thus lead to serious harms. Moreover, Bridges (2017) explains that – in the US context at least – there is the additional problem of poorer people experiencing a *de facto* weaker protection of their privacy rights.

14 Another possibility is that, for instance, poorer people will be targeted with advertisements for short-term high-interest credits, which might end up harming them more than benefiting them.

The predicament sketched above is also the point of departure of the theory of privacy that adds a fundamentally new perspective: Nissenbaum's (2010) theory of privacy as contextual integrity. Nissenbaum observes that we are surrounded by all kinds of information flows that, on the one hand, may threaten our privacy, but, on the other hand, are also necessary for many essential or useful services. Instead of trying to argue that the flow and use of information should be controlled by individuals as much as possible, Nissenbaum argues that information should flow in *appropriate* ways. The appropriateness of a particular flow of information can be determined by analysing whether 'context-relative informational norms' are respected (Nissenbaum 2010, 129). Society is made up of various social contexts such as the educational contexts, the healthcare context, and the commercial marketplace. In each context different goals, ends, and purposes are at stake; or, put differently, each context is structured around a different set of values. These values inform the norms of a context that determine how activities within the context can be conducted in an appropriate manner. Privacy is respected when information flows without breaching context-relative informational norms (Nissenbaum 2010, 129-157). By shifting focus away from individual control and towards contextual norms, Nissenbaum's theory is an attempt to theorize privacy's function and value for a time where (personal) data are generated, disseminated, and analysed at such a rapid pace that (complete) individual control over data seems no longer attainable.

3.5.3 Challenges to decisional privacy

Because data are a salient feature of many contemporary technological developments, it is unsurprising that informational privacy is the primary analytical frame often used. However, the relevance of decisional privacy as source of relevant ethical norms should not be overlooked. The datafication of our life world through the emergence of big data, the Internet of things, social media, the quantified self, and smart cities also increases the potential of technology to influence us, persuade us, or even manipulate our behaviour (Spahn 2012). Consider Yeung's (2017) concept of 'hypernudge'. Hypernudges are nudges[15]

15 The term 'nudge' was popularized by Thaler and Sunstein (2008) and refers to 'any aspect of the choice architecture that alters people's behavior in a predictable way without forbidding any options or significantly changing their economic incentive' (Thaler and Sunstein 2008, 6). The basic idea is to exploit known cognitive biases to help people make better decisions. *The Cafeteria* is the best known-example: by placing the salad in an easier-to-reach place than the less healthy lasagna, significantly more people will end up choosing the salad. This outcome can be predicted and explained by insights from behavioural economics.

supercharged with big data technology. As Yeung explains 'Big Data-driven nudges make it possible for enforcement to take place *dynamically* (Degli Esposti 2014), with both the standard and its execution being continuously updated and refined within a networked environment that enables real-time data feeds which, crucially, can be used to *personalize* algorithmic outputs' (Yeung 2017: 122). Because hypernudges can be personalized on the basis of personal profiles, they are expected to be much more effective in terms of influencing our behaviour. An important question is whether we want to allow this type of access to our decisional sphere. Consider another example that has come to be known as the Facebook emotional manipulation study. In a large-scale experiment, for which they later apologized, Facebook tried to manipulate the emotions of its users, by showing either more positive or more negative content to users and check whether and how it influenced the behaviour of these users (Hill 2014). The researchers where indeed able to measure significant effects. The study seems to suggest that a large platform can influence how we feel and possibly how we act as a result of those invoked feelings. This raises, again, questions on decisional privacy. Do we deem it acceptable for platforms to enter out decisional sphere behind our backs, trying to manipulate what we do by adjusting our 'psychological levers', possibly 'away from their ideal setting' (Noggle 1996, 47)? This chapter is not the right place to answer such a question, but it is important to point out that the growing potential for manipulation, and the importance of norms within informational as well as decisional privacy for preventing manipulation, should not be overlooked when discussing contemporary challenges to privacy.

3.6 Conclusion

This chapter has presented many different theories on privacy's value, as well as critiques of these theories. The reader might ask herself what to do with such a multiplicity of perspectives. Is one to just pick and choose between these perspectives, based on personal preference?

One way of understanding the meaning and practical use of these different perspectives, is to acknowledge that there is a common thread running through all the theories discussed and the discussions between theories. This common thread can already be found in Warren and Brandeis' foundational text on the right to privacy. As both society and technology are constantly developing and changing, we are also confronted with a constant reconfiguration of norms that regulate what we may know of each other, what we may see of each other, what places we may enter, what information we may share, and what private decisions we may (try to) influence. Many of the theories discussed are an attempt to (1) make sense of these shifting norms, and (2) suggest how we *should*, ideally, understand and enforce privacy norms.

There are, of course, persistent and fundamental disagreements as to how we *should* understand and enforce privacy norms. Different theories build on, and promote, different values, and those values can clash. Most of the time, however, different theories focusing on different developments and different aspects of privacy can, when taken and understood together, complement each other and allow for a richer understanding of the privacy challenge at hand. The hope is that this chapter provides the reader with a rich toolbox filled with normative and conceptual tools that help the reader understand and theorize how privacy *should* take shape, now and in the (near) future.

Further reading

Liberal theory and the value of privacy to the individual

Benn, S.I. (1984). 'Privacy, Freedom, and Respect for Persons' in F.D. Schoeman (ed.), *Philosophical Dimensions of Privacy: An Anthology*. Cambridge: Cambridge University Press, 225-243.

Bloustein, E.J. (1964). 'Privacy as an Aspect of Human Dignity: An Answer to Dean Prosser'. *New York University Law Review* 39, 962-1007.

Cohen, J.L. (1992). 'Redescribing Privacy: Identity, Differences and the Abortion Controversy'. *Columbia Journal of Gender and Law* 3(1), 43-117.

Fried, C. (1984). 'Privacy: A Moral Analysis' in F.D. Schoeman (ed.), *Philosophical Dimensions of Privacy: An Anthology*. Cambridge: Cambridge University Press, 203-222.

Marmor. A. (2015). 'What Is the Right to Privacy?' *Philosophy and Public Affairs* 43(1), 3-26.

Moore, A.D. (2010). *Privacy Rights: Moral and Legal Foundations*. University Park: The Pennsylvania State University Press.

Reiman, J. (1976). 'Privacy, Intimacy, and Personhood'. *Philosophy and Public Affairs* 6(1), 26-44.

Reiman, J. (1995). 'Driving to the Panopticon: A Philosophical Exploration of the Risks to Privacy Posed by the Highway Technology of the Future'. *Santa Clara High Technology Law Journal* 11(1), 27-44.

Richards, N.M. (2015). *Intellectual Privacy: Rethinking Civil Liberties in the Digital Age*. Oxford: Oxford University Press.

Roessler, B. (2005). *The Value of Privacy*. Cambridge: Polity Press.

Schoeman, F. D. (Ed.) (1984a). *Philosophical Dimensions of Privacy: An Anthology*. Cambridge: Cambridge University Press.

Schoeman, F.D. (1984b). 'Privacy and Intimate Information' in F.D. Schoeman (ed.), *Philosophical Dimensions of Privacy: An Anthology*. Cambridge: Cambridge University Press, 403-418.

Warren, S.D. and L.D. Brandeis. (1890). 'The Right to Privacy'. *Harvard Law Review* 4(5), 193-220.

Westin, A.F. (1967). *Privacy and Freedom*. New York: Atheneum.

The social and democratic value of privacy

DeCew, J.W. (2015). 'The Feminist Critique of Privacy: Past Arguments and New Social Understandings' in B. Roessler and D.M. Mokrosinska (eds.), *Social Dimensions of Privacy: Interdisciplinary Perspectives*. Cambridge: Cambridge University Press, 403-418.

Gavison, R. (1980). 'Privacy and the Limits of Law'. *The Yale Law Journal* 89(3), 421-471.

Goold, B.J. (2009). 'Surveillance and the Political Value of Privacy'. *Amsterdam Law Forum* 1(4), 3-6

Goold, B.J. (2010). 'How Much Surveillance Is Too Much? Some Thoughts on Surveillance, Democracy, and the Political Value of Privacy' in D.W. Schartum (ed.), *Overvåking i en rettsstat* [*Surveillance in a Constitutional Government*]. Bergen: Fagbokforlaget, 403-418.

Hughes, K. (2015). 'The Social Value of Privacy, the Value of Privacy to Society and Human Rights Discourse' in B. Roessler and D.M. Mokrosinska (eds.), *Social Dimensions of Privacy: Interdisciplinary Perspectives*. Cambridge: Cambridge University Press, 403-418.

Lever, A. (2006). 'Privacy Rights and Democracy: A Contradiction in Terms?' *Contemporary Political Theory* 5(2), 142-162.

Lever, A. (2015). 'Privacy, Democracy and Freedom of Expression' in B. Roessler and D.M. Mokrosinska (eds.), *Social Dimensions of Privacy: Interdisciplinary Perspectives*. Cambridge: Cambridge University Press, 162-180.

Nissenbaum, H. (2010). *Privacy in Context: Technology, Policy, and the Integrity of Social Life*. Stanford: Stanford University Press.

Rachels, J. (1975). 'Why Privacy Is Important'. *Philosophy and Public Affairs* 4(4), 323-333.

Regan, P. (1995). *Legislating Privacy: Technology, Social Values, and Public Policy*. Chapel Hill: University of North Carolina Press.

Richards, N.M. (2015). *Intellectual Privacy: Rethinking Civil Liberties in the Digital Age*. Oxford: Oxford University Press.

Roessler, B. and D.M. Mokrosinska. (2013). 'Privacy and Social Interaction'. *Philosophy & Social Criticism* 39(8), 771-791.

Schoeman, F.D. (ed.) (1984a). *Philosophical Dimensions of Privacy: An Anthology*. Cambridge: Cambridge University Press.

Simitis, S. (1987). 'Reviewing Privacy in the Information Society'. *University of Pennsylvania Law Review* 135(3), 707-746.

Solove, D.J. (2008). *Understanding Privacy*. Cambridge, MA: Harvard University Press.

Solove, D.J. (2015). 'The Meaning and Value of Privacy' in B. Roessler and D.M. Mokrosinska (eds.), *Social Dimensions of Privacy: Interdisciplinary Perspectives*. Cambridge: Cambridge University Press, 71-82.

Steeves, V. (2009). 'Reclaiming the Social Value of Privacy' in I. Kerr, V. Steeves, and C. Lucock (eds.), *Privacy, Identity, and Anonymity in a Networked World: Lessons from the Identity Trail*. New York: Oxford University Press, 191-208.

Critiques of privacy

Allen, A. (1988). *Uneasy Access: Privacy for Women in a Free Society*. Totowa: Rowman and Littlefield.

Bridges, K.M. (2017). *The Poverty of Privacy Rights*. Stanford: Stanford University Press.

DeCew, J.W. (1997). *In Pursuit of Privacy: Law, Ethics, and the Rise of Technology*. Ithaca: Cornell University Press.

DeCew, J.W. (2015). 'The Feminist Critique of Privacy: Past Arguments and New Social Understandings' in B. Roessler and D.M. Mokrosinska (eds.), *Social Dimensions of Privacy: Interdisciplinary Perspectives*. Cambridge: Cambridge University Press, 85-103.

Etzioni, A. (1999). *The Limits of Privacy*. New York: Basic Books.

Gavison, R. (1992). 'Feminism and the Public/Private Distinction'. *Stanford Law Review* 45(1), 1-45.

MacKinnon, C. (1989). *Toward a Feminist Theory of the State*. Cambridge, MA: Harvard University Press.

Pateman, C. (1989). *The Disorder of Women: Democracy, Feminism, and Political Theory*. Stanford: Stanford University Press.

Thomson, J.J. (1975). 'The Right to Privacy'. *Philosophy and Public Affairs* 4(4), 295-314.

New challenges and topical discussions

boyd, d. (2010). 'Social Network Sites as Networked Publics: Affordances, Dynamics, and Implications' in Z. Papacharissi (ed.), *A Networked Self: Identity, Community and Culture on Social Network Sites*. New York: Routledge, 39-58.

boyd, d. (2014). *It's Complicated*. New Haven: Yale University Press.

Brey, P. (2005). 'Freedom and Privacy in Ambient Intelligence'. *Ethics and Information Technology* 7(3), 157-166.

De Vries, K. (2010). 'Identity, Profiling Algorithms and a World of Ambient Intelligence'. *Ethics and Information Technology* 12(1), 71-85.

Etzioni, A. and O. Etzioni. (2016). 'AI Assisted Ethics'. *Ethics and Information Technology* 18(2), 149-156.

Ezrachi, A. and M.E. Stucke. (2016). 'Is Your Digital Assistant Devious?' *Oxford Legal Studies Research Paper* 52/2016. Available at https://ssrn.com/abstract=2828117

Marwick, A.E. (2013). *Status Update: Celebrity, Publicity, and Branding in the Social Media Age*. New Haven: Yale University Press

Nissenbaum, H. (2010). *Privacy in Context: Technology, Policy, and the Integrity of Social Life*. Stanford: Stanford University Press.

Spahn, A. (2012) 'And Lead Us (Not) Into Persuasion...? Persuasive Technology and the Ethics of Communication'. *Science and Engineering Ethics* 18(4), 633-650.

Stucke, M.E. and A. Ezrachi. (2017). 'How Your Digital Helper May Undermine Your Welfare, and Our Democracy'. *Berkeley Technology Law Journal*, forthcoming. Available at https://ssrn.com/abstract=2957960

Van Dijk, N. (2010). 'Property, Privacy and Personhood in a World of Ambient Intelligence'. *Ethics and Information Technology* 12(1), 57-69.

Yeung, K. (2017). '"Hypernudge": Big Data as a Mode of Regulation by Design'. *Information, Communication & Society* 20(1), 118-136.

Zuboff, S. (2015). 'Big Other: Surveillance Capitalism and the Prospects of an Information Civilization'. *Journal of Information Technology* 30(1), 75-89.

References

Allen, A. (1988). *Uneasy Access: Privacy for Women in a Free Society*. Totowa: Rowman and Littlefield.

Allen, A. (2011). *Unpopular Privacy: What Must We Hide?* Oxford: Oxford University Press.

Benn, S.I. (1984). 'Privacy, Freedom, and Respect for Persons' in F. D. Schoeman (ed.), *Philosophical Dimensions of Privacy: An Anthology*. Cambridge: Cambridge University Press, 223-244.

Bennett, C., and C. Raab. (2006). *The Governance of Privacy: Policy Instruments in Global Perspective*. Cambridge, MA: MIT Press.

boyd, d. (2010). 'Social Network Sites as Networked Publics: Affordances, Dynamics, and Implications' in Z. Papacharissi (ed.), *A Networked Self: Identity, Community and Culture on Social Network Sites*. New York: Routledge, 39-58.

Brey, P. (2005). 'Freedom and Privacy in Ambient Intelligence'. *Ethics and Information Technology* 7(3), 157-166.

Bloustein, E.J. (1964). 'Privacy as an Aspect of Human Dignity: An Answer to Dean Prosser'. *New York University Law Review* 39, 962-1007.

Bridges, K.M. (2017). *The Poverty of Privacy Rights*. Stanford: Stanford University Press.

Cohen, J.L. (1992). 'Redescribing Privacy: Identity, Differences and the Abortion Controversy'. *Columbia Journal of Gender and Law* 3(1), 43-117.

Cohen, J.L. (2002). *Regulation Intimacy: A New Legal Paradigm*. Princeton: Princeton University Press.

Cohen, J.E. (2008). 'Privacy, Visibility, Transparency, and Exposure'. *The University of Chicago Law Review* 75(1), 181-201.

DeCew, J.W. (1997). *In Pursuit of Privacy: Law, Ethics, and the Rise of Technology*. Ithaca: Cornell University Press.

DeCew, J.W. (2015). 'The Feminist Critique of Privacy: Past Arguments and New Social Understandings' in B. Roessler and D.M. Mokrosinska (eds.), *Social Dimensions of Privacy: Interdisciplinary Perspectives*. Cambridge: Cambridge University Press, 85-103.

Degli Esposti, S. (2014). 'When Big Data Meets Surveillance: The Hidden Side of Analytics'. *Surveillance & Society* 12(2), 209-225.

De Vries, K. (2010). 'Identity, Profiling Algorithms and a World of Ambient Intelligence'. *Ethics and Information Technology* 12(1), 71-85.

Ember, S. (2016). 'See That Billboard? It May See You, Too'. *The New York Times*, 28 February. Available at https://www.nytimes.com/2016/02/29/business/media/see-that-billboard-it-may-see-you-too.html. Accessed 5 December 2017.

Etzioni, A. (1999). *The Limits of Privacy*. New York: Basic Books.

Etzioni, A., and O. Etzioni. (2016). 'AI Assisted Ethics'. *Ethics and Information Technology* 18(2), 149-156.

Ezrachi, A., and M.E. Stucke. (2016). 'Is Your Digital Assistant Devious?' *Oxford Legal Studies Research Paper* 52/2016. Available at https://ssrn.com/abstract=2828117

Fried, C. (1984). 'Privacy: A Moral Analysis' in F.D. Schoeman (ed.), *Philosophical Dimensions of Privacy: An Anthology*. Cambridge: Cambridge University Press, 203-222.

Gavison, R. (1980). 'Privacy and the Limits of Law'. *The Yale Law Journal* 89(3), 421-471.

Gavison, R. (1992). 'Feminism and the Public/Private Distinction'. *Stanford Law Review* 45(1), 1-45.

Gibbs, S. (2016). 'Shops Can Track You Via Your Smartphone, Privacy Watchdog Warns'. *The Guardian*, 21 January. Available at https://www.theguardian.com/technology/2016/jan/21/shops-track-smartphone-uk-privacy-watchdog-warns. Accessed 5 December 2017.

Goffman, E. (1959). *The Presentation of Self in Everyday Life*. New York: Anchor Books.

Goold, B.J. (2009). 'Surveillance and the Political Value of Privacy'. *Amsterdam Law Forum* 1(4), 3-6

Goold, B.J. (2010). 'How Much Surveillance Is Too Much? Some Thoughts on Surveillance, Democracy, and the Political Value of Privacy' in D.W. Schartum (ed.), *Overvåking i en rettsstat* [*Surveillance in a Constitutional Government*]. Bergen: Fagbokforlaget, 38-48.

Hill, K. (2014). 'Facebook Manipulated 689,003 Users' Emotions for Science'. *Forbes*, 28 June. Available at https://www.forbes.com/sites/kashmirhill/2014/06/28/facebook-manipulated-689003-users-emotions-for-science/#701febd3197c. Accessed 5 December 2017.

Hughes, K. (2015). 'The Social Value of Privacy, the Value of Privacy to Society and Human Rights Discourse' in B. Roessler and D.M. Mokrosinska (eds.), *Social Dimensions of Privacy: Interdisciplinary Perspectives*. Cambridge: Cambridge University Press, 225-243.

Inness, J. (1992). *Privacy, Intimacy and Isolation*. Oxford: Oxford University Press.

Kirkpatrick, M. (2010). 'Facebook's Zuckerberg Says the Age of Privacy Is Over'. *The New York Times*, 10 January. Available at www.nytimes.com/external/readwriteweb/2010/01/10/10readwrite-web-facebooks-zuckerberg-says-the-age-of-privac-82963.html. Accessed 5 December 2017.

Lever, A. (2006). 'Privacy Rights and Democracy: A Contradiction in Terms?' *Contemporary Political Theory* 5(2), 142-162.

Lever, A. (2015). 'Privacy, Democracy and Freedom of Expression'. In B. Roessler and D.M. Mokrosinska (eds.), *Social Dimensions of Privacy: Interdisciplinary Perspectives*. Cambridge: Cambridge University Press, 225-243.

MacIntyre, A. (1981). *After Virtue*. Notre Dame: University of Notre Dame Press.

MacKinnon, C. (1989). *Toward a Feminist Theory of the State*. Cambridge, MA: Harvard University Press.

Marmor. A. (2015). 'What Is the Right to Privacy?' *Philosophy and Public Affairs* 43(1), 3-26.

Mayer-Schönberger, V., and K. Cukier. (2013). *Big Data: A Revolution That Will Transform How We Live, Work and Think*. London: John Murray Publishers.

Mill, J.S. (1991 [1859]). *On Liberty and Other Essays*. Oxford: Oxford University Press.

Miller, V. (2016) *The Crisis of Presence in Contemporary Culture: Ethics, Privacy and Speech in Mediated Social Life*. London: SAGE.

Moore, A.D. (2010). *Privacy Rights: Moral and Legal Foundations*. University Park: The Pennsylvania State University Press.

Nissenbaum, H. (2010). *Privacy in Context: Technology, Policy, and the Integrity of Social Life*. Stanford: Stanford University Press.

Noggle, R. (1996). 'Manipulative Actions: A Conceptual and Moral Analysis'. *American Philosophical Quarterly* 33(1), 43-55.

Pateman, C. (1989). *The Disorder of Women: Democracy, Feminism, and Political Theory*. Stanford: Stanford University Press.

Pierce, D. (2018). 'Let's Hope CES 2018 Brings Some Sanity to Personal Tech'. *Wired*. Available at https://www.wired.com/story/ces-2018-what-to-expect/

Prosser, W.L. (1960). 'Privacy'. *California Law Review* 48, 383-423.

Rachels, J. (1975). 'Why Privacy Is Important'. *Philosophy and Public Affairs* 4(4), 323-333.

Regan, P. (1995). *Legislating Privacy: Technology, Social Values, and Public Policy*. Chapel Hill: University of North Carolina Press.

Reiman, J. (1976). 'Privacy, Intimacy, and Personhood'. *Philosophy and Public Affairs* 6(1), 26-44.

Reiman, J. (1995). 'Driving to the Panopticon: A Philosophical Exploration of the Risks to Privacy Posed by the Highway Technology of the Future'. *Santa Clara High Technology Law Journal* 11(1), 27-44.

Richards, N.M. (2015). *Intellectual Privacy: Rethinking Civil Liberties in the Digital Age*. Oxford: Oxford University Press.

Roessler, B. (2005). *The Value of Privacy*. Cambridge: Polity Press.

Roessler, B., and D.M. Mokrosinska. (2013). 'Privacy and Social Interaction'. *Philosophy & Social Criticism* 39(8), 771-791.

Roessler, B., & D.M. Mokrosinska. (eds.) (2015). *Social Dimensions of Privacy: Interdisciplinary Perspectives*. Cambridge: Cambridge University Press.

Roux, B., and M. Falgoust (2013). 'Information Ethics in the Context of Smart Devices'. *Ethics and Information Technology* 15(3), 183-194.

Sandel, M. (1981). *Liberalism and the Limits of Justice*. Cambridge: Cambridge University Press.

Sax, M. (2016). Big Data: Finders Keepers, Losers Weepers?. *Ethics and Information Technology* 18(1), 25-31.

Scanlon, T. (1975). 'Thomson on Privacy'. *Philosophy and Public Affairs* 4(4), 315-322.

Schoeman, F.D. (ed.) (1984a). *Philosophical Dimensions of Privacy: An Anthology*. Cambridge: Cambridge University Press.

Schoeman, F.D. (1984b). 'Privacy and Intimate Information' in F.D. Schoeman (ed.), *Philosophical Dimensions of Privacy: An Anthology*. Cambridge: Cambridge University Press, 225-243.

Simitis, S. (1987). 'Reviewing Privacy in the Information Society'. *University of Pennsylvania Law Review* 135(3), 707-746.

Sloot, B. van der (2017). 'Editorial: 10 Questions on Balancing'. *European Data Protection Law Review* 3(1), 1-12.

Solove, D.J. (2008). *Understanding Privacy*. Cambridge, MA: Harvard University Press.

Solove, D.J. (2015). 'The Meaning and Value of Privacy' in B. Roessler and D.M. Mokrosinska (eds.), *Social Dimensions of Privacy: Interdisciplinary Perspectives*. Cambridge: Cambridge University Press, 225-243.

Spahn, A. (2012) 'And Lead Us (Not) Into Persuasion...? Persuasive Technology and the Ethics of Communication'. *Science and Engineering Ethics* 18(4), 633-650.

Steeves, V. (2009). 'Reclaiming the Social Value of Privacy' in I. Kerr, V. Steeves, and C. Lucock (eds.), *Privacy, Identity, and Anonymity in a Networked World: Lessons from the Identity Trail*. New York: Oxford University Press, 225-243.

Stucke, M.E. and A. Ezrachi. (2017). 'How Your Digital Helper May Undermine Your Welfare, and Our Democracy'. *Berkeley Technology Law Journal*, forthcoming. Available at https://ssrn.com/abstract=2957960

Taylor, C. (1989). *Sources of the Self: The Making of Modern Identity*. Cambridge: Cambridge University Press.

Thaler, R.H. and C.R. Sunstein. (2008). *Nudge: Improving Decisions about Health, Wealth and Happiness*. New Haven: Yale University Press.

Thomson, J.J. (1975). 'The Right to Privacy'. *Philosophy and Public Affairs* 4(4), 295-314.

Turow, J. (2011). *The Daily You: How the New Advertising Industry Is Defining Your Identity and Your Worth*. New Haven: Yale University Press.

Van Dijck, J. (2014). 'Datafication, Dataism, Dataveillance: Big Data Between Scientific Paradigm and Ideology'. *Surveillance & Society* 12(2), 197-208.

Van Dijk, N. (2010). 'Property, Privacy and Personhood in a World of Ambient Intelligence'. *Ethics and Information Technology* 12(1), 57-69.

Waldron, J. (2003). 'Security and Liberty: The Image of Balance'. *The Journal of Political Philosophy* 11(2), 191-210.

Walzer, M. (1983). *Spheres of Justice*. Oxford: Blackwell.

Warren, S.D. and L.D. Brandeis. (1890). 'The Right to Privacy'. *Harvard Law Review* 4(5), 193-220.

Westin, A.F. (1967). *Privacy and Freedom*. New York: Atheneum.

Wittgenstein, L. (1953). *Philosophical Investigations*, translated by G.E.M. Anscombe. Oxford: Basil Blackwell.

Yeung, K. (2017). '"Hypernudge": Big Data as a Mode of Regulation by Design'. *Information, Communication & Society* 20(1), 118-136.

Zuboff, S. (2015). 'Big Other: Surveillance Capitalism and the Prospects of an Information Civilization'. *Journal of Information Technology* 30(1), 75-89.

Nudging: A Very Short Guide[1]

Cass R. Sunstein[2]

This brief essay offers a general introduction to the idea of nudging, along with a list of ten of the most important 'nudges.' It also provides a short discussion of the question whether to create some kind of separate 'behavioral insights unit,' capable of conducting its own research, or instead to rely on existing institutions.

I. Liberty-preserving approaches

Some policies take the form of *mandates* and *bans*. For example, the criminal law forbids theft and assault. Other policies take the form of *economic incentives* (including disincentives), such as subsidies for renewable fuels, fees for engaging in certain activities, or taxes on gasoline and tobacco products. Still other policies take the form of *nudges* – liberty-preserving approaches that steer people in particular directions, but that also allow them to go their own way. In recent years, both private and public institutions have shown mounting interest in the use of nudges, because they generally cost little and have the potential to promote economic and other goals (including public health).

In daily life, a GPS is an example of a nudge; so is an 'app' that tells people how many calories they ate during the previous day; so is a text message, informing customers that a bill is due or that a doctor's appointment is scheduled for the next day; so is an alarm clock; so is automatic enrollment in a pension plan; so are the default settings on computers and cell phones; so is a system for automatic payment of credit card bills and mortgages. In government, nudges include graphic warnings for cigarettes; labels for energy efficiency or fuel economy; 'nutrition facts' panels on food; the 'Food Plate,' which provides a simple guide for healthy eating (see choosemyplate.gov); default rules for public assistance programs (as in 'direct certification' of the eligibility of poor children for free school meals); a website like data.gov or data.gov.uk, which makes a large number of data sets available to the public; and even the design of government websites, which list certain items first and in large fonts.

1 This essay has been published in 37J. Consumer Pol'y 583 (2014).

2 Robert Walmsley University Professor, Harvard University. Special thanks to Lucia Reisch, Maya Shankar, and Richard Thaler for valuable comments and suggestions, and to Thaler for many years of collaboration on these questions; none of them should be held responsible for any errors or infelicities here.

A. Nudges maintain freedom of choice

It is important to see that the goal of many nudges is to make life simpler, safer, or easier for people to navigate. Consider road signs, speed bumps, disclosure of health-related or finance-related information, educational campaigns, paperwork reduction, and public warnings. When officials reduce or eliminate paperwork requirements, and when they promote simplicity and transparency, they are reducing people's burdens. Some products (such as cell phones and tablets) are intuitive and straightforward to use. Similarly, many nudges are intended to ensure that people do not struggle when they seek to interact with government or to achieve their goals.

It is true that some nudges are properly described as a form of 'soft paternalism,' because they steer people in a certain direction. But even when this is so, nudges are specifically designed to preserve full freedom of choice. A GPS steers people in a certain direction, but people are at liberty to select their own route instead. And it is important to emphasize that some kind of social environment (or 'choice architecture'), influencing people's choices, is always in place. New nudges typically replace preexisting ones; they do not introduce nudging where it did not exist before.

B. Transparency and effectiveness

Any official nudging should be transparent and open rather than hidden and covert. Indeed, transparency should be built into the basic practice. Suppose that a government (or a private employer) adopts a program that automatically enrolls people in a pension program, or suppose that a large institution (say, a chain of private stores, or those who run cafeterias in government buildings) decides to make healthy foods more visible and accessible. In either case, the relevant action should not be hidden in any way. Government decisions in particular should be subject to public scrutiny and review. A principal advantage of nudges, as opposed to mandates and bans, is that they avoid coercion. Even so, they should never take the form of manipulation or trickery. The public should be able to review and scrutinize nudges no less than government actions of any other kind.

All over the world, nations have become keenly interested in nudges. To take two of many examples, the United Kingdom has a Behavioral Insights Team (sometimes called the 'Nudge Unit'), and the United States has a White House Social and Behavioral Sciences Team. The growing interest in nudges stems from the fact that they usually impose low (or no) costs, because they sometimes deliver prompt results (including significant economic savings), because they maintain freedom, and because they can be highly effective. In some cases, nudges have a larger impact than more expensive and more coercive tools. For example, default rules, simplification, and uses of social norms have sometimes been found to have even larger impacts than significant economic incentives.

In the context of retirement planning, automatic enrollment has proved exceedingly effective in promoting and increasing savings. In the context of consumer behavior, disclosure requirements and default rules have protected consumers against serious economic harm, saving many millions of dollars. Simplification of financial aid forms can have the same beneficial effect in increasing college attendance as thousands of dollars in additional aid (per student). Informing people about their electricity use, and how it compares to that of their neighbors, can produce the same increases in conservation as a significant spike in the cost of electricity. If properly devised, disclosure of information can save both money and lives. Openness in government, disclosing both data and performance, can combat inefficiency and even corruption.

C. The need for evidence and testing

For all policies, including nudges, it is exceedingly important to rely on evidence rather than intuitions, anecdotes, wishful thinking, or dogmas. The most effective nudges tend to draw on the most valuable work in behavioral science (including behavioral economics), and hence reflect a realistic understanding of how people will respond to government initiatives. But some policies, including some nudges, seem promising in the abstract, but turn out to fail in practice. Empirical tests, including randomized controlled trials, are indispensable. Bad surprises certainly are possible, including unintended adverse consequences, and sensible policymakers must try to anticipate such surprises in advance (and to fix them if they arise). Sometimes empirical tests reveal that the planned reform will indeed work – but that some variation on it, or some alternative, will work even better.

Experimentation, with careful controls, is a primary goal of the nudge enterprise. Fortunately, many nudge-type experiments can be run rapidly and at low cost, and in a fashion that allows for continuous measurement and improvement. The reason is that such experiments sometimes involve small changes to existing programs, and those changes can be incorporated into current initiatives with relatively little expense or effort. If, for example, officials currently send out a letter to encourage people to pay delinquent taxes, they might send out variations on the current letter and test whether the variations are more effective.

II. Ten important nudges

Nudges span an exceedingly wide range, and their number and variety are constantly growing. Here is a catalogue of ten important nudges – very possibly, the most important for purposes of policy – along with a few explanatory comments.

(1) **default rules** (e.g., automatic enrollment in programs, including education, health, savings)

Comment: Default rules may well be the most effective nudges. If people are automatically enrolled in retirement plans, their savings can increase significantly. Automatic enrollment in health care plans, or in programs designed to improve health, can have significant effects. Default rules of various sorts (say, double-sided printing) can promote environmental protection. Note that unless *active choosing* (also a nudge) is involved, some kind of default rule is essentially inevitable, and hence it is a mistake to object to default rules as such. True, it might make sense to ask people to make an active choice, rather than relying on a default rule. But in many contexts, default rules are indispensable, because it is too burdensome and time-consuming to require people to choose.

(2) **simplification** (in part to promote take-up of existing programs)

Comment: In both rich and poor countries, complexity is a serious problem, in part because it causes confusion (and potentially violations of the law), in part because it can increase expense (potentially reducing economic growth), and in part because it deters participation in important programs. Many programs fail, or succeed less than they might, because of undue complexity. As a general rule, programs should be easily navigable, even intuitive. In many nations, simplification of forms and regulations should be a high priority. The effects of simplification are easy to underestimate. In many nations, the benefits of important programs (involving education, health, finance, poverty, and employment) are greatly reduced because of undue complexity.

(3) **uses of social norms** (emphasizing what most people do, e.g., 'most people plan to vote' or 'most people pay their taxes on time' or 'nine out of ten hotel guests reuse their towels')

Comment: One of the most effective nudges is to inform people that most others are engaged in certain behavior. Such information is often most powerful when it is as local and specific as possible ('the overwhelming majority of people in your community pay their taxes on time'). Use of social norms can reduce criminal behavior and also behavior that is harmful whether or not it is criminal (such as alcohol abuse, smoking, and discrimination). It is true that sometimes most or many people are engaging in undesirable behavior. In such cases, it can be helpful to highlight not what most people actually do, but instead what most people *think* people should do (as in, '90 percent of people in Ireland believe that people should pay their taxes on time').

(4) **increases in ease and convenience** (e.g., making low-cost options or healthy foods visible)

Comment: People often make the easy choice, and hence a good slogan is this: 'make it easy.' If the goal is to encourage certain behavior, reducing various barriers (including the time that it takes to understand what to do) is often helpful. Resistance to change is often a product not of disagreement or of skepticism, but of perceived difficulty – or of ambiguity. A supplemental point: If the easy choice is also fun, people are more likely to make it.

(5) **disclosure** (for example, the economic or environmental costs associated with energy use, or the full cost of certain credit cards – or large amounts of data, as in the cases of data.gov and the Open Government Partnership, see opengovernmentpartnership.org)

Comment: The American Supreme Court Justice Louis Brandeis said that 'sunlight is the best of disinfectants,' and disclosure can make both markets and governments much 'cleaner.' For consumers, disclosure policies can be highly effective, at least if the information is both comprehensible and accessible. Simplicity is exceedingly important. (More detailed and fuller disclosure might be made available online for those who are interested in it.) In some settings, disclosure can operate as a check on private or public inattention, negligence, incompetence, wrongdoing, and corruption. The Open Government Partnership, now involving sixty-four nations, reflects a worldwide effort to use openness as a tool for promoting substantive reform. (6) **warnings, graphic or otherwise** (as for cigarettes)

Comment: If serious risks are involved, the best nudge might be a private or public warning. Large fonts, bold letters, and bright colors can be effective in triggering people's attention. A central point is that attention is a scarce resource, and warnings are attentive to that fact. One virtue of warnings is that they can counteract the natural human tendency toward unrealistic optimism and simultaneously increase the likelihood that people will pay attention to the long-term. There is a risk, however, that people will respond to warnings by discounting them ('I will be fine'), in which case it would make sense to experiment with more positive messages (providing, for example, some kind of reward for the preferred behavior, even if the reward is nonmonetary, as in apps that offer simple counts and congratulations). Research also shows that people are far less likely to discount a warning when it is accompanied by a description of the concrete steps that people can take to reduce the relevant risk ('you can do X and Y to lower your risk').

(7) **precommitment strategies** (by which people commit to a certain course of action)

Comment: Often people have certain goals (for example, to stop drinking or smoking, to engage in productive activity, or to save money), but their behavior falls short of those goals. If people precommit to engaging in certain action – such as a smoking cessation program – they are more likely to act in accordance with their goals. Notably, committing to a specific action at a *precise* future moment in time better motivates action and reduces procrastination.

(8) **reminders** (for example, by email or text message, as for overdue bills and coming obligations or appointments)

Comment: People tend to have a great deal on their minds, and when they do not engage in certain conduct (for example, paying bills, taking medicines, or making a doctor's appointment), the reason might be some combination of inertia, procrastination, competing obligations, and simple forgetfulness. A reminder can have a significant impact. For reminders, timing greatly matters; making sure that people can act immediately on the information is critical (especially in light of the occasional tendency to forgetfulness). A closely related approach is 'prompted choice,' by which people are not required to choose, but asked whether they want to choose (for example, clean energy or a new energy provider, a privacy setting on their computer, or to be organ donors).

(9) **eliciting implementation intentions** ('do you plan to vote?')

Comment: People are more likely to engage in activity if someone elicits their implementation intentions. With respect to health-related behavior, a simple question about future conduct ('do you plan to vaccinate your child?') can have significant consequences. Emphasizing people's identity can also be effective ('you are a voter, as your past practices suggest').

(10) **informing people of the nature and consequences of their own past choices** ('smart disclosure' in the US and the 'midata project' in the UK)

Comment: Private and public institutions often have a great deal of information about people's own past choices – for example, their expenditures on health care or on their electric bills. The problem is that individuals often lack that information. If people obtain it, their behavior can shift, often making markets work better (and saving a lot of money).

III. Institutionalizing nudges: two approaches

What is the best method for implementing nudges? It is certainly possible to rely entirely on existing institutions. We could imagine a system in which an understanding of nudges is used by current officials and institutions, including leaders at the highest levels. For example, the relevant research could be enlisted by those involved in promoting competitiveness, environmental protection, public safety, consumer protection, and economic growth – or in reducing private and public corruption and combating poverty, infectious diseases, and obesity. Focusing on concrete problems rather than abstract theories, officials with well-established positions might be expected to use that research, at least on occasion.

If the relevant officials have both knowledge and genuine authority, they might be able to produce significant reforms, simply because they are not akin to a mere research arm or a think-tank. (Even a single person, if given the appropriate authority and mission, could have a large impact.) On one model, the relevant officials would not engage in new research, or at least not in a great deal of it. They would build on what is already known (and perhaps have formal or informal partnerships with those in the private sector who work on these issues). In an important sense, this approach is the simplest, because it does not require new offices or significant additional funding, but only attention to the relevant issues and a focus on the right appointments. In the United States, this kind of approach has proved highly successful, with the adoption of numerous nudges.

A quite different approach would be to create a new institution – such as a behavioral insights team or a 'nudge unit' of some sort (as in the United Kingdom, the United States, and increasingly many nations). Such an institution could be organized in different ways, and it could have many different forms and sizes. On a minimalist model, it would have a small group of knowledgeable people (say, five), bringing relevant findings to bear and perhaps engaging in, or spurring, research on their own. On a more ambitious model, the team could be larger (say, thirty or more), engaging in a wide range of relevant research. A behavioral insights team could be created as a formal part of government (the preferred model, to ensure real impact) or could have a purely advisory role.

Whatever its precise form, the advantage of such an approach is that it would involve a dedicated and specialized team, highly informed and specifically devoted to the relevant work, and with expertise in the design of experiments. If the team could work with others to conduct its own research, including randomized controlled trials, it might be able to produce important findings (as has in fact been done in the United Kingdom and the United States, and similar efforts are occurring elsewhere). The risk is that such a team would be akin to an academic adjunct, a kind of outsider, without the ability to power or ability initiate real reform. Authority greatly matters. The United Kingdom has had the most experience with this kind of approach, and it has succeeded in part because it has enjoyed high-level support and access.

In this domain, one size does not fit all, but it is noteworthy that a growing number of nations have concluded that it is worthwhile to have a dedicated team. Of course the two approaches might prove complementary.

4. Privacy from an Economic Perspective

Edo Roos Lindgreen

4.1 Introduction

Elsewhere in this book, it has been made clear that privacy is a multidisciplinary field that can and should be viewed through many different lenses – e.g. social, legal, psychological, political, philosophical, ethical, technological, and economic (Hui and Png 2005). If privacy is studied through an *economic* lens, a multitude of intriguing questions arises. What are the economic trade-offs when it comes to privacy, both on the individual level and on the policy level? Is there a way to determine the economic value of privacy? Is there a difference between the real and perceived value of privacy? What are the individual, organizational, and societal costs and benefits of maintaining or giving up privacy? In the field of privacy economics, researchers are looking for the answers to these and similar questions, which recent technological and social developments have made more relevant than ever.

This section will introduce some basic terms and concepts in economics that are relevant for this chapter; it will discuss the field of privacy economics; and it will touch upon the social-economic impact of current technological developments.

4.1.1 What is economics?

While most people will have an intuitive understanding of what the long-standing social science of economics is about, Backhouse and Medema (2009) point out that, in fact, the word *economics* has many definitions and interpretations. To one, it is the study of economies, both at the individual level and for society as a whole (Krugman and Wells 2004); to others, it is the study of how society manages its scarce resources (Mankiw 2001). The definition and interpretation of economics may vary from scholar to scholar. This chapter adopts the Oxford Dictionary's definition: the branch of knowledge concerned with the production, consumption, and transfer of wealth.

Note on terminology: this chapter will follow the economic tradition to distinguish the individual agents in an economic transaction by using terms that clarify their role, such as *consumer* and *merchant*. These terms may pertain to individuals or organizations, depending on the context.

Within the field of economics, a distinction is made between macroeconomics and microeconomics. *Macroeconomics* studies the behaviour of economies and the influence of economic policies on an aggregated level, addressing themes like growth, inflation, and employment. *Microeconomics*, on the other hand, studies the economic behaviour of individuals and companies. This chapter will discuss privacy through a microeconomic lens.

Central to microeconomics is the idea that the *price* of products and services is established in a competitive *market* where demand meets supply. As we will see below, this assumption does not hold for privacy.

Two important factors that influence the price of a product are utility and cost.

> *Utility* – Utility is a measure of the usefulness, benefit, or satisfaction that a consumer obtains from a good, a service, or a transaction. A consumer's *willingness to pay* a certain amount of money for a product or service is often used as a measure for utility. In this chapter, the terms utility and benefit will be used interchangeably. In section 2, we will attempt to analyse the utility of privacy.
>
> *Cost* – The cost of a product is largely determined by the price of the resources required to produce it and bring it to the market. *Opportunity cost* is defined as the loss of potential gain from other alternatives when one alternative is chosen; it will play a role in section 2 of this chapter.

Benefits and costs that can be quantified and can be attributed to an identifiable asset are called *tangible*. Likewise, *intangible* benefits and costs are subjective and cannot be measured directly in monetary terms. Examples of intangibles for individual consumers, organizations, and society at large are, respectively: well-being, safety, reputation, freedom of choice, happiness (individual consumers), customer goodwill, employee morale, corporate reputation (organizations), and societal well-being, resilience, safety, social security, freedom (societies).

Lastly, a word on *rationality*. Traditional economics assumes that the agents in an economic transaction base their decision on rational considerations of the cost versus the expected utility of that transaction. Behavioural economists have shown conclusively that such a *homo economicus* does not exist. Simon (1972) introduces the term *bounded rationality* to describe the decision maker's cognitive limitations of both knowledge and computational capacity. Kahneman and Tversky (2002) introduce *prospect theory*, which describes the way people make choices between probabilistic alternatives that involve risk, and conclude that the *rational agent* is a figment of our imagination. Ariely (2009) exposes the predictably irrational behaviour of people making decisions and even makes the case for the end of rational economics. Thanks to these scholars, it is now widely accepted that people make decisions based on incomplete information, psychological biases, and irrational considerations, even if all ingredients for a rational analysis are present. More on this in sections 2 and 3.

4.1.2 The economics of privacy

In the study of economics, privacy was never more than an incidental guest. The past decade, however, the economics of privacy – or *privacy economics*, for short – has become a discipline in itself.

Privacy economics studies the economic trade-offs people make when confronted with privacy-related decisions. Such trade-offs are made by individuals, by organizations, and by society at large. For example, an individual may decide to disclose some personal data to obtain a discount; a company may decide to collect personal data to increase its advertising revenue; and, at the policy level, a government may decide to adopt and implement costly regulation to protect the privacy of its citizens. In order to study these and other trade-offs involving privacy, it is necessary to study the aforementioned properties of utility and cost. When studying these, one cannot escape studying privacy's opposite: the uncontrolled *disclosure* of personal data.

Starting as early as the 1960s, research in privacy economics has developed in roughly three stages, to be touched upon in section 3. The last stage has produced the now dominant school of thinking, which will be discussed in section 4.

Privacy economics is a complex area of research. Acquisti et al. (2016) point out that the economic parameters of privacy are highly dependent on the context and the actors involved and pose three observations:

> *No single theory* – A single unified economic theory of privacy economics seems infeasible, given the diversity of contexts in which the issue arises.
>
> *Positive and negative effects* – Protecting privacy may have positive and negative economic effects, not only for individuals, but for society as a whole.
>
> *Incomplete information* – It is near impossible for consumers to make *informed* decisions on privacy, simply because they do not know which

data is being collected, for what purposes, and what the consequences might be, in an ecosystem where companies are systematically collecting vast amounts of personal data with substantial economic value.

Indeed, the more one studies privacy economics, the more complex the subject seems to become. A fundamental reason for this, is that privacy is so much more than a simple economic good or service. Instead, privacy, in its meaning of the right and ability of an individual to control the protection and selective disclosure of his or her personal data, is perhaps the ultimate example of an *intangible asset* as discussed above. Intangibles play a very important role in privacy economics. Of course, there are many examples of benefits and costs that are utterly tangible, such as discounts, special offers and tailor-made news for consumers, or higher conversion rates, increased sales and lower costs for merchants. But not everything of value can be expressed directly in economic terms, and many other aspects of privacy clearly extend beyond the tangible. Examples of intangibles include the adverse psychological effects of being monitored or manipulated, the social exclusion of those not using social media, or the public reputation of companies known to use personal data as a core element of their revenue model. More on this in section 2.

Section 1.1 discussed the general economic premise that the price of goods is established in a market where supply meets demand. As has been pointed out by many authors, there is no clear, transparent, open market for personal data. It is true that personal data are traded on an enormous scale, but its market is far from transparent and certainly not open to everyone – especially not to the subject of the data. Granted, individuals do 'sell' their personal data to companies, but usually implicitly, as the by-product of using a specific service, such as a search function or a social platform. Given the absence of an open market, an accurate and fair valuation of personal data and hence privacy is – by the principles of economics – impossible.

4.1.3 Privacy economics in the digital age

As Nissenbaum (2009) has stated, the notion of privacy is an oversimplification, a catchword conveniently used to denote a very complex and delicate system, a social fabric of assumptions, norms, conventions, and the like regarding the disclosure and protection of personal data – that is, data relating to an identifiable person – in many different contexts, evolved over centuries. The widespread adoption of digital technology in the past decades has made deep cuts in this fabric. And the end is not in sight; technological developments seem to be accelerating rather than slowing down. Thus, the economics of privacy cannot be addressed without considering the influence of technological developments on privacy itself. Are these developments changing the privacy trade-offs we are studying in privacy economics? Do they impact the real and perceived value of privacy?

One can debate whether the digital age started with the world's first computer programme, written in 1843 by Ada Lovelace (Fuegi and Francis 2003; Koetsier 2001), with the first electronic computer, built in 1936 by Conrad Zuse (Rojas 1997), or with the use of integrated circuits in computer systems (Moore 1965); see frame. More important is that today, information technology has permeated nearly all aspects of people's lives. In this environment, data plays a pivotal role. For example, organizations use data to profile customers, predict their preferences, and so increase sales volume and customer satisfaction. Or to analyse markets, geographic areas, and demographic strata, yielding insights that assist in strategic decisions. The ubiquitous use of data, made possible by technology, has serious consequences for privacy.

Below, a brief analysis of the technology market is presented, to identify the companies involved in collecting, processing, enriching, and using all that data, and to get a grasp of the sheer size of the market collectively formed by these companies.

The technology sector is currently valuated at hundreds of trillions of dollars. In this sector, economic power is concentrated in a few well-known companies. Based in Silicon Valley, publicly listed on the Nasdaq stock exchange, and commonly known as The Big Five, their total market capitalization exceeds three trillion dollars (table 1). Billions of people use their products and services on a daily basis: smartphones, tablets, and notebooks; the operating systems running on those devices; the networks, servers and datacentres that are used to provide their services. Thus, these companies have full control over the software and data that these billions of people use to live their daily lives – to communicate, to socialize, to search, to study, to work, to write, to buy, to sell, to trade, to apply for a job, to share, to watch, to present and identify themselves. In addition, serving as intermediaries, they control the advertising platforms that other companies must use to reach their customers.

The value of these companies is largely determined by investors' expectations about their future performance. These expectations are based on the expected sales of products and services and the accompanying margins, which, in turn, are based on the perceived value of the data they collect. So, ultimately, the value of these companies is determined by the value of their data. Every day, ever more data is being collected, stored, analysed, and enriched, data that completely describes its subjects and can predict their preferences and actions better than anybody or anything else. Indeed, personal data has been identified as a new asset class in itself (Schwab et al, 2011).

Table 4.1: Big technology companies

Company	Main products and services	Business model	Mcap *	Users **
Apple	Smartphones, computers, accessories, software, cloud services	Products, licenses	>825	>1.000
Alphabet	Search (Google), video sharing (YouTube), mail (gmail), operating systems (Android), cloud services, navigation	Advertising, licences	>700	>2000
Amazon	Retail, cloud services	Retail, licences	>500	>300
Facebook	Social media (Facebook, Instagram), messaging (WhatsApp, Messenger)	Advertising, licences	>500	>2000
Microsoft	Business software, cloud services, social media (LinkedIn), messaging (Skype), search (Bing)	Licences	>600	>1000

* Mcap = market capitalization in billions of dollars. ** Users = number of active users in millions of users. Data from public sources and annual reports, 2017.

The companies in table 1 do not operate in isolation, but are highly interconnected, both financially and functionally. They were all founded on Silicon Valley venture capital and use each other's platforms to accelerate the growth of their user population and revenue. For example, Facebook, Instagram, and Whatsapp are the most popular apps driving the sales of today's smartphones, including Apple's iPhone and all non-Apple smartphones, where Google's Android is the leading operating system; smartphones which you can order through Amazon, for which you can use your smartphone, etc. So it is a tightly knit ecosystem, where only Microsoft is a bit of an outcast, dominating the business space and owning LinkedIn.

According to their annual financial reports, both Google and Facebook have a business model that is almost exclusively based on collecting, enriching, and monetizing personal data by selling narrowly targeted advertising space to the highest bidder. Facebook's Custom Audience and Google's Adwords programmes allow clients to buy advertising space for highly specific categories of users. The current market capitalization of these companies is sometimes used as a yardstick to measure the value of personal data. One could argue, for example, that the value of a Facebook profile equals the company's market capitalization divided by its number of active users (which, at the time of writing this chapter, would yield a value of approximately $250), but such an estimate would be very speculative.

Apple, Microsoft, and Amazon collect and use data too, but these companies have other primary revenue drivers (selling smartphones, business software, and nearly everything respectively) and use personal data to increase their own revenue instead of selling targeted advertising space to others.

Note that besides the Big 5, a legion of other companies is active in the same space; they range from very small to extremely large. Effectively, in the digital age, tens of thousands of companies are systematically harvesting and monetizing ever-increasing amounts of personal data. In addition to its direct commercial utility, the data collected by these technology companies of enormous value to intelligence agencies for surveillance purposes.

To summarize, in less than a decade, full control over inconceivable amounts of personal data has been transferred from individual citizens to a complex ecosystem of private companies and government organizations, where the bulk of this data – and thus economic power – is concentrated in a handful of companies that totally dominate their respective markets. The use of this data for targeted advertising, influencing, and surveillance has a huge social-economic impact, and a huge impact on privacy. But how does it influence privacy economics?

4.1.4 Outline of this chapter

The outline of the remainder of this chapter is as follows. In section 2, the meaning and function of privacy economics will be discussed. Section 3 will discuss classic texts and authors. Section 4 describes the prevalent schools of thought and current debates regarding the economics of privacy. Section 5 will discuss new challenges and topical discussions in the field of privacy economics. Section 6 will present the conclusions of this chapter and give suggestions for further reading.

4.2 Meaning and function of privacy

Consumers may reap economic benefits by sharing specific personal data, but may also experience disadvantages when sharing other personal data (Varian 1997). This section will give a brief introduction to the meaning of privacy in the domain of economics by systematically analysing the potential economic benefit and cost of privacy for individuals, for organizations, and for society at large (see table 2). Lastly, this section focuses on the privacy trade-offs made by these actors.

Table 4.2: Economic benefit and cost for individuals, organizations, and society at large

	Individual	Organization	Society
Benefit	Better negotiation Reduced vulnerability Improved well-being	Protection of reputation Prevention of fines	Protecting human rights
Cost	Opportunity cost	Opportunity cost Cost of control	Opportunity cost Cost of surveillance Stagnation
	microeconomic	microeconomic	macroeconomic

4.2.1 The economic benefits of privacy

For parties engaged in an economic transaction, being able to control the disclosure of personal data may have direct economic benefits. Below, we distinguish between benefits for individuals, organizations, and society at large.

4.2.1.1 Economic benefits for individuals

The key economic benefits of privacy for individuals are twofold: improving one's negotiation position and reducing one's vulnerability.

Improving negotiation position – In an economic transaction, actors intend to improve their position by obtaining as much relevant information on the other party as possible, while keeping their own cards close to their chest. Thus, privacy – controlling the sharing of personal data – can be beneficial to one's negotiation position (Varian 1997). If you have information on the other party in a transaction, you can use that information for many purposes; e.g. to come up with a good proposition, to optimize your negotiation strategy, to get the best price, or to cut the best deal. Conversely, if the other party has relevant information on you, your economic position may be weakened in a similar way. It follows that protecting your personal data – or, to be more precise, controlling the disclosure of your personal data – can be beneficial in economic transactions. Conversely, protecting the other person's privacy would benefit him or her, but would not improve your economic position.

Reducing vulnerability – Controlling the disclosure of personal data may also reduce the risks one is exposed to. For example, being discrete about one's financial position, good or bad, can make one less vulnerable to parties who want to profit from it, for example by offering loans or investment opportunities at unfavourable conditions, or subjecting one to criminal activities, such as theft, robbery or extortion (Stigler 1980). Being not too open about one's lifestyle and behaviour at parties may reduce the risk of being rejected at a job application. Keeping silent about an unhealthy lifestyle may prevent a raise in one's health insurance fee.

Besides these more-or-less direct economic benefits, there are many other examples of utility that are obtained from protecting one's privacy; examples include one's well-being or the well-being of others, such as family members. For example, it has been shown that a certain level of privacy is necessary to offer children and adolescents an environment to develop a sense of self, personal responsibility, autonomy, and intimacy in human relations (Van Manen and Levering 1996). As another example, preventing the disclosure of information pertaining to activities or properties that are considered unacceptable or shameful in the subject's societal context can prevent shaming, social exclusion, or worse (Solove 2007a).

The above illustrates that for the individual, there are direct and indirect economic benefits to keeping at least some of one's personal data to oneself. Quantifying these benefits, however, is far from easy.

4.2.1.2 *Economic benefits for organizations*

From a strictly economic viewpoint, privacy in itself does not bring direct benefits to public and private organizations. On the contrary: it is in the economic interest of an organization to collect as much relevant personal data as possible. Data on consumers, suppliers, employees, competitors, and other individuals can be used to its advantage, e.g. by increasing sales, improving one's competitive position, market share, or service levels. From a strictly economic point of view, respecting and protecting the privacy of individuals does not directly benefit a private company. The same argument applies to organizations in the public sector, such as law enforcement, government, healthcare, and education. Although long-term profitability is not a strategic objective of such organizations, they will benefit in other ways by collecting personal data and using it to their advantage.

There is one clear benefit for a company to protect personal data after it has been collected: protecting its reputation and preventing fines. If a company fails to protect the data collected and an inadvertent disclosure is disclosed itself, its reputation will be damaged or fines may be imposed. Also, openly using personal data for purposes that are considered unethical by the public may be detrimental to the company's image and reduce consumer trust and consumer spending. For example, in 2014, the Dutch bank ING announced plans to use their customer data for commercial purposes, which resulted in nationwide negative publicity and probably loss of a few clients (Munsterman 2014).

There is an inherent tension between the economic value of personal data for organizations on the one hand and the high utility of privacy for individuals on the other hand. This tension – and all the privacy violations that have been caused by it – have led to increasing levels of privacy regulation, such as Europe's General Data Protection Regulation (GDPR). It is no coincidence that this regulation uses financial levers of control to enforce compliance; GDPR, for example, imposes sanctions on privacy violations in the form of high fines, which may be up to 4% of the company's revenue. By placing a financial incentive on privacy, the regulator has made it economically beneficial for organizations to protect the privacy of its customers and employees. For a more extensive description, see the legal chapter in this book by Bart van der Sloot.

4.2.1.3 Economic benefits for society

From society's perspective the utility of privacy is high enough to warrant its global acceptance as a fundamental right, stipulated by article 12 of the Universal Declaration of Human Rights of the United Nations: 'No one shall be subjected to arbitrary interference with his privacy, family, home or correspondence, nor to attacks upon his honour and reputation. Everyone has the right to the protection of the law against such interference or attacks' (UN 1948). Following this article, privacy is contained in international and national legislation, like the GDPR in the European Union and the national laws implementing it. Please note that fundamental rights and economic driving forces are not necessarily aligned. Indeed, violations of these rights are often motivated by the desire to obtain economic benefit, and privacy is no exception. The question remains if privacy's utility to society can be expressed in immediate economic benefits that can be quantified in one way or another. Apparently, this question is quite difficult to answer. No answer can be found in existing literature; more on this in section 4.

4.2.2 The cost of privacy

Protecting one's personal data does not only bring benefits, but also incurs costs. Below, cost factors are analysed, again from an individual, an organizational, and a societal perspective.

4.2.2.1 Costs for the individual

For individual consumers, privacy can be expensive because it generates *opportunity costs.* Indeed, the reasons to share personal data are often direct or indirect tangible economic benefits, such as discounts, convenience, or access to services (Varian 1997). It follows that not sharing this data – in other words, protecting one's privacy – will incur direct or indirect opportunity costs. Examples include less discount on personalized offers, the inconvenience and waste of time caused by receiving information that does not fit one's needs or preferences, reduced opportunities on the labour market for not having a LinkedIn profile, or receiving inadequate medical treatment because one's patient information is not readily available.

A surprising source of intangible opportunity costs is of a psychological nature. Tamir and Mitchell (2012) note that, on average, people spend 30-40% of their communication capacity to disclosing facts and stories about themselves. They describe an experiment where subjects are given a fee to answer questions; subjects are willing to forego a premium of 17% to answer questions about themselves rather than about other people. The authors conclude that people see an intrinsic utility in sharing personal information. In the context of this section, this means that restraining the disclosure of personal data may lead to intangible opportunity costs simply because it prevents people from talking about themselves.

4.2.2.2 Costs for organizations

The costs of privacy for private companies (and, probably to a lesser extent, for public organizations) fall into two categories: opportunity cost and cost of control.

> *Opportunity cost* – Less information on potential clients means less opportunities for tailored propositions, less opportunities for price discrimination (Zuiderveen Borgesius 2015), and hence lower margins. Goldfarb and Tucker (2013) calculate the effect of privacy regulation on advertising, stating that the decreased effectiveness of advertising due to harsher privacy regulation have led to a 2.85 times higher advertising spend in Europe compared to the US. Though tangible, these costs are hard to quantify.

Cost of control – The cost of control becomes relevant when privacy regulation is tightened; it comprises the costs of implementing the regulation (including legal advice, information systems and infrastructure) and the costs of maintaining the ensuing framework of controls and demonstrating regulatory compliance. For cost of control, it is possible to make reasonable estimates based on time and materials spent.

4.2.2.3 Costs for society

From a societal perspective, protecting the privacy of individuals may incur costs as well. The Chicago School, among others represented by Posner (1981), argues that privacy reduces the efficiency of the market place by increasing information asymmetry, thus increasing costs and reducing value, and even suggests that privacy regulation might lead to lower wages, higher unemployment, and higher interest rates. On the other hand, Shapiro and Varian (1997) pose that controlling the dissemination of personal data actually leads to a more efficient market, with positive macroeconomic effects.

Below, three potential cost factors for society at large are discussed: opportunity costs, the blocking of innovation, and cost of surveillance.

Opportunity costs – Sharing personal data may yield societally beneficial results that are unattainable otherwise; not sharing this data will incur opportunity costs. For example, trusted reviews by verified users give valuable information on the price and quality of products and services, which allows consumers to make better choices, increases the quality/price ratio, and thus benefits society at large (e.g. Calzolari and Pavan 2006). Choosing not to share and use this data incurs indirect opportunity costs. Another example are insights into the condition of patients and the effectiveness of medical treatments brought forward by using electronic medical records. Today, medical research is strongly dependent on collecting and aggregating medical information, deepening the gap between privacy and research goals (Konnoth 2015). Miller and Tucker (2011a) find that an increase of 10% in the use of electronic medical records reduces neonatal mortality rates by 3%. Although difficult to express in monetary terms, the digitization of healthcare incurs substantial benefits, and choosing not to share personal data may prohibit some of them. Other examples include the aggregation of online searches, which may yield insight into interactions between medications (White et al., 2013) or early warnings for epidemics (Dugas et al., 2012).

Blocking innovation – It is a complaint often heard by startups and corporates alike: we have fantastic ideas and opportunities for innovation and growth, but privacy laws prevent us from implementing them. But is it true? Goldfarb and Tucker (2012) investigate the relationship between innovation and privacy for advertising, healthcare, and operational efficiency. They conclude that privacy regulation will affect the direction of innovation, which appears to be a euphemism for slowing it down, and that there is an inherent tension between the economic value of personal data and the need to safeguard the privacy of consumers; the authors argue that protecting personal data will prevent its value from being unlocked and utilized. They also assert that privacy is interlinked with innovation and economic growth and note the tension between the economic value of using personal data and the need to safeguard privacy.

Cost of surveillance – There are significant economic effects of privacy in the context of justice, law enforcement, intelligence, and national security. It is often claimed that implementing massive surveillance systems – ranging from CCTV systems to bulk interception of Internet traffic – may lead to lower crime rates and hence lower costs for society. Such claims are presented as self-evident, but the factual evidence supporting them is thin. The privacy effects of using electronically collected personal data for surveillance purposes has been the source of a heated debate that started at least half a century ago (Westin 1967) and continued in the decades that followed (e.g. Solove 2007b), focusing on the legal authorization and practical capabilities for law enforcement and intelligence to collect, process, and use personal data, and the necessity to impose limitations on them. The debate is conducted from many different perspectives, including law, ethics, national security, social sciences, and so on; see elsewhere in this book. Less attention is paid, however, to the economics of using personal data for law enforcement and intelligence. In private and public communications, professionals in these disciplines unanimously state that they see privacy regulation as an obstacle that prevents them from doing their work effectively and efficiently. Such claims, however, are seldom backed by objective evidence. A popular but rather speculative contrary notion among privacy advocates nowadays is that using too much personal data for law enforcement and intelligence is not only disproportional, but actually leads to inefficiency for trying to find the same needles in a much bigger haystack. It seems safe to conclude that the apparent utility of collecting personal data for law enforcement and intelligence purposes implies that privacy will incur opportunity costs for society; the amount of these costs, however, will most likely remain impossible to calculate.

4.2.3 Conclusion

Having analysed the benefit and cost factors of privacy for individuals, organizations, and society at large, the main lesson is that even tangible, direct benefits and costs of privacy may be very hard or even impossible to quantify and express in monetary terms; let alone the indirect intangibles. The impact of this lesson will be discussed in section 4.

4.3 Classic texts and authors

This section will give a brief introduction into the classic texts and authors on privacy in the domain of economics.

A comprehensive treatment of scholarly papers on the economics of privacy is presented by Acquisti et al. (2016). According to the authors, research on the economics of privacy has come in three waves, addressed below.

4.3.1 First wave: market efficiency

The first wave of research is generally considered part of the Chicago School of economic reasoning, which evolved at the University of Chicago in the mid-1950s. A central theme in the reasoning of this school is the presumed natural tendency of ecosystems to gravitate to an economic optimum by bargaining (Coase 1937). The Coase Theorem argues that, given sufficiently low transaction costs, institutions evolve to a state of Pareto efficiency, meaning that it is impossible for one member of an ecosystem to obtain a better position without worsening the position of another member. Another central theme in the Chicago School is the limited or even adverse effect of government policymaking. Most famously, Friedman and Schwartz (1963) argue that the Great Depression of the 1930s was not caused by the crash of the New York stock exchange, but by the monetary policy of the US government at the time.

In the late 1970s, Chicago scholars projected a number of their central themes on privacy and privacy regulation. For instance, Posner (1981) argues that privacy regulation will create inefficiencies in the marketplace by creating information inequalities, thus reducing transparency. The argument is based on an example of job seekers applying for vacant positions, where protecting personal data on the applications would negatively affect a firm's hiring decisions.

In the spirit of Friedman, Stigler (1980) argues that privacy regulation will have negative effects on market efficiency. The argument is based on the assumption that individuals have a tendency to disclose positive information and to hide negative information about themselves. In this light, even the protection of personal data gives information, since it can be an indicator of a negative trait. The author argues that privacy regulation – aimed at blocking the flow of personal data – would lead to market inefficiencies, since it removes information from the marketplace.

Hirshleifer (1980) disputes the rational agent model underlying the privacy models of the Chicago School. He argues that purely rational agents may be driven to collect too much personal data, reducing rather than increasing efficiencies. An interesting perspective is offered by (Spence 1973) in the context of job market signalling; the author argues that the aggregate cost of collecting personal data may well exceed the benefits obtained from it. Gottlieb and Smetters (2011) argue that systems where no personal data is disclosed at all may work more efficiently than systems where personal data are disclosed. In the latter, they state, a lot of additional effort will be spent on the optimization and presentation of this data by the data subject, and the interpretation of this data by the data user.

Building on the Coase Theorem, several authors have argued that personal data will be redistributed based on its value as perceived by the actors in a system, and reach an equilibrium that is independent of the initial allocation of rights (Kahn et al. 2000), adding that upfront investments and difficulties in making commitments can hamper the bargaining process.

4.3.2 Second wave: technological developments

In the second wave, research is focused on the impact of technological developments, especially the rise of personal computing and the Internet. Varian (1997) is the first to argue that consumers may reap economic benefits by sharing specific personal data, but may experience disadvantages when sharing other personal data. The author also discusses the implications of the secondary use of personal data, and points out that consumers have no clue who will be using their data, when, and for what purpose.

In response to the observation that individuals are not in control of their personal data, Laudon (1996) proposes the creation of an information market where individuals may sell the rights to their personal data and receive fair compensation for the use of information about themselves. This is an interesting line of thought that has been explored a number of times, for example in the context of Facebook (Dhar 2012) but has never made it to reality, probably due to its limited practical feasibility. Moreover, Acquisti et al. (2016) argue that a market where individuals would be free to sell their personal data would spiral to an equilibrium that would benefit a monopolist rather than the consumers.

4.3.3 Third wave: informational privacy

The third wave of research in privacy economics was ignited by the technological developments described in section 1 of this chapter. The rise of online consumer activity and the resulting explosion of personal data that is collected, analysed, and used for commercial and other purposes gave birth to a notable increase in privacy-related research. In the third wave, the number of research projects and scholarly papers increased with an order of magnitude, and a number of dedicated, multidisciplinary academic privacy conferences saw the light, such as the highly esteemed Amsterdam Privacy Conference, which was established in 2012 and draws an international crowd of over 500 academics

Privacy economics still makes up only a relatively small percentage of total privacy research. Current research is more based on formal economic theories and models than the previous waves of research. It is often focused on issues surrounding specific technological developments such as those sketched in section 1.

Classic authors in this wave are (Acquisti et al. 2016). The authors provide an encompassing description of extant research; their paper is a proper starting point for any research in this field.

The next section will highlight some of the key topics and traditional debates in current research.

4.4 Traditional debates and dominant school

This section will give a brief introduction into the dominant school in the field of privacy economics research and the current debates.

4.4.1 Dominant school

As mentioned in the previous section, the dominant school of thought in privacy economics emerged in the late 1990s. It focuses on the economic value of privacy and the economic consequences of protecting and disclosing personal data from an informational point of view, paying special attention to the trade-offs and decisions made by consumers in a highly digital environment.

Compared to the preceding decades, there are now more players in the field, which has become more fragmented as a result (Acquisti et al, 2016). Below, a highlight of recent research topics is given.

4.4.1.1 Price discrimination

One important research topic in privacy economics is price discrimination (Zuiderveen Borgesius 2015): the differentiation of prices for goods and services between different categories of customers or even individual customers based on personal data. The personal data involved can be data on past purchases and other online activity, such as cookies, Facebook likes, and online searches; but it can also include a consumer's location, or a consumer's psychological profile distilled by an intermediary from a myriad of online interactions.

Price discrimination can be used as an instrument for many purposes, the most prominent being margin improvement, for example by offering higher prices to customers who can afford it or who have expressed a higher willingness to pay. Other objectives can be acquisition or poaching of customers, for example by offering lower prices to new customers (Villas-Boas 2004; Fudenberg and Tirole 2000); or to obtain information on clients, for example by offering discounts in exchange for personal data (Chen and Zhang 2009).

An interesting finding from this stream of research is that price discrimination can backfire if consumers find out about it; for example, consumers may choose to defer a purchase to avoid being branded a regular customer and being charged higher prices in the future, or avoid the merchant altogether (Villas-Boas 2004). For this reason, merchants may choose to commit themselves openly to refraining from price discrimination.

4.4.1.2 Data intermediaries

A second research topic is the role of data intermediaries, who build up customer profiles and sell targeted advertising space to merchants. De Cornière (2017) shows that such an ecosystem may lead to lower prices, lower search costs, and a better match between supply and demand; these advantages may be offset, however, by the costs of the intermediary itself. Bergemann and Bonatti (2015) demonstrate that data intermediaries can decide to reduce the precision of customer information in order to sell more data and increase revenue. De Cornière and Nijs (2014) show that, in ecosystems where merchants place bids on targeted advertisements, higher prices will result. Acquisti et al. (2016) conclude that merchants have no incentive to attain an optimum match between supply and customer demand – the less effective their targeting, the more merchants have to spend to reach their target audience.

4.4.1.3 *Marketing techniques*

A third category of existing research in privacy economics is related to various marketing techniques, including unsolicited e-mail (spam) and targeted advertising.

Although it is often said that the use of e-mail is declining in favour of more direct forms of communication, data suggests that both forms of electronic communication are on the rise, with 86% of European Internet users using e-mail, irrespective of age group (Eurostat 2018). Hann et al. (2008) depicts a 'spam arms race' by showing that the more consumers protect themselves from unsolicited mail, the more merchants will spend to reach them, and so on.

The opposite of spam is targeted advertising (Taylor 2014), addressed elsewhere in this chapter. Interestingly, several authors find that targeted advertising may have an adverse effect on consumer spending due to privacy concerns (White et al, 2008; Goldfarb and Tucker, 2011).

4.4.2 Traditional debates

Section 2 analysed and discussed the tangible and intangible benefits and costs of privacy. One of the main conclusions was that, in many cases, these are hard to quantify, if possible at all. This brings us to the central debate in privacy economics: given these largely unquantifiable benefits and costs, how do parties engaged in economic transactions make their privacy trade-offs? As it turns out, there are many other factors complicating such trade-offs. Elements of this debate are appearing in many influential research papers and are treated below, distinguishing between individual consumers and organizations on the one hand, and society on the other.

4.4.2.1 *Privacy trade-offs for individuals and organizations*

Regarding the privacy trade-offs made by individuals and organizations, the following topics are encountered in literature.

Subjective utility – The utility of both privacy and personal data is highly subjective and context-dependent; see also (Nissenbaum 2010; Varian 1997; Acquisti et al. 2016). What is valuable for one person or organization, can be worthless for another; what is important in one context, may be completely irrelevant in another.

Incomplete information – When contemplating the economic impact of a privacy decision, most people are confronted with information that is incomplete and incorrect at best. For example, they have no idea who will use their data in the future, or for what purpose (Varian 1997). As a contemporary example: many people choose to use free apps with advertisements and banners. What they probably don't know, is that the banners in the apps send their phone's location to a marketing company every time they switch on their phone. This way, the marketing company collects thousands of data points. Every data point consists of a unique code identifying one's telephone, the location, and the date and time. The marketing company sells this data to parties that analyse it and use the results for a variety of purposes. The codes in the data points themselves cannot be traced back to individual telephones – except when it is known where the owner of that telephone usually resides: for example, at home, at the office, and at the gym. In that case, it is easy for an employee at the marketing company to identify the person behind the data, and from there, it is easy to determine where this person has been the past few years.

Temporal aspects of utility and cost – Benefits and costs of privacy decisions may change over time, reversing the economic balance multiple times in their course. For example, disclosing personal data may give immediate benefits, such as a discount provided by the merchant, but may also incur costs in the long term, because the merchant builds up a profile that he can use to get more out of his customer's wallet in the years to come. Additionally, the merchant may decide to sell the profile data to other parties, so that a consumer will never know who will control his personal data in the future, and for what purposes (Varian 1997). From construal level theory (Trope and Liberman 2010), we know that people clearly see the short-term, concrete effects of an action but have difficulty creating a clear mental picture of the long-term, more abstract effects. Thus, also when it comes to privacy matters, people tend to choose for the short-term benefit, and ignore the long-term cost, or vice-versa (Demmers 2017). In the example mentioned above, even if people knew about the secondary use of personal data, they would focus on the short-term benefit, ignore any potential long-term risks, and would still install the free app.

Irrational agent – In section 1, it was argued that the rational agent does not exist, and that economic decisions are always influenced by incomplete information, psychological biases and irrational behaviour. Assuming that the same holds for privacy-related decisions, the consequence is that there is no such thing as a clinical 'privacy calculus'; privacy decisions are never based on a purely rational trade-off of economic benefits and costs (Demmers 2017). Rather, choices are made unconsciously, or at best based on perceptions of net value that are influenced by many uncontrollable factors and may be a far cry from reality.

Reflection effect – According to what Kahneman (2002) has termed the reflection effect, people tend to estimate the negative value of loss higher than the positive value of gain. The reflection effect offers an explanation for one of the many paradoxes in privacy: when it comes to privacy, people seldom put their money where their mouth is. According to many studies, respondents will express concerns over their privacy when asked. There is a large gap, however, between people's privacy concern and the willingness to pay for a remediation of that concern. This is popularly known as the *privacy paradox* (Rainie et al. 2013). For example, according to a study by Beresford et al. (2012), participants, assumedly valuing their privacy, predominantly were willing to provide information about their monthly income and date of birth for a mere discount of one euro when purchasing DVDs online. The reflection effect and construal level theory offer an explanation for this phenomenon; consumers value the immediate opportunity loss higher than the potential long-term privacy gain.

Lack of choice – A trade-off implies that the decision maker actually has different options to choose from. Taking all immediate and long-term benefits and costs into account, the privacy trade-off boils down to the decision: do I disclose some of my personal data or do I keep it to myself? In the digital age, opting out is often no realistic option. Those who opt out from the digital economy not only face immediate economic disadvantages, but also less tangible disadvantages, such as social exclusion or reduced opportunities on the labour market. For all but a few consumers, the opportunity costs of not using big tech are prohibitive, leaving not much to trade off.

From the above, it follows that, in general, it is very difficult, if not impossible for individuals to make an informed economic privacy-trade-off.

4.4.2.2 *Privacy trade-offs for society*

Given the inherent difficulty of making the privacy trade-off for individuals, how are things for society? Do governments make trade-offs? The answer appears to be negative. Big policy decisions on privacy are seldom based on meaningful economic analyses. By its nature, it seems, privacy simply does not lend itself to quantitative impact analysis. As of to date, little to no comprehensive analysis of the costs and benefits of privacy regulation has been conducted and it is safe to say that privacy regulation is conceived and implemented without even the most basic form of impact analysis. Some feeble attempts have been made. The Conference Board of Canada (2012), for example, identifies a number of benefits of privacy regulation, including creating necessary conditions for economic growth; these benefits, however, are not quantified. The report also identifies a number of potential cost sources, including the cost of compliance, estimated at CAD 5142 per employee per year, and the impact on investment and innovation, adding up to total administrative costs of CAD 3.8 billion per year for Canadian companies. According to another report, privacy regulation in Europe has led to a reduction in venture capital investment in European online advertising companies of around USD or CAD? 249 million in nearly nine years (Lerner 2012). Results like these are incomplete at best and highly speculative at worst and provide too thin a basis for a well-informed and rational decision. Worst, their pretence of precision runs the risk of decisions being misinformed, irrational, or both.

4.4.3 Towards a new dominant school

Given the above, it must be concluded that the notion of individuals and organizations, let alone societies, rationally contemplating and calculating the economic, financial impact of their privacy-related decisions, is a complete and utter illusion.

One implication of this conclusion is the expectation that privacy economics itself will be transitioning towards a fourth wave that is not so much dominated by a focus on information economics, economic models, and digital technology but on human behaviour – much like Kahneman c.s. have transformed economics into behavioural economics. More on this in the next section.

4.5 New challenges and topical discussions

This section will discuss new challenges in the field of privacy economics, focusing on the following topics: the economic benefits and costs of privacy for society and the economics of privacy and trust, as has been argued elsewhere in this chapter.

4.5.1 Societal benefits of privacy: prosperity, growth, and well-being

In section 2, it has been argued that the immediate economic benefits of privacy for society at large requires more research.

A first angle to study the societal benefits of privacy would be to look at the direct effects of privacy on economic prosperity or even economic growth. Indeed, it is sometimes said that an adequate level of privacy is necessary to create the conditions for investment and economic growth (e.g. Descôteaux and Szoka 2013). Such claims, however, are seldom backed by solid evidence. It might be worthwhile to study the relationship between privacy and economic growth or prosperity – not only from a macroeconomic theoretical perspective, but also based on hard evidence, comparing different economies on both dimensions.

Another angle to study the positive economic effects of privacy for society at large may be found in the relation between privacy and subjective well-being (Diener et al. 1999). If such a relation exists, it would have an economic effect for at least two reasons: (a) well-being can be seen as having economic utility in itself, and (b) well-being is not only related to economic growth, but in fact should be seen as the ultimate goal of economic activity (Stutzer and Frey 2010). Intuitively, one would expect that the level of privacy in a society is in some way related to the subjective well-being of its members, but existing literature does not provide evidence to support or falsify this hypothesis. Dolan et al. (2008) give an extensive overview of academic research related to subjective well-being in the following categories: (1) income; (2) personal characteristics; (3) socially developed characteristics; (4) how we spend our time; (5) attitudes and beliefs towards self/others/life; (6) relationships; and (7) the wider economic, social, and political environment. In their review, privacy is not mentioned; trust, however, is, and the relation between privacy and trust will be touched upon briefly below.

4.5.2 Societal costs of privacy: bridging the gap between privacy and the common good

In section 2, the costs of privacy to society at large were briefly addressed, and it was concluded that privacy regulation may incur significant (although unquantifiable) opportunity costs, for example in the medical domain, where research is increasingly based on the collection and analysis of large amounts of clinical and biometric data residing in national or even international bio-banks. Blocking this flow of data through regulation may be desirable from a privacy standpoint but will also impair the speed and quality of medical research and thus harm society at large. Several authors have suggested possible solutions to close the widening gap between privacy and societal benefit. Konnoth (2015), for example, proposes to view the collection of medical data as a form of taxation – not monetary, but informational. In doing so, collecting medical data can be seen as a collective endeavour, in which all citizens participate.

Like the economic benefits, the opportunity costs of privacy and privacy policy deserve deeper research.

4.5.3 The economics of privacy and trust

A strictly economic view on privacy tends to overlook other dimensions of privacy found in scholarly literature, such as anonymity, secrecy, autonomy, freedom, solitude, etc. It could be worthwhile to take a closer look at the economics of these more personal and social dimensions of privacy. As noted above, one of these dimensions is the notion of *trust*. The economic importance of trust is undisputed and it is generally assumed that a high level of trust has a positive effect on economic growth. For example, it has been shown that social trust increases economic growth rates (Bjornskoff 2012), that there is a causal relation between the level of trust between international trade partners and the volume of trade flows between them (Den Butter and Mosch 2003), and that in low-trust environments, investment rates are reduced (Zak and Knack 2001).

For this reason, it would be interesting to gain insight into the relation between trust and privacy. One would expect privacy and trust to go hand in hand, given that they are both perceived as positive qualities, at least in the Western world.

Intuitively, the relationship seems simple. If you trust someone, there is no need to know one's deeds, whereabouts or other personal details. On the other hand, the less you trust the other party, the more information you will likely need to build a trust relation. So, in interpersonal relationships, trust inspires privacy. Respecting a person's privacy can be seen as an expression of trust, and disrespecting a person's privacy can be seen as an expression of distrust.

But the link is not symmetric. In a relationship between two entities, trust is a predictor for trustworthy future behaviour based on past evidence. The more information each partner has about the identity and the past of the other,

Fig.4.1: Privacy and trust: an asymmetrical relationship

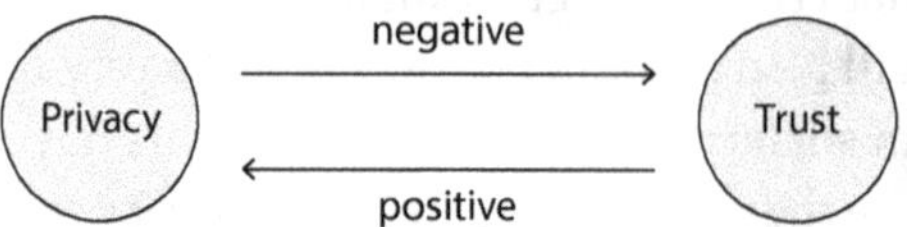

the higher the level of mutual trust can be. Apparently, in this case, privacy is not congruent with trust, but can be traded for it (Seigneur and Damgaard Jensen 2004). Indeed, it is privacy – perceived as secrecy – that inspires distrust, and transparency – or a *lack* of privacy – that inspires trust (see also Posner 1981). Apparently, the connection between privacy and trust in individual relationships is more complicated than we think and deserves a closer study.

It would be equally interesting to investigate the relation between privacy and trust on a societal scale. There is some evidence that societies with a relatively low level of privacy – and therefore a high level of transparency – exhibit a high level of trust: Friedman and Resnick (2001), for example, point out that in communities where people use pseudonyms rather than real identities, trust building will be low because people cannot be held accountable. On the other hand, we might find that communities where people trust each other have a high level of privacy. The relation between privacy and trust in individual and societal settings seems a worthwhile subject for future research.

4.6 Conclusion

In this chapter, we have identified and clarified various factors of influence on the economics of privacy in the digital age. As it turns out, it is relatively easy to identify positive economic factors (benefits) and negative economic factors (costs) of privacy for individuals, organizations, and society at large. For individuals, controlling the disclosure of personal data has significant direct benefits, but also leads to opportunity costs: the indirect costs of not being able to enjoy other benefits. For private and public organizations, collecting and using personal data leads to significant economic benefits; prohibiting them from doing so will erode their competitive advantage and incur opportunity costs. For society at large, the situation is quite unclear.

It is extremely difficult to quantify the above factors in a meaningful way, due to a lack of empirical data and the large influence of subjectivity and context on the valuation of privacy and personal data. As a result, the generally accepted notions of 'privacy trade-off' need to be revisited, as privacy trade-offs cannot be made objectively and rationally, let alone quantitatively.

Further reading

The reader is strongly encouraged to read the material listed below

Acquisti et al. (2016) give a reasonably complete, but rather terse overview of existing research and provide a good starting point.
Nissenbaum (2012) is a must-read for those interested in the impact of digital technology.
Ariely (2009) describes the end of rational economics in a rather entertaining way.
The work of Varian (1997 and later) provides excellent insights into the basics of privacy economics.
Posner (1981) and Stigler (1980) give a good insight into the early ruminations on privacy by the Chicago School.
Zuiderveen Borgesius (2015) gives insight into the relationship between privacy economics and price discrimination.
Goldfarb and Tucker (2012) dive into the subject of privacy and innovation.

References

Acquisti, A., C. Taylor, and K. Wagman, K. (2016). 'The Economics of Privacy'. *Journal of Economic Literature* 54ª2), 442-492.

Ariely, D. (2009). 'The End of Rational Economics'. *Harvard Business Review* July-August.

Backhouse, R.E., and S.G. Medema. (2009). 'Retrospectives: On the Definition of Economics'. *The Journal of Economic Perspectives* 23(1), 221-234.

Beresford, A.R., D. Kübler, and S. Preibuscha. (2012). 'Unwillingness to Pay for Privacy: A Field Experiment'. *Economics Letters* 117, 25-27.

Bergemann, D. and A. Bonatti. (2015). 'Selling Cookies'. *American Economic Journal: Microeconomics* 7(3), 259-294.

Bjornskoff, C. (2012). 'How Does Social Trust Affect Economic Growth?' *Southern Economic Journal* 78(4), 1346-1368.

Calzolari, G. and A. Pavan. (2006). 'On the Optimality of Privacy in Sequential Contracting'. *Journal of Economic Theory* 130(1), 168-204.

Chen, Y. and Z.J. Zhang. (2009). 'Dynamic Targeted Pricing with Strategic Consumers'. *International Journal of Industrial Organization* 27(1), 43-50.

Coase, R.H. (1937). 'The Nature of the Firm'. *Economica* 4, 386-405.

Cohen, J.A. (2005). *Intangible Assets – Valuation and Economic Benefit.* City: Wiley.

Conference Board of Canada. (2012). *Exploring the Iceberg: The Economic Impact of Privacy Policy, Laws and Regulations on Commercial Activity.* City: Publisher.

De Cornière, A. (2016). 'Search Advertising'. *American Economic Journal: Microeconomics* 8(3), 156-188.

De Cornière, A. and R. De Nijs, R. (2016). 'Online advertising and privacy'. *RAND Journal of Economics* 47(1), 48-72.

Demmers, J. (2017). *Consumers and their data – when and why they share it.* Amsterdam: University of Amsterdam. PhD Dissertation.

Den Butter, F.A.G. and R.H.J. Mosch. (2003). 'Trade, Trust and Transaction Cost'. Tinbergen Institute Discussion Paper 03-082/3.

Descôteaux, D. and B. Szoka. (2013). *Protecting Personal Data: The Economic Impact of Regulating the Internet.* Montreal: Montreal Economic Institute.

Dhar, V. (2012). 'Get Paid for Your Data on Facebook'. *Wired.com* 6 November. Accessed on date.

Diener, E., E.M. Suh, R.E. Lucas, and H.L. Smith. (1999). 'Subjective well-being: Three Decades of Progress'. *Psychological Review* 125, 276-302.

Dolan, P., T. Peasgood, and M. White. (2008). 'Do We Really Know What Makes Us Happy? A Review of the Economic Literature on the Factors Associated with subjective well-being'. *Journal of Economic Psychology* 29, 94-122.

Dugas, A.F. et al. (2012). 'Google Flu Trends: Correlation with Emergency Department Influenza Rates and Crowding Metrics'. *Clinical Infectious Diseases* 54(4), 463-469.

Eurostat. (2018). *Internet Access and Use Statistics – Households and Individuals.* ec.europa.eu.

Friedman, M. and A.J. Schwartz. (1963). *A Monetary History of the United States, 1867-1960.* National Bureau of Economic Research Publications. Princeton: Princeton University Press.

Fudenberg, D. and J. Tirole. (2000). 'Poaching and Brand Switching'. *The RAND Journal of Economics* 31(4), 634-657.

Fuegi, J., and J. Francis. (2003). 'Lovelace & Babbage and the creation of the 1843 "notes"'. *Annals of the History of Computing* IEEE 25(4), 16-26. doi: 10.1109/MAHC.2003.1253887.

Goldfarb, A. and C. Tucker. (2012). 'Privacy and Innovation'. *Innovation Policy and the Economy* 12(1), 65-89.

Gottlieb, D. and K. Smetters. (2011). 'Grade Non-disclosure'. National Bureau of Economic Research Working Paper 17465.

Greenwald, G. (2013). 'NSA Prism Program Taps in to User Data of Apple, Google and Others'. https://www.theguardian.com/world/2013/jun/06/us-tech-giants-nsa-data.

Hirshleifer, J. (1980). 'Privacy: Its Origin, Function, and Future'. *Journal of Legal Studies* 9(4), 649-664.

Kahn, C.M., J. McAndrews, and W. Roberds. (2000). 'A Theory of Transactions Privacy'. FRB Atlanta Working Paper 2000-22.

Kahneman, D. (2002). *Maps of Bounded Rationality: a Perspective on Intuitive Judgment and choice.* Nobel Prize Lecture.

Koetsier, T. (2001). On the Prehistory of Programmable Machines: Musical Automata, Looms, Calculators. Mechanism and Machine Theory. *Elsevier* 36(5), 589-603. doi: 10.1016/S0094-114X(01)00005-2.s

Konnoth, C. (2016). 'Classification and Standards for Health Information: Ethical and Practical Approaches'. *Washington and Lee Law Review Online* 72(3), 397-408.

Krugman, P. and R. Wells. (2004). *Microeconomics.* New York: Worth.

Laudon, K.C. (1996). 'Markets and Privacy'. *Communications of the ACM* 39(9), 92-104.

Mankiw, N.G. (2001). *Principles of Economics.* 2nd Ed. Harcourt. Fort Worth: Harcourt.

Miller, A.R., and C. Tucker. (2009). 'Privacy Protection and Technology Diffusion: The Case of Electronic Medical Records'. *Management Science.* 55(7), 1077-1093.

Moore, G.E. (1965). 'Cramming More Transistors onto Integrated Circuits'. *Electronics* 38(8), 56-59.

Mueller, R. (2018). Internet Research Agency Indictment. Case 1:18-cr-00032-DLF.

Munsterman, R. (2014). 'ING stopt Big-Data plan terug in de kooi'. *Follow The Money* March 17. https://www.ftm.nl/artikelen/ing-trekt-big-data-plan-terug

Nissenbaum, H. (2010). *Privacy in Context – Technology, Policy, and the Integrity of Social Life.* Stanford Law Books. Stanford: Stanford University Press.

Posner, R.A. (1981). 'The Economics of Privacy'. *The American Economic Review* 71(2), 405-409. New York: American Economic Association, 405-409.

Rojas, R. (1997). 'Konrad Zuse's Legacy: The Architecture of the Z1 and Z3'. *IEEE Annals of the History of Computing* 19(2), 5-16.

Schwab, K., A. Marcus, J.R. Oyola, W. Hoffman, and M. Luzi. (2011). *Personal Data – The Emergence of a New Asset Class*. City: World Economic Forum.

Seigneur, J.M. and C. Damsgaard Jensen. (2004). *Trading Privacy for Trust. Proceedings of the International Conference on Trust Management. Lecture Notes in Computer Science*. Heidelberg: Springer, 95-107.

Simon, H. (1991). 'Bounded Rationality and Organizational Learning'. *Organizational Science* 2(1), 125-134.

Solove, D.J. (2007a). *The Future of Reputation: Gossip, Rumor, and Privacy on the Internet*. Yale: Yale University Press.

Solove, D.J. (2007b). '"I've Got Nothing to Hide" and Other Misunderstandings of Privacy'. *San Diego Law Review* 44, 1-23.

Spence, M. (1973). 'Job Market Signaling'. *Quarterly Journal of Economics* 87(3), 355-374.

Stutzer, A. and B.S. Frey. (2010). 'Recent Advances in the Economics of Individual Subjective Well-Being'. *Social Research* 77(2), 679-714.

Tamir, D.I. and J.P. Mitchell. (2012). 'Disclosing Information about the Self is Intrinsically Rewarding'. *Proceedings of the National Academy of Science* 109(21), 8038-8034.

Taylor, C.R. and L. Wagman. (2014). 'Consumer Privacy in Oligopolistic Markets: Winners, Losers, and Welfare'. *International Journal of Industrial Organization* 34, 80-84.

Trope, Y., and N. Liberman. (2010). 'Construal-Level Theory of Psychological Distance'. *Psychological Review* 117(2), 440-463.

UN (1948). *Universal Declaration of Human Rights*. New York: United Nations.

UN (2017). *World Happiness Report*. New York: United Nations.

van Manen, M., and B. Levering. (1996). *Childhood's Secrets: Intimacy, Privacy, and the Self Reconsidered*. New York: Teachers College Press.

Varian, H.R. (1997). *Economic Aspects of Personal Privacy*. Privacy and Self-Regulation in the Information Age. Boston: US Department of Commerce.

Varian H.R. (2009). 'Economic Aspects of Personal Privacy' in W. Lehr, and L. Pupillo L. (eds.), *Internet Policy and Economics*. Boston: Springer.

Villas-Boas, J. Miguel. (2004). 'Price Cycles in Markets with Customer Recognition'. *RAND Journal of Economics* 35(3), 486-501.

Westin, A.F. (1967). 'Legal Safeguards to Insure Privacy in a Computer Society'. *Communications of the ACM* 10(9), 533-537.

Westin, A.F. (1968). 'Privacy And Freedom. Book Review'. *Washington & Lee Law Revie* 25(1), 166-167.

White, T.B., D.L. Zahay, H. Thorbjørnsen, and S. Shavitt. (2008). Getting too Personal: Reactance to Highly Personalized Email Solicitations. *Marketing Letters* 19(1), 39-50.

White, R.W., N.P. Tatonetti, N.H. Shah, R.B. Altman, and E. Horvitz. (2013). 'Web-Scale Pharmacovigilance: Listening to Signals from the Crowd'. *Journal of the American Medical Informatics Association* 20, 404-408.

Zak, P.J. and S. Knack. (2001). 'Trust and Growth'. *The Economic Journal* 111(470), 295-321.

Zuiderveen Borgesius, F. (2015). Online Price Discrimination: Is and Should It Be Allowed? Amsterdam: Amsterdam Privacy Conference.

Security, Privacy, and the Internet of Things (IoT)

Mikko Hypponen

The IoT revolution is happening, whether we like it or not. And the reason is simple: Cost versus benefits. It's becoming very cheap to add Internet connectivity to appliances and things. When connectivity is cheap, the benefits don't have to be very large for vendors to adopt it.

In many cases, devices won't go online to benefit the consumer; rather, the benefit will be for the manufacturer. For example, home appliances can collect analytics about how and when they're used, or about customers' physical locations. Information like this is extremely valuable to vendors. This means that even the most mundane of machines, like toasters, will eventually go online – to collect data. Because data is the new oil.

That's not to say that the IoT won't offer consumer benefits – of course it will. Imagine the convenience of being able to fire up your coffee maker while you're in bed or switch on your washing machine while you're at work. Smart homes also offer improved safety – think of a security system that alerts you when it detects something suspicious. Energy efficiency is another benefit, one that translates to cost savings – take the example of a thermostat that optimizes performance based on your behaviours. And a host of other IoT innovations promise to boost our quality of life.

Do a Google image search for 'smart home' and you're bombarded with visuals of sleek, polished living spaces in ultramodern white. But the problem with this whole picture, as attractive as it is, is that cyber security is too often missing from the design.

Cyber security, you see, is not a selling point for something like a washing machine. Selling points for washing machines are size, colour, price, load capacity, and wash programmes. Because security is not a selling point, appliance vendors can't invest a lot in it. This leads to insecure appliances. And we've already seen where that leads.

We've already seen several botnets targeting IoT, the biggest of them being the Mirai botnet. There were more than 100,000 hacked systems in the original Mirai attack network, and none of them were computers – they were all IoT devices. In other words, they were all appliances from our homes.

The existence of the Mirai botnet was possible because of the use of default login credentials on those devices. Hackers wrote a malicious piece of software that tried out manufacturers' known default username and password combinations against devices found on the Internet. As we know, too many devices were using default credentials. The consequences were far-reaching.

This is just one of the IoT security concerns that needs more attention. Other problematic issues are infrequent or non-existent software updates, indiscriminate data collection, and lack of proper data encryption.

As IoT appliance vendors want to collect more and more data, even the 'traditional' or 'stupid' home appliances will eventually be online. Once the costs of the IoT chipsets plummets to a few cents per unit, everything will go online. And they will go online so the vendors can get the data they want.

The adoption of the technology is being driven by businesses eager to gain valuable data from citizens, with little concern for their privacy or the protection of that data.

The IoT has profound implications for us in terms of surveillance, privacy, and consumer rights. Without rights and protections, we are at risk of becoming a component of the IoT. So instead of being in control of the technology, we may end up impotent, left to the mercy of the sensors, the databases, the servers, and the analytical software engines and algorithms that now roam the Internet.

All of these companies have been to analyst briefings and they've been told over and over again that data is the new oil and they look at Google and Facebook and see them making billions out of analytics, so they want to collect analytics. It's clear that some of it is useful to companies because they know physically where their customers are, and when they are using their products

Many consumers have not yet even considered the implications of this one-way flow of data from their homes.

The Mobile Ecosystem Forum did a poll for 5,000 mobile users and found that globally 62% were concerned about their privacy and 54% were worried about threats to their home security. In the US, the figure rises to 70% and it stands at 69% in France. Meanwhile a survey released by Gartner found that almost two thirds of consumers are worried about IoT devices in their homes eavesdropping on their conversations.

Perhaps even more disturbingly for the technology industry, the survey also found that most people were not convinced that they needed a smart home. Many of the benefits of a smart home, such as automating tasks around the house such as dimming and turning off lights, controlling heating systems, and carrying out other household tasks left people cold, with 75% of the 10,000 people contacted responding that they would rather do those things themselves than have an IoT device do them.

The frustration and confusion that the population feels regarding technology is a very real issue. There is now a need to increase awareness of cyber security issues among people who feel disenfranchised by technology.

The Internet wasn't built for security or privacy. We built it first, and then we realized we needed to play catch-up to secure it. We're still working on that all the time. Unfortunately, the Internet of Things is not being built for security either. But it's not too late.

We need to take IoT security seriously, now. Before the problems caused by neglecting it become too difficult to handle.

The IoT revolution is happening, whether we like it or not.

5. Privacy from an Informatics Perspective

Matthijs Koot & Cees de Laat

5.1 Introduction

Both 'privacy' and 'informatics' are semantically overloaded concepts; no broad consensus exists on a single definition of either. This chapter has the following objectives:

- to provide an intuition of 'privacy' and of 'informatics';
- to provide an understanding of relations between privacy and informatics;
- to provide references to academic and other authoritative sources for further research.

Elaboration is provided on selected topics in this theme. For topics that are already described and discussed in existing sources, references are provided.

5.1.1 An intuition of privacy

At the risk of minor overlap with other chapters, a short characterization of privacy follows to keep this chapter self-contained. It is adapted from earlier work.[1]

Privacy entails some desire to hide one's characteristics, choices, behaviour, and communication from scrutiny by others. A corollary is that privacy entails some desire to exercise control over the use of personal information, for example to prevent future misuse. Phrases commonly associated with privacy include[2] 'the right to be let alone', meaning freedom of interference by others; 'the selective control of access to the self or to one's group', meaning the ability to seek or avoid interaction in accordance with the privacy level desired at a particular time; and 'informational self-determination', meaning the ability to exercise control over disclosure of information about oneself.

Contrary to what some believe, the rise of social media and ubiquitous computing does not imply the 'end' or 'death' of privacy. Rather, as Evgeny

1 Koot 2012.

2 Warren 1890; Altman 1975.

Morozov paraphrased from Helen Nissenbaum's book[3] on contextual integrity in *The Times Literary Supplement* of 12 March 2010: 'the information revolution has been so disruptive and happened so fast (...) that the minuscule and mostly imperceptible changes that digital technology has brought to our lives may not have properly registered on the social radar'. In her two and a half-year ethnographic study of American teens' engagement with social network sites, danah boyd observed[4] that teens 'developed potent strategies for managing the complexities of and social awkwardness incurred by these sites'. So, rather than privacy being irrelevant to them, the teens found a way to *work around* the lack of built-in privacy. In conclusion: privacy is not dead. At worst, it is in intensive care, beaten up by overzealous and careless use of technology. It can return to good health as policymakers, technologists, and consumers learn why, what, where, when, and how to define privacy objectives.

Privacy can also be conceived of as a means of personal security: by controlling disclosure of one's own personal information, one can self-protect against known and unknown threats stemming from potential (future) uses of that information, such as identity fraud or yet-unforeseen uses of profiling.

Now that a broad intuition of privacy has been given, an intuition of informatics follows. Further on, the relation between privacy and informatics will be defined in terms of the importance of information security to privacy.

5.1.2 An intuition of informatics

In this chapter, 'informatics' is meant in the sense of 'Information and Communication Technology' (ICT): the hardware and software that spawn from science, technology, engineering, and mathematics (STEM) and enable storage, processing, and communication of data. Relevant academic disciplines include, inter alia, computer science, electrical engineering, information science, and logic.

For two reasons, this chapter does not focus on a single STEM discipline, but on applications of their, often joint, outcomes. First, legibility must be maintained for readers that have no background in STEM disciplines. Second, privacy issues are often not yet sufficiently clear in the course of practising any single discipline without considering specific applications. For instance, design of computer networking and wireless communication

3 Nissenbaum 2010.

4 boyd, 2008. (Note: boyd spells her Christian name and surname in lowercase, as explained here: http://www.danah.org/name.html.)

protocols may focus firstly on achieving robust and efficient means of communication, and not always take security and privacy requirements into account that emerge in their use in certain application domains. Similarly, the fundamentals of artificial intelligence are purely mathematical, and not until the mathematics are applied to specific domains (healthcare, public security, insurance, and so on), specific security and privacy risks start to become clear.

The design and use of ICT for the processing of personal data by definition relates to privacy. The use of technology results in increased frequency and size of collection, retention and use of personal data, and generates forms of personal data that did not exist before: for instance, sensors inside personal devices that make measurements about the user and/or the user's environment, such as the pedometer, gyroscope, location-related sensors based on the Global Positioning System[5] (GPS), and data trails due to Wi-Fi, Bluetooth, ZigBee, and so on. These measurements are not a privacy problem per se, but the relation between the measurements, the user's identity, and other data results in new potential privacy hotspots, depending on who can access the data. This is especially relevant when devices are tethered to a service provider or corporate environment where the user's real, verified identity is already known, such as in the case of personal devices tethered to Apple or Google, or enrolled in a corporate Mobile Device Management (MDM) environment.

ICT functions can be grouped into three areas:
- **storage**: solid-state disks, hard disks, etc.;
- **networking:** network equipment, communication protocols, etc.;
- **computation**: Central Processing Units (CPUs), Field Programmable Gate Arrays (FPGAs), Systems-on-Chip (SoCs), algorithms, etc.

These functions respectively map to three main states of data:[6]
- **data at rest**: data while stored;
- **data in transit**: data while transferred over computer networks;
- **data in use**: data while calculations are performed on it.

Software applications run on devices and communicate via network infrastructures to provide functionality to end-users. The distinction between

5 Or based non-US alternatives to GPS such as Galileo (EU), BeiDou (China), and Glonass (Russia).

6 The three-states model is useful to provide an understanding of ICT, but is not formally defined.

the three states of data is not apparent to the end-user, but does matter for those who want to understand data protection from a technological perspective. There is no single mechanism that protects data in all states: the mechanisms to protect data in transit are different from mechanisms to protect data at rest, and so on; although basic building blocks can serve purposes in more than one data state, such as cryptographic algorithms.

When a smartphone user takes a photo and shares it via Facebook's mobile app, for instance, what happens can be approximated in simplified terms as follows:

- First, the image sensor ('camera') of the phone generates data, which is then processed by the CPU (data in use) and finally stored on the phone (data at rest).
- Second, the Facebook app reads the photo from disk (the photo then becomes data in use) to send it to Facebook's data centre (data in transit).
- Third, in Facebook's data centre, the photo is processed while being received (data in use) and then stored on disk in Facebook's data centres (data at rest).

Being aware of these three states helps grasp data and communications privacy from an informatics perspective, including potential threats to privacy and countermeasures to protect against such threats. A selection of available protective measures in each state will be discussed shortly, after first introducing basic security and privacy controls which can be a part of those protections.

Digital privacy requires digital security. Security is a systems property: all components must be secure in order for the system as a whole to be secure and by extension to protect user privacy: hardware, operating systems, and applications. If the security of one component fails, other components can fail, undermining security and as a result potentially undermining privacy; for instance when the vulnerabilities result in data breach. Vulnerabilities in software and hardware are still a fact of life. For that reason, an elaboration on digital security follows in the next section.

Whereas the concept of 'privacy' is not well defined in informatics, a proposal for common definitions of 'anonymity' and related concepts exists in the area of anonymity research due to Pfitzmann and Hansen.[7] A simplified explanation of 'anonymity':

- a **subject** can be said to be sufficiently anonymous;
- from the perspective of an **observer**;

7 Pfitzmann and Hansen 2010.

- with regard to an **item of interest**;
- if the **observer** cannot link the **item of interest** to the **subject** with sufficiently high probability to be useful to the observer's objective.

The subject is a person, the item of interest is an activity or data (e.g. an online transaction, a database record, or knowledge of the subject's social network), and the observer is an entity from which the subject seeks to hide its link to the item of interest ('unlinkability'). Depending on context, potential observers may include untrusted peers on a shared system or network, Internet providers, or a so-called 'global passive observer' who is attributed the ability to eavesdrop on large parts of global Internet traffic (e.g. multinational cooperation between intelligence agencies, CloudFlare, and so on).

Informatics affects privacy of personal information, privacy of personal behaviour, and privacy of personal communications; and with the emergence of wearables, millimetre wave body scanners, and e-health devices, also privacy of the person ('bodily privacy'). The use of technology such as mobile apps generates a continuous stream of 'items of interest' that are, from the perspective of its creators, linkable to an identified or identifiable subject. The latter certainly applies to mobile apps that require the user to register via a social media account ('social login').

5.1.2.1 Security and privacy controls

A characterization of information security that gained popularity since its conception at NASA in the 1970-1980s, is the so-called 'CIA triad':

1. **confidentiality**: protecting data against unauthorized read access. Example measures: logical access control (make sure only user X or group Y can read a file or a certain record in a database), physical access control (access to server rooms), encryption (make sure only users who have the right cryptographic key can access data);
2. **integrity**: protecting data against unauthorized write access. Example measures: cryptographic signatures, logical access controls;
3. **availability**: making sure data is available to authorized users. Example measures: redundant data storage and connectivity, making backups of data.

Privacy can be a motivating factor for deciding on these controls. While the CIA triad, in its simplicity, is still widely present in expert publications, it has been argued that these three controls alone are insufficient for the proper understanding of reality and advancing security.[8] For instance, authentication, authorization, and non-repudiation have been suggested to be included as separate controls, rather than implied to be part of the three traditional controls.

A popular approach to threat modelling named STRIDE[9] captures this. Threat modelling can help detect security threats (or privacy threats[10]) that may exist despite security controls, or due to a lack of security controls.[11] STRIDE was created by Microsoft in 1999, and is an acronym for six types of threats, each of which has an associated security control to counter it:

– **Spoofing**: possibility to impersonate a user
  Security control: **authentication**
– **Tampering:** possibility to perform unauthorized changes
  Security control: **integrity**
– **Repudiation:** possibility to deny that an action was performed
  Security control: **non-repudiation**
– **Information disclosure** (data breach): possibility to access/obtain data
  Security control: **confidentiality**
– **Denial of service:** possibility to render a service unavailable to legitimate users
  Security control: **availability**
– **Elevation of privilege**: possibility to obtain more or higher privileges
  Security control: **authorization**

Distinguishing six security controls and types of threats, rather than three, provides a more fine-grained way to identify potential threats and decide on countermeasures.

The STRIDE threat modelling process is informal and, at a minimum, consists of drawing a high-level diagram about a system or infrastructure, and subsequently identifying 'trust boundaries'. For an internet-facing web application, for instance, a trust boundary exists at least between the web application and its end-users: systems should never trust user input to conform to what the application (implicitly) expects. Failing to do so

8 Ross 2016.

9 Shostack 2014.

10 Threat modelling can also be applied to privacy. For instance, see Adam Shostack, 19 February 2018: 'Threat Modeling the Privacy of Seattle Residents'. Available at https://seattleprivacy.org/threat-modeling-the-privacy-of-seattle-residents/

11 Threat modelling can also be applied to privacy. For instance, see Adam Shostack, 19 February 2018: 'Threat Modeling the Privacy of Seattle Residents'. Available at https://seattleprivacy.org/threat-modeling-the-privacy-of-seattle-residents/

may result in vulnerabilities that can be exploited to gain access to the system, the data and/or or underlying infrastructure. Everywhere a data flow crosses a trust boundary, the STRIDE elements can be considered to determine which threats are relevant and necessitate protective controls.

Which protective controls should be implemented is a context-specific matter and depends on risk management and the economics of information security and privacy. It is important to note that technologies that provide confidentiality, integrity, availability, authentication, authorization, and non-repudiation can serve security objectives and privacy objectives simultaneously.

It is important to validate whether security controls are implemented adequately. This is usually done through mandatory compliance requirements. Ideally, these are not merely approached as a 'checkbox exercise' that should be passed with the least possible effort, but embraced by upper management as critical to values. Requirements can include operational security testing such as subjecting ICT infrastructure or applications to (authorized) penetration tests, social engineering, and so on. This provides insight into the vulnerabilities in technology, procedures, and human behaviour. Security testing is already mandatory for certain categories of ICT: for instance, systems that offer their users a login via the Dutch national authentication scheme DigiD must be subjected to such testing every year. This is in accordance with a norm[12] issued by the Dutch government. Similar requirements exist or may emerge in other domains.

As long as vulnerabilities in software and hardware exist, there is a potential risk to security and privacy. The 'legacy problem' exacerbates this: organizations that keep business-critical systems that contain known vulnerabilities operational because no patches, upgrades, or less vulnerable alternatives are readily available. The legacy problem can also exist at the level of individuals: not all vendors of personal devices provide patches for the entire expected device lifetime, not all users know how to install the patches, and not all can afford to buy newer, less vulnerable models; so individuals, too, can keep vulnerable devices in use.

Data protection regulation requires data controllers to ensure that personal data has 'appropriate security'. It does not make a distinction between states of data. To assess what 'appropriate' means, threats must be

12 Specifically, norm elements C.03 and C.04 of the 'Norm ICT-beveiligingsassessments DigiD' versio 2.0, issued by Logius, a body of the Ministry of the Interior. Available at https://www.logius.nl/fileadmin/logius/ns/diensten/digid/assessments/20161215_norm_V2_ict-beveiligingsassessments_digid.pdf

identified while taking into account available methods for digital security. The next sections provide an understanding of how data can be protected in its various states: at rest, in transit, and in use.

5.1.2.2 *Protecting data at rest*

Protection of data at rest can consist of physical, procedural, and logical measures. Logical measures include applying encryption, keeping encryption keys secure, and applying access controls (authentication and authorization) to disk storage (filesystem permissions) and to end-user applications (application access permissions) which can access data from storage. This holds for any computer: standalone computers at home, in-house corporate file servers, shared infrastructure in data centres,[13] and so on.

For data stored in a data centre, physical protection involves technical and procedural measures to prevent, detect, and (insofar possible) repress unauthorized physical access to the data centre and within the data centre itself (compartmentalization; customers should not be able to physically access equipment of other customers). Besides fences, security cameras, burglar alarms, and physical presence of security personnel, authorized persons should be trained to be mindful of attackers attempting to gain access through social engineering. For instance, attackers may attempt to impersonate an ICT vendor, cleaning company, elevator repair person, a customer, as well as leveraging tricks to distract or manipulate security personnel to gain access. Social engineering may also involve bribery or blackmail of authorized persons. Personnel at high-privilege positions, such as security personnel themselves, may need periodic screening for potential vulnerability to enticement by criminals or foreign states via for instance Money (bribery), Ideology (strong political or religious views), Coercion (blackmail), or Ego (e.g. self-importance or revenge) (MICE) or other angles.[14]

5.1.2.3 *Protecting data in transit*

Protection of data in transit, too, consists of physical, procedural, and logical measures. Internet exchanges are organizations that route network traffic between Internet providers and eventually, via Internet access providers, to end-users, including corporate consumers and individual consumers at home. The exchanges have to cope with risks that are similar to that of data

13 The term 'data centre' is used throughout this chapter. 'Cloud computing' is a marketing term that designates data centres: at all times, data is stored on real equipment, accessible by real operators, in a real jurisdiction.
14 Burkett 2013.

Fig 5.1: The seven-layer OSI model of data communication.

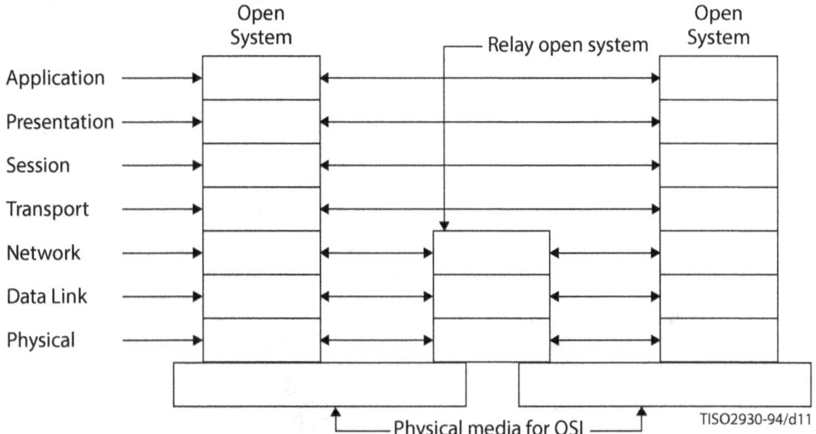

centres: the networking equipment should be protected against unauthorized physical or logical access. This is a responsibility of these exchanges.

The Open Systems Interconnection (OSI) model laid down in 1994 in ISO/IEC 7498-1 is a reference model used to characterize, design, and engineer protocols for communication between devices and applications running on those devices, including internet protocols (a topic that will be returned to later in this chapter). A basic understanding of the OSI model helps understand the protection of data in transit. The OSI model is a reference model for communication protocols. The picture below depicts a core part of the OSI model, namely the distinction of seven functional layers.

Here, 'Open System' refers to a device that participates as a sender or (final) receiver in the communication, for instance a smartphone, a laptop, or a web server in a data centre. 'Relay open system' refers to what is commonly referred to as a router. In a connection between two open systems on the global Internet, a packet travels across a series of intermediate routers, informally referred to as 'hops'. When browsing the web directly from a home computer, the home router is the first hop.

The OSI models specifies seven functional layers, seen at the left and right 'towers' in fig 5.1. Many common Internet protocols do not strictly fit in a single layer, but the model does serve a shared vocabulary in, firstly, engineering communities. The model can be (very) roughly simplified to four parts:

1. **application** + **presentation** + **session**: e.g. HTTP (web), SMTP (email), DNS ('the Internet's phonebook');

   intuition: a letter is typed by a user;

2. **transport**: e.g. TCP, UDP;
   intuition: the letter is put inside an envelope;
3. **network**: e.g. IPv4, IPv6;
   intuition: the recipient address is written on an envelope and the envelope is handed over to a postal service;
4. **physical** + **data link**: e.g. Ethernet over optic-fibre cables, Wi-Fi/Bluetooth over radio;
   intuition: the postal service hands over the envelope to an intermediate postal service, which hands it over to another intermediate postal service, and so on, until the envelope is delivered by the recipient's own postal service.

To make sure the letter in the envelope (example: an HTTP request sent by a browser, an email message, a DNS lookup) is delivered at the intended recipient, and that postal employees cannot read or change the letter or the envelope, measures can be taken at various layers. For instance, at the top layer, the sender and recipient may agree on a certain method and/or code for secret writing, so that the letter is only legible by them, unless an attacker has compromised the method or code. This is 'end-to-end encryption'.

In addition to that, the envelope can be sealed and tamper-evident. This can be done through SSL/TLS, as seen in e.g. HTTPS[15] and SMTPS.[16] Also, the postal vehicle can be armoured and protected against unauthorized road diversions: IP layer encryption may be used, DNSSEC (to protect attackers from tricking the phonebook into giving users a wrong number), and at the IP resource level through, i.a. Resource Public Key Infrastructure (RPKI).

The use of SSL/TLS, best known in relation to HTTP, where its use is referred to as HTTPS or informally 'the padlock in the browser', can provide confidentiality and integrity for communication. It provides confidentiality of communication against snooping by whoever is able to access a communication link between two communicating devices. For instance between a smartphone that runs a web browser and the web server that it connects to, or between servers in two data centres. It provides integrity through the use of cryptographic signatures over the contents of the communication: the sender cryptographically signs the communication content, and the receiver verifies this signature. If the verification fails, the data may have been tampered with, and the receiving system will reject the data. Both integrity and confidentiality are provided through cryptography and public

15 HTTP + SSL/TLS = HTTPS.
16 SMTP + SSL/TLS = SMTPS.

key infrastructure (PKI). For critical perspective on the latter topics, readers are referred to Asghari (2012) and Durumeric (2013).

SSL/TLS can be said to provide privacy in that the confidentiality it brings protects users against behavioural profiling by ISPs. A well-known example of (planned) snooping by ISPs is found in the UK around 2008: three ISPs considered deploying the Phorm Webwise system,[17] which would allow the ISPs to monetize on subscriber's web traffic through targeted advertising based on profiling built using keyword searches in individual users' web traffic. The plans led to public outcry, and were subsequently withdrawn. The Webwise system involved technology that is referred to as Deep Packet Inspection (DPI). HTTPS can help protect against such techniques.

5.1.2.4 Protecting data in use

Protection of data in use is a relatively state-of-the-art topic, and involves the use of novel cryptography to perform operations on encrypted data. That means that data never has to exist in unencrypted form on the system that performs calculations on it. Specifically, this involves 'fully homomorphic encryption'.[18]

At all times, the fundamental underlying question is: where is the data, and how does it need to be protected? One way to examine this properly is through the use of threat modelling:[19] an informal but structured approach to model threats and defences to data flows at any level of abstraction. This method can be applied to discover threats and decide on defences for data flows across the world, inside a single organization, inside a single device, or inside single application. The latter, for instance, is relevant when data is processed by a mobile app, and the mobile app must be robust against other apps running on the same mobile device.

Recognizing the three basic states of data is important to understand what data protection entails from a technical perspective. Data may be protected while transferred over the network using SSL/TLS, but be stored unencrypted on servers. Proper protection takes into account the entire lifecycle of data, from the moment it enters the system (from a sensor or from user input) until it is definitively removed.

Now that an intuition of both privacy and informatics is provided, the next section constructs two perspectives on the relation between both.

17 Clayton 2008.
18 Gentry 2009; Dulek 2017.
19 Shostack 2014.

5.2 Meaning and function of privacy

In simplified terms, the relation between privacy and ICT can be understood from two perspectives:

- ICT poses privacy challenges;
- privacy poses ICT challenges.

The first perspective gives examples of how the adoption of Internet technology – and vulnerabilities that come with it – gives rise to security needs at businesses and governments, and how the fulfilment of those needs can affect privacy. The second perspective focuses on how the need for privacy, whether expressed in policy and laws or expressed by individuals and groups, gives rise to requirements that technologists generally were not used to take (sufficiently) into account. Both perspectives are discussed next.

5.2.1 Perspective: 'ICT as a privacy challenge'

The first perspective, 'ICT as privacy challenge', pertains to the ever-increasing scale of computation, storage, and network power and use of that power in the private and public sector exacerbates existing privacy challenges. Examples include:

- private companies performing checks on social media. Besides legitimate uses, such as identifying insurance and welfare fraud, arguably less legitimate uses exist, such as screening and retaining employees' opinions expressed on social media that are not related to their job;
- privacy companies 'taking in' social media for commercial objectives, including marketing;
- the use of big data for safety and security, crowd control, behavioural analytics and prediction;
- automated facial recognition against public security camera footage;
- Automatic Number Plate Recognition (ANPR) on highways, but also in urban areas;
- Internet of Things (IoT): an increasing number of devices at home, at work, and/or worn by users are connected to the Internet. These may be built to provide convenience and functionality, not to protect their owner's privacy.

New means of ICT can generate personal data that did not exist before, or at least was not systematically stored and used. Humans are at all times connected to a time and place, and that connection is increasingly captured by sensors and transactions (e.g. payments that require physical presence of a phone or credit card). The automotive industry introduces odometry sensors, ultrasonic sensors, front and back cameras, and light detection and ranging ('lidar') sensors generate data, as does car navigation equipment. The data may be non-personal data when considered in isolation, but longitudinal measurements that can be associated with a car owner become personal data. The data generated by sensors might be stored on the car for maintenance or insurance purposes; and may be 'phoned home' to the car manufacturer or measured by devices placed above or around highways. Additionally, individual movements may be tracked through electronic emissions from personal devices, which often emit information that is intended or can be repurposed as (partially or uniquely) identifying information. Physical characteristics of emissions themselves, both wired and radio, can be used for fingerprinting[20] with varying degrees of accuracy, reliability, and practicality. Physical characteristics may also be used to identify[21] rogue devices, for instance to detect cloned devices or illegal transmitters.

Whereas electrical appliances are subject to a mandatory (self-)certification scheme regarding safety, health, and environmental protection (the 'CE' marking for appliances traded within the EU), no such scheme exists in general for software or hardware with regard to security or privacy requirements. This is left up to the vendors. For specific domains, such as point-of-sale systems and payment cards, rigorous compliance tests are imposed, for instance by Mastercard. Whether mandatory certification can apply to software and hardware vendors in other domains, what tests should be part of such certification, and whether such certification should be carried out by the vendors themselves (self-certification; as is the case with CE markings) or by independent certification bodies, remain open questions.

5.2.1.1 Protection against digital threats can affect privacy

New categories of technologies come with new categories of threats and vulnerabilities, and countermeasures against those can affect privacy. A logical consequence of how Internet technology is designed and the rapid growth in global coverage and adoption is the emergence of botnets and phishing attacks. To protect against new phenomena that pose a risk to national security, such as the use of Internet by terrorists, organized crime,

20 Gerdes 2012; Shi 2011.

21 Hou 2014; Wang 2016.

and hostile nations, new methods and technologies are continuously being developed. These can involve big-data systems storing data that, at least in raw form, constitutes personal data. For instance DNS requests, that by definition describe an IP address performing an 'Internet phonebook lookup' for an Internet domain name when a user accesses a website. An example of a big-data system that collects DNS request data for the purpose of protecting against certain categories of new threats is SIDN's ENTRADA system. ENTRADA[22] is an experimental system that stores DNS requests received by the two authoritative name servers for the .nl top domain. Some 15,000 DNS requests per second are observed, and if stored with full IP and Ethernet headers, some 60GB[23] of data is added per day. The processing of such data can help detect botnet activity, and website spoofing; there have been court rulings[24] in the Netherlands on scammers setting up fake webshops that mimic real webshops for well-known brands. The data processed is obviously privacy-sensitive; SIDN itself took the initiative to establish an enforceable privacy framework that addresses privacy concerns associated with this data processing. This supports public trust in SIDN as maintainer of the .nl domain.

5.2.1.2 Digital espionage

Software that can be used for digitally spying on others is commercially available to individuals, or can be crafted by tech-savvy individuals. One recent example in the US is the case of Phillip Durachinsky, an American citizen who used malware dubbed 'Fruitfly' to spy on Americans. On 10 January 2018, Reuters reported[25] that the indictment states that Durachinsky collected data from thousands of computers belonging to individuals, companies, schools, a police department, and the US Department of Energy, from 2003 through early 2017. That would constitute no less than some thirteen years of computer hacking and spying without getting caught. The sensitive nature of digital espionage software becomes clear when realizing that such software, when only available to governments (as opposed to being available for the general public,

22 Wullink 2016.

23 Jansen 2016.

24 In the 'Meiberg' case, for instance, the Public Prosecution Office demanded up to three years imprisonment for large-scale scams involving falsified webshops. In Dutch: https://www.om.nl/@101212/eisen-3-jaar-cel/ (29 November 2017). The court ruling shows the defendants received between 48 and 146 weeks imprisonment: https://www.rechtspraak.nl/Organisatie-en-contact/Organisatie/Rechtbanken/Rechtbank-Den-Haag/Nieuws/Paginas/Gevangenisstraffen-voor-internetoplichting.aspx (22 December 2017).

25 Reuters, 10 January 2018: 'Ohio man indicted for using "Fruitfly" malware to spy on Americans'. Available at https://www.reuters.com/article/us-usa-justice-malware/ohio-man-indicted-for-using-fruitfly-malware-to-spy-on-americans-idUSKBN1EZ2KO

whether for free or paid), is subject to export controls under the Wassenaar Arrangement; 'intrusion software' was added to the List of Dual-Use Goods and Technologies in December 2013.[26] The purpose of the Wassenaar Arrangement is to support international peace by preventing military and 'dual-use' equipment, including hardware and software, from ending up in the hands of, for instance, governments that do not subscribe to nuclear non-proliferation treaties or that are known to abuse human rights. The addition of 'intrusion software' to this list was an initiative of Dutch MEP Marietje Schaake.

5.2.1.3 *Cryptography vs. cryptanalysis and 'breaking' cryptography*

With regard to cryptography, it is important to note that cryptographic algorithms tend to be broken over time. The typical lifetime of many cryptographic methods in the early days of the Internet was just about ten years. Advances in mathematics and cryptanalysis, and increases in computational resources made breaking encryption feasible. For certain classes of cryptographic algorithms, quantum computing may be able to break encryption using, for instance, Shor's algorithm[27] or Bernstein et al.'s GEECM.[28] Data that is encrypted and captured *today* may thus become decryptable in the near future. In some cases, existing methods may have longer lifetimes by imposing extended key-length requirements and/or key renewal schemes. In short, 'hygiene' with regard to the use of cryptographic methods and keys, such as timely re-encrypting data at rest with new algorithms or longer keys when necessary, is an important technical and procedural challenge to privacy.

5.2.2 Perspective: 'privacy as an ICT challenge'

A second perspective on the relationship between privacy and ICT is: 'privacy poses ICT challenges'. That is, ICT can mitigate or redress privacy challenges brought forth by ICT, or provide privacy where no privacy was possible before.

A well-known aphorism in Internet law is 'code is law'[29], attributed to Lawrence Lessig. This refers to the observation that the way hardware and software are designed and programmed ('coded') form a de facto regulatory

26 Matthijs R. Koot's Notebook, 12 December 2013, '"Intrusion software" now export-controlled as "dual-use" under Wassenaar Arrangement'. Available at https://blog.cyberwar.nl/2013/12/intrusion-software-now-export-controlled-as-dual-use-under-wassenaar-arrangement/.

27 Shor 1997.

28 Bernstein 2017.

29 Lessig 1999.

framework for cyberspace. John Borking contends[30] that this development is undesirable and undemocratic. Borking suggests that 'privacy law is code' is preferable, with privacy requirements laid down in legislation as (mandatory) guidelines to be followed by those who dream up and implement ICT. This relates to 'privacy by design'.[31] As stated earlier, privacy requires security. Besides privacy by design, there is the older notion of 'security by design'. The latter does not necessarily support privacy objectives. Rather, privacy by design and security by design are paradigms that can both be practised to pursue systems that are both reasonably secure and reasonably privacy-friendly.

Furthermore, the emergence of the General Data Protection Regulation (GDPR) in the EU motivates the organization of new academic events, in addition to existing recurring events, to advance privacy in ICT; one example being the IEEE International Workshop on Privacy Engineering (IWPE) (http://iwpe.info/), which has been co-hosted at the long-standing IEEE Symposium on Security & Privacy.

In 1994, a report[32] commissioned by the European Council, informally referred to as the 'Bangemann report', already identified personal data protection as a critical factor for consumer trust in the information society:

> The Group believes that without the legal security of a Union-wide approach, lack of consumer confidence will certainly undermine the rapid development of the information society. Given the importance and sensitivity of the privacy issue, a fast decision from Member States is required on the Commission's proposed Directive setting out general principles of data protection.

In other words: user confidence in the information society may suffer if 'the privacy issue', in the sense of data protection, is not properly dealt with. Regulatory points of view are discussed in other chapters in this book, for instance the chapter by Bart van der Sloot.

5.3 Classic texts and authors

The Internet era started some three decades ago, and developments have been so rapid and diverse that work published in the early days has often

30 Borking 2010.
31 Cavoukian 2009.
32 Bangemann 1994.

been superseded by new insights. A full historiography of computers, cryptology,[33] and digital security is beyond the scope of this chapter. Some insights described in early work however still apply today, or demonstrate that privacy and security challenges discussed today have existed before. Three topics are discussed below: the Ware report (a seminal work in the history of information security), the advent of public-key cryptography (notably RSA), and the creation of Pretty Good Privacy (PGP).

5.3.1 1970: The Ware report

One seminal work in computer security is due to the US Defense Science Board's Task Force on Computer Security which in 1970 released its report 'Security Controls For Computer Systems', also known as the 'Ware report', after its writer, Willis H. Ware. Prior to the task force and its report, Ware organized the 1967 Spring Joint Computer Conference session that discussed challenges that led to the establishment of the Task Force. The report, which has been characterized[34] as 'the paper that started it all, first raising computer security as a problem', states:

> Thus, the security problem of specific computer systems must, at this point in time, be solved on a case-by-case basis, employing the best judgment of a team consisting of system programmers, technical hardware and communication specialists, and security experts.

The report was written prior to the emergence of Internet, during early conceptualizations and advancements in computing and networking that eventually led to the Internet.

Now, close to 50 years after this report, that statement still applies, as do its seven conclusions:

1. Providing satisfactory security controls in a computer system is in itself a system design problem. A combination of hardware, software, communication, physical, personnel, and administrative-procedural safeguards is required for comprehensive security. In particular, software safeguards alone are not sufficient.

33 Macrakis 2010; De Leeuw 2015; Budiansky 2016.

34 Cited from the 'Seminal Papers' page of U.C. Davis' security lab, maintained by computer security scholar Matt Bishop. Available at http://seclab.cs.ucdavis.edu/projects/history/seminal.html

2. Contemporary technology can provide a secure system acceptably resistant to external attack, accidental disclosures, internal subversion, and denial of use to legitimate users for a *closed environment* (cleared users working with classified information at physically protected consoles connected to the system by protected communication circuits).
3. Contemporary technology cannot provide a secure system in an open environment, which includes uncleared users working at physically unprotected consoles connected to the system by unprotected communications.
4. It is unwise to incorporate classified or sensitive information in a system functioning in an open environment unless a significant risk of accidental disclosure can be accepted.
5. Acceptable procedures and safeguards exist and can be implemented so that a system can function alternately in a closed environment and in an open environment.
6. Designers of secure systems are still on the steep part of the learning curve and much insight and operational experience with such systems is needed.
7. Substantial improvement (e.g., cost, performance) in security controlling systems can be expected if certain research areas can be successfully pursued.

These findings were made in the context of (government) systems processing classified or otherwise sensitive information, but it is easy to see that the findings also largely apply to contemporary computer systems; one only needs to interpret 'open environment' as 'internet-connected'. Readers interested in lessons that can be learned from the Ware report regarding security certification of technology are referred to Murdoch (2012).

5.3.2 1976, 1978: advent of public-key cryptography (RSA)

One of the challenges in cryptography is key distribution. Before the advent of public-key cryptography, parties that want to communicate securely need to share a secret key. This is referred to as 'symmetric encryption', where 'symmetric' refers to the fact that parties use a single, shared secret key. To communicate a secret key, you need to have a secure channel, or rely on out-of-band methods, such as physical exchange via couriers. This changed with the introduction of public-key cryptography, which is also referred to as 'asymmetric encryption'. In public-key cryptography, each communicating party has two keys: a public key and a private key, derived at the same time via a mathematical algorithm. The public key can only be used to encrypt and to verify cryptographic signatures, and thus does not need to be kept secret (hence, 'public' key). The private key can only be used to decrypt and to generate cryptographic signatures, and must be kept secret by its owner. Under assumptions of certain 'hard problems' in mathematics, deriving a private key from its associated public key is intractable. To communicate securely, parties only need to exchange their public key, which can be done via open channels.

The first published work that introduces the idea of public-key crypto systems is due to Diffie and Hellman[35] in 1976, under influence of Merkle who subsequently published[36] a seminal work in 1978. In that same year, Rivest, Shamir, and Adleman introduced[37] a crypto system that has since been known as 'RSA', an acronym of the authors' last names. The RSA system builds on the assumption expressed by Euler's theorem, which dates back to the 1700s, which essentially boils down to the assumption that it is very hard to factorize large prime numbers. RSA remains in widespread use today, for instance in SSL/TLS, and in PGP, the next topic.

5.3.3 Zimmerman (1991): Pretty Good Privacy (PGP)

In 1991, the US Senate drafted an anti-crime bill[38] that included the following clause, that would essentially require providers of encrypted communication services and manufacturers of encrypted communications equipment to place backdoors in their systems to allow the government to access plain-text (i.e. unencrypted) communications:

> SEC. 2201. COOPERATION OF TELECOMMUNICATIONS PROVIDERS WITH LAW ENFORCEMENT.
> It is the sense of Congress that providers of electronic communications services and manufacturers of electronic communications service equipment shall ensure that communications systems permit the government to obtain the plain text contents of voice, data, and other communications when appropriately authorized by law.

35 Diffie 1976.

36 Merkle 1978.

37 Rivest 1978.

38 'S.266 – Comprehensive Counter-Terrorism Act of 1991', 102nd US Congress. Available at https://www.congress.gov/bill/102nd-congress/senate-bill/266/text

This led US-based software engineer Phil Zimmermann to create software he dubbed 'Pretty Good Privacy' (PGP) and make it available to the general public via an Internet-connected file exchange server in that same year. PGP was the first publicly available software that implemented a public-key cryptography system: RSA. At the time, strong cryptography was considered to be subject to US Arms Export Control Act, but the PGP software nonetheless ended up outside the US.

Current versions of PGP, notably the open-source software GnuPG, remain in use today in a variety of high-security contexts, including communication with CERTs about incidents and vulnerabilities, and communication between journalists and their sources.

5.4 Traditional debates and dominant schools

The development of ICT has mostly taken place in politics-agnostic environments, and many technologists' attitude was, and remains, one of 'technology is neutral'. This neutrality is suspect when the rationale and funding for R&D have roots in organizations with a political agenda, and cannot always be seen as politics-agnostic. ICT exists in a habitat that is not isolated from personal choices, market forces, and government decisions, all of which are to some extent political. The development of Internet standards by communities of engineers is an example; also recall Lessig's 'code is law' and Borking's 'privacy law is code'.

A brief reflection on the history of standardization of Internet protocols follows, to illustrate that the 'technology-is-neutral' point of view is, for better or worse, no longer upheld, or at least faces increases opposing voices within some Internet engineering communities. Simply put, the below shows how privacy (and security) by design, notions that are not inherently politics-agnostic, gain presence in these communities.

How computers 'talk' to each other on the Internet is largely laid down in technical Internet standards. An Internet standard starts with an idea for change or new functionality. Under the umbrella of the Internet Engineering Task Force (IETF) that idea is further developed into a 'Request for Comments' (RFC) document. These are presently published on the IETF Datatracker.[39] This process is completely open: anyone who has relevant knowledge and insights can join IETF discussions. When an idea reaches a draft status, and sometimes earlier than that, ICT vendors implement

39 IETF's Datatracker is available at https://datatracker.ietf.org/.

the idea. Possibly after minor changes or corrections, and with sufficient adoption by industry, the idea can reach maturity and is promoted to the status of 'Internet Standard'. Roughly put this is how Internet technology has developed from the 1980s into what it is today. Examples of Internet standards include the protocol used for communication between web browsers and web servers (HTTP and HTTP/2,) email (SMTP), a protocol intended to protect the confidentiality and integrity of such communications (TLS), and the 'Internet address book' that resolves domain names to IP addresses (DNS).

The predecessor of the Internet, ARPANET, and the early Internet, were networks that consisted solely of parties that had some trust relation. Because of that, Internet protocols designed during the early Internet (1980s and early 1990s) did not take security or privacy into account. Concerns about inadequate security arose when the Internet expanded further and commercialized, and it was decided in 1993 that new RFCs must contain a 'Security Considerations' paragraph. This is laid down in RFC 1543.[40] The paragraph must contain a discussion about possible threats and attacks on the protocol described in a new standard. After several years of (sometimes bad) experiences with writing such paragraphs, it was clarified in 2003 what exactly should be in that section; this is laid down in RFC 3552.[41] This section should describe which digital attacks are relevant to the protocol, which are not, and why. For relevant attacks, it must describe whether the protocol protects against them. Among other things, it is mandatory to pay attention to eavesdropping (confidentiality), to the injection, modification, or removal of data (integrity), and to denial-of-service attacks that may interfere with services that use the protocol (availability). Such a paragraph will never be perfect, but requiring protocol designers to think about security properties should lead to improvement of security on the Internet. In addition, RFCs are 'living documents', in that updates and errata can be published.

Snowden's revelations have shown that intelligence services, especially the NSA (US) and GCHQ (UK), are actively gathering intelligence on the Internet on a large scale, using a wide variety of methods and techniques. These revelations, in conjunction with cases of ethically doubtful behaviour by nongovernment entities, eventually led to a rough consensus within the IETF that 'pervasive monitoring' should be considered to be an 'attack' that

40 RFC 1543: Instructions to RFC Authors, October 1993. Available at https://tools.ietf.org/html/rfc1543

41 RFC 3552: Guidelines for Writing RFC Text on Security Considerations, July 2003. Available at https://tools.ietf.org/html/rfc3552

designers of new internet protocols should take into account. Pervasive monitoring is defined as follows:

> Pervasive Monitoring (PM) is widespread (and often covert) surveillance through intrusive gathering of protocol artefacts, including application content, or protocol metadata such as headers. Active or passive wiretaps and traffic analysis, (e.g., correlation, timing or measuring packet sizes), or subverting the cryptographic keys used to secure protocols can also be used as part of pervasive monitoring. PM is distinguished by being indiscriminate and very large scale, rather than by introducing new types of technical compromise.

Furthermore: 'The motivation for [pervasive monitoring] can range from non-targeted nation-state surveillance, to legal but privacy-unfriendly purposes by commercial enterprises, to illegal actions by criminals'.

The consensus that pervasive monitoring should be considered to be an 'attack' was laid down in 2014 in RFC 7258 by Stephen Farrell, research fellow at the school of Computer Science and Statistics at Trinity College Dublin, and Hannes Tschofenig, a senior engineer at microprocessor manufacturer ARM Limited. It has the status of 'Best Current Practice' (BCP), and promotes mitigation of pervasive monitoring in new protocols. It should be noted that the BCP does not mandate prevention of monitoring by motivated attackers, which may include law enforcement and intelligence services. Rather, the BCP states the following: '"Mitigation" is a technical term that does not imply an ability to completely prevent or thwart an attack. Protocols that mitigate PM will not prevent the attack but can significantly change the threat.'

Adherence to the BCP is expected to result in better privacy-by-default in new Internet protocols. Readers interested in matters of privacy and ethics in Internet protocol design are also referred to RFC 8280[42] and RFC 6973.[43] In short, the aphorism 'architecture is politics', attributed to Mitchell Kapor, applies to the digital realm as well. Interested readers are also referred to Milan (2017) which provides a Science and Technology Studies (STS) perspective on policy related to the Internet architecture and infrastructure.

As a final example: governments may seek to influence standardization bodies for Internet protocols to protect national security interests;

42 RFC 8280: Research into Human Rights Protocol Considerations, October 2017. Available at https://tools.ietf.org/html/rfc8280

43 RFC 6973: Privacy Considerations for Internet Protocols, July 2013. Available at https://tools.ietf.org/html/rfc6973

classified documents leaked via Edward Snowden indicate the existence of government programmes that pursue this: NSA's Bullrun programme and GCHQ's Edgehill programme. A famous example of the alleged weakening of cryptography by government actors related to the cryptographic algorithm 'Dual_EC_DRBG', which turned out to contain a vulnerability that has the characteristics of an intentional backdoor crafted by cryptologist-mathematicians. Between 2006 and 2014, the US NIST agency recommended 'Dual_EC_DRBG' for use; and it was widely in use due to RSA Security products using that algorithm by default. Interested readers are referred to https://projectbullrun.org.

5.5 New challenges and topical discussions

In addition to the challenges regarding nternet standards as laid out in the previous section, current and new challenges include:[44]

- ethics of big data and artificial intelligence;
- ubiquitous identification and surveillability;
- privacy-enhancing technologies (PETs);
- digital vulnerabilities in current and emerging technology.

These are discussed in the next subsections.

5.5.1 Ethics of big data and artificial intelligence

Big data holds the promise of filtering out human cognitive bias in data analysis, but it isstill humans who programme algorithms and Interpret their outcomes. As such, logical fallacies must still be taken into account. Skepticism toward overzealous and questionable uses of big data, while avoiding techno-panic[45] and threat inflation, remains relevant. For instance, a digital vulnerability hitting mainstream news may indicate that a vulnerability of that statute occurs infrequently; media attention exacerbates perception of risk, which on the hand can at times be qualified as spreading 'Fear, Uncertainty, and Doubt' (FUD), but on the other hand can reinforce public

44 This list is necessarily incomplete. A plethora of other privacy challenges and topics exist, notably in specific application domains, such as healthcare, personal finance, law enforcement, and intelligence. The topics discussed in this chapter were selected on the basis of having relevance beyond a single application domain.

45 Thierer 2013.

awareness of the reality of technological fallibility and promote adoption of privacy-by-design and security-by-design by makers and buyers of ICT goods and services.

One recommended resource about fallibilities in big data and artificial intelligence is a Spring 2017 course taught at the University of Washington named 'Calling Bullshit: Data Reasoning in a Digital World', created by mathematical biologist Carl T. Bergstrom and data scientist Jevin West. The course aims 'to teach you how to think critically about the data and models that constitute evidence in the social and natural sciences'. From the website:[46]

> *Bullshit* involves language, statistical figures, data graphics, and other forms of presentation intended to persuade by impressing and overwhelming a reader or listener, with a blatant disregard for truth and logical coherence;
> *Calling bullshit* is a performative utterance, a speech act in which one publicly repudiates something objectionable. The scope of targets is broader than bullshit alone. You can call bullshit on bullshit, but you can also call bullshit on lies, treachery, trickery, or injustice.

This calls for awareness of the possibility of false positives and flaws in profile-building, both of which may unjustly result in unjust harms to privacy of individuals and groups. A toy example to explain the phenomenon of false positives: suppose that the algorithms have an 99% accuracy level, and one out of 100,000 people is a true threat. With 99% accuracy, there is 1% inaccuracy, i.e. unjustly indicating a person as a threat. Hence, a false positive. For every 100,000 persons, this will yield 1000 false positives, yielding a 0.1% overall false positive rate. Safeguards may be needed to prevent and redress the impact that 'false flagging' can have on an individual.

It can be noted that the Dutch legislator already recognizes this issue in the context of the Dutch intelligence services: the Memorandum of Explanation of the new Dutch intelligence and security services law explicitly[47] forbids the services from promoting or taking measures towards a person based on outcomes of automated data analysis alone. Human decision-making must augment automated data analysis[48].

46 Available at http://callingbullshit.org/ (includes course materials).
47 'Wet op de inlichtingen- en veiligheidsdiensten 2017' (Wiv2017), Memor. of Explanation, pp. 175-176.
48 This of course begs the question how human analysts interpret the outcomes of automated data analysis.

One way forward in addressing ethical questions in big data and artificial intelligence is algorithmic transparency and accountability. In January 2017, the US Public Policy Council of the Association for Computing Machinery (USACM) released[49] a statement that included a list of principles that support algorithmic transparency[50] and accountability: awareness, access and redress, accountability, explanation, data provenance, auditability, and validation and testing. In March 2018, the same council released[51] a statement on the importance of preserving personal privacy, in the context of big data and the Internet of Things. Interested readers are also referred to a survey[52] exploring potential malicious uses of artificial intelligence, published in February 2018.

Also worth noting are initiatives for codes of ethics in informatics. For instance, a programmers' equivalent to the Hippocratic Oath was proposed[53] in early 2018 by software developer Nick Johnstone, in a joint effort with other developers:

> As a programmer, I swear to fulfill these tenets:
> - I will only undertake honest and moral work. I will stand firm against any requirement that exploits or harms people.
> - I will respect the lessons learned by those who came before me, and will share what I learn with those to come.
> - I will remember that programming is art as well as science, and that warmth, empathy and understanding may outweigh a clever algorithm or technical argument.
> - I will not be ashamed to say 'I don't know', and I will ask for help when I am stuck.
> - I will respect the privacy of my users, for their information is not disclosed to me that the world may know.
> - I will tread most carefully in matters of life or death. I will be humble and recognize that I will make mistakes.

49 ACM US Public Policy Council (USACM), 'Statement on Algorithmic Transparency and Accountability', 12 January 2017. Available at https://www.acm.org/binaries/content/assets/public-policy/2017_usacm_statement_algorithms.pdf

50 'Algorithmic transparency' does not entail public disclosure of source code. Although such disclosure would provide a strong safeguard, other interests may be prohibitive to such disclosure, for instance protection of intellectual and business interests.

51 ACM US Public Policy Council (USACM), 'Statement on the Importance of Preserving Personal Privacy', 1 March 2018. Available at https://www.acm.org/binaries/content/assets/public-policy/2018_usacm_statement_preservingpersonalprivacy.pdf

52 Brundage 2018.

53 See https://github.com/Widdershin/programmers-oath.

- I will remember that I do not write code for computers, but for people.
- I will consider the possible consequences of my code and actions. I will respect the difficulties of both social and technical problems.
- I will be diligent and take pride in my work.
- I will recognize that I can and will be wrong. I will keep an open mind, and listen to others carefully and with respect.

Not much is known about the effects and (in)effectiveness of such ethics codes in informatics, however. Similar proposals have been seen in the past in the realm of system administrators[54] and database administrators who, due to the nature of their job, often have highly privileged access to systems and data. Administrators can be confronted with requests related to investigations fraud of incidents.

5.5.2 Ubiquitous identifiability and surveillability

New technology and increased connectivity come with new possibilities to identify, and subsequently track, devices and users. This topic can be illustrated in terms of the OSI model.

At the Data Link layer (OSI layer 2), protocols such as Bluetooth and Wi-Fi render personal devices identifiable via the Media Access Control (MAC) address[55] associated with a network interface (a Bluetooth interface or a Wi-Fi interface). MAC addresses are a key part in the mechanism that enables communication between devices over some wired or wireless physical medium (radio, copper, fibre); which is the whole idea of protocols at this layer. Although MAC addresses are usually not *globally* unique, they are, by intent, *locally* unique within smaller scopes; and may be unique within, for instance, a single country. The risk of ubiquitous surveillability via MAC tracking is addressed through 'MAC randomization', variations of which are already implemented in recent versions of Android and iOS. Flaws[56] in design or code may thwart this protection and still allow tracking. And even if the design and code are flawless, surveillability remains: if Bluetooth beacons

54 For instance, see https://www.usenix.org/system-administrators-code-ethics.

55 In the OSI model, MAC addresses reside within the Data Link layer. Wi-Fi, Bluetooth, and Ethernet are examples of protocols that provide functions at that OSI layer and implement MAC addresses.

56 Matte 2016; Matte 2017; Martin 2017. Also see The Guardian: 'MAC randomisation: A massive failure that leaves iPhones, Android mobes open to tracking' (by Thomas Claburn), 10 March 2017. Available at https://www.theregister.co.uk/2017/03/10/mac_address_randomization/

become widespread, incentives emerge to nudge[57] users into installing an app that requires Bluetooth pairing with a beacon, or with some different Bluetooth device controlled by the same company. Mobile phones may allow apps that are granted the Bluetooth permission to communicate over Bluetooth also when the app is not in use; communication can take place without the user being aware of being tracked.

At the Network layer (OSI layer 3) all the way up to the Application layer (OSI layer 7), protocol behaviour and artefacts can be found that allow web tracking. At the Application layer, every web visit discloses some technical information to one or more websites. Not only to the website the user knowingly visits, but also to any third parties from which that website includes content, such as systems controlled by online advertising brokers. A user can be tracked[58] on the web by (combination of) their IP address (Network layer), cookies, browser/device fingerprinting,[59] and other recurring patterns in observable device or user behaviour. Websites that contain, for instance, a 'Like' button (Facebook) or 'Tweet' button (Twitter) cause web browsers to load content from third-party servers. If a user makes an online purchase and discloses their real identity, address, and other information to a web shop, that web shop knows which real identity is associated with a certain unique combination of technical information. Depending on jurisdiction and terms of service, the web shop may monetize that data, for instance by selling (access to) it to third parties, who can leverage the data to enhance behavioural targeting.

Furthermore, if a website includes code from a third-party system and the website does not include proper security instructions for browsers,[60] the user is exposed to the possibility of malicious code being loaded if the third

57 For instance by offering a service or discount only via an app that requires the user to grant Bluetooth permission and enable Bluetooth; in addition to making it less easy to fully disable Bluetooth communication, as observed in a change made between iOS 10 and iOS 11. Also see The Guardian: 'iOS 11: toggling wifi and Bluetooth in Control Centre doesn't actually turn them off' (by Samuel Gibbs), 21 September 2017. Available at https://www.theguardian.com/technology/2017/sep/21/ios-11-apple-toggling-wifi-bluetooth-control-centre-doesnt-turn-them-off. Furthermore, Apple removed the audio jack from new iPhone models, requiring users to either purchase a Lightning-to-audio adapter, or use a Bluetooth headphone. The latter may increase the number of users that have Bluetooth enabled by default.

58 EFF's Panopticlick website allows visitors to test how uniquely identifiable their browser is. It was launched in 2010, received a significant update in 2015. Available at https://panopticlick.eff.org/ (suggestion: do the test both from a normal browser and from Tor Browser and see the difference in uniqueness, expressed in bits of entropy). The methodology is explained in the About page at https://panopticlick.eff.org/about.

59 Eckersley 2010; Mowery 2012; Acar 2014.

60 For instance by using 'Content-Security-Policy' (CSP), 'Subresource Integrity' (SRI), or 'Confinement with Origin Web Labels' (COWL).

party is compromised.[61] This risk is especially relevant to web applications that allow authenticated users to access sensitive data (e.g. personal data) or functions (e.g. security management).

The risk of web tracking can, to some extent, be mitigated through Tor Browser, a web browser that provides users with some degree of privacy while browsing the web. Tor Browser is implemented such that its users, by default, 'blend into the crowd' with other users by suppressing or generalizing information it emits and that would otherwise allow observers to 'zoom in' on a certain part of the users to link a web request to its real source. In addition to the digital footprint of Tor Browser being less identifying, it hides the user's IP address by routing web traffic via the Tor network ('dark web'), where the last hop, a so-called 'exit node', submits the web request to the web server, thus acting as a proxy. The Tor network is a decentralized network consisting of nodes, often volunteer-operated, physically spread around the world (though the highest-bandwidth exit nodes tend to be located in Western countries).

Web tracking is also mentioned[62] as an issue affecting the protection of the covert identity of intelligence agents deployed abroad. Tor Browser has some use in such contexts by hiding the digital exhaust from at least local, low-resourced[63] eavesdroppers. Tor Browser itself, while 'hardened', should be expected to remain vulnerable to 0-days, i.e. vulnerabilities that are found but not disclosed to the vendor so that the vulnerabilities can be exploited. Tor Browser is based on Firefox ESR and security vulnerabilities in Firefox ESR may also apply to Tor Browser. (0-days are one of the means law enforcement and intelligence services can deploy in attempt to deanonymize users. This has for instance been done by the FBI in Operation Pacifier, which targeted users of an onion service used to exchange child sex abuse imagery).

Another example of surveillability is users' DNS lookups. Each time the user visits a website with a browser (Internet Explorer, Firefox, Chrome, etc.), or sends a message via an email application running on the user's system (Thunderbird, Outlook, etc.), the user's system emits a DNS request that looks up information about the website's domain or email recipient's domain. Due

61 The Register, 11 February 2018: 'UK ICO, USCourts.gov... Thousands of websites hijacked by hidden crypto-mining code after popular plugin pwned', The Register, 11 February 2018. Available at https://www.theregister.co.uk/2018/02/11/browsealoud_compromised_coinhive/.

62 Dujmovic 2018.

63 Tor is not designed to protect against the so-called 'global passive adversary': this type of adversary is explicitly excluded from Tor's original design. It is assumed that an attacker who is able to simultaneously intercept the first link and last link in the three-hop, thus five-link, connection that Tor builds can deanonymize Tor users. Such capabilities would likely require multinational signals intelligence efforts; that topic is beyond the scope of this chapter.

to the hierarchical structure of the DNS ecosystem and absence of encryption, this traffic is observable at various systems and networks on the Internet. This includes: 1) the user's ISP,[64] 2) the operator of authoritative name servers for the top-level domain (example: '.nl' is operated by SIDN), and 3) the operator of the authoritative name servers for the second-level domain (example: lookups for 'uva.nl' are sent to 'dns-prod1a.uva.nl', 'dns-prod2a.uva.nl', or 'dns-prod3a.uva.nl', operated by the University of Amsterdam itself). To protect end-user privacy, various methods have been proposed that provide varying protection against surveillance by eavesdropping on network links, including[65] DNSCrypt, DNSCurve, DNS-over-HTTPS (DOH), QNAME minimization,[66] and Oblivious DNS[67] (ODNS). These methods could be characterized as privacy-enhancing technologies, the topic of the next section.

Default privacy and security settings in technology standards (RFCs, ISO norms, and so on), operating systems, applications, and communication providers determine the privacy and security settings that apply to most users: most users do not know or care to change these settings. An example of possible consequences can be found in a 'heat map' data visualization published in 2018 by the company that made a fitness tracking app called Strava: the map inadvertently revealed locations of secret US military bases abroad. This situation can be attributed to Strava's default setting regarding user location data, which was by default not set to 'private'.[68] It turned out that personnel at US military bases in Syria, Afghanistan, and Antarctica used the app in its default settings. The media reports about this also resulted in a question[69] raised by a Dutch member of parliament.

64 Or if a public hotspot is used, the operator of that hotspot, as well as *its* upstream ISP.

65 DNSSEC is not listed because it does not encrypt DNS lookups. DNSSEC provides authentication and integrity, not confidentiality.

66 QNAME minimization is laid down in RFC 7816, 'DNS Query Name Minimisation to Improve Privacy' (March 2016, still a draft), available at https://datatracker.ietf.org/doc/rfc7816/. QNAME minimization does not provide encryption, but does reduce unnecessary leakage of DNS lookups that is due to DNS resolvers directly communicating full domain names (e.g. 'www.google.com') to the authoritative root servers. With QNAME minimization, systems can traverse the DNS hierarchy in a step-by-step approach where the full domain name is only communicated to the DNS server that is authoritative for that particular domain.

67 See https://odns.cs.princeton.edu/ and https://freedom-to-tinker.com/2018/04/02/a-privacy-preserving-approach-to-dns/.

68 *The Guardian*, 28 January 2018: 'Fitness tracking app Strava gives away location of secret US army bases'. Available at https://www.theguardian.com/world/2018/jan/28/fitness-tracking-app-gives-away-location-of-secret-us-army-bases.

69 Transcript of the 45th meeting of the House of Representatives 2017-2018 that took place on 30 January 2018. Available in Dutch at https://www.tweedekamer.nl/kamerstukken/plenaire_verslagen/detail?vj=2017-2018&nr=45&version=2.

Providers of social media have been known to change default privacy settings, so as to increase web visits via accessible user-generated content and thus generate more ad revenue. Privacy-enhancing settings are often opt-in rather than an opt-out. In 2010, Matt McKeon illustrated erosion in Facebook's default privacy settings between 2005 and 2010 in a series of pictures; fig. 5.3 depicts this series.[70] And more recently, a German court in 2018 ruled[71] against Facebook in a court case brought by Verbraucherzentrale Bundesverband, the federation of German consumer organizations, over Facebook's default privacy settings. Facebook profiles are by default indexed by search engines, and its default settings, the mobile Facebook app shares users' location data. Furthermore, prior to the revelations[72] surrounding Cambridge Analytica, the Facebook API allowed third-party apps to obtain not only information of Facebook-enabled app users, but also of the *friends* of those users. In short, default settings remain an essential topic in the discussion and assessment of privacy (and by extension, security).

Another emerging topic in surveillability is Mobile Device Management (MDM) software. MDM software is used by organizations to allow employees to use mobile devices to access confidential corporate data while providing the organization controls to cope with threats such as device loss and mobile malware. Various MDM vendors exist, and their products differ in terms of potential impact on privacy of employees. If the device is a personal device, owned by the employee, and the employee enrols in the MDM solution, the employee grants the organization a certain degree of control over their device and data on the device. Functionality available to MDM administrators can include the ability to track the physical location of devices (and hence track the person who is carrying it), enumerate mobile apps installed on a device (which may include dating apps, medical apps, and so on), access the mobile browser history, or inspect the user's live web traffic by routing web traffic through a corporate web proxy.

Readers interested in surveillance and privacy from an intelligence standpoint are referred to the chapter by Willemijn Aerdts and Giliam de

70 McKeon 2010. Figures used with permission of the author.

71 ZDNet, 13 February 2018: 'Facebook is breaking law in how it collects your personal data, court rules'. Available at http://www.zdnet.com/article/facebook-is-breaking-law-in-how-it-collects-your-personal-data-court-rules/.

72 *The Guardian*, 17 March 2018: 'Revealed: 50 million Facebook profiles harvested for Cambridge Analytica in major data breach'. A copy of the old Facebook Developers API involved, taken offline by Facebook in 2015, can be found at the Internet Archive: https://web.archive.org/web/20131218130854/https://developers.facebook.com/docs/reference/login/extended-profile-properties/.

Fig. 5.2: Diminishing default privacy settings on Facebook from 2005 till 2010.

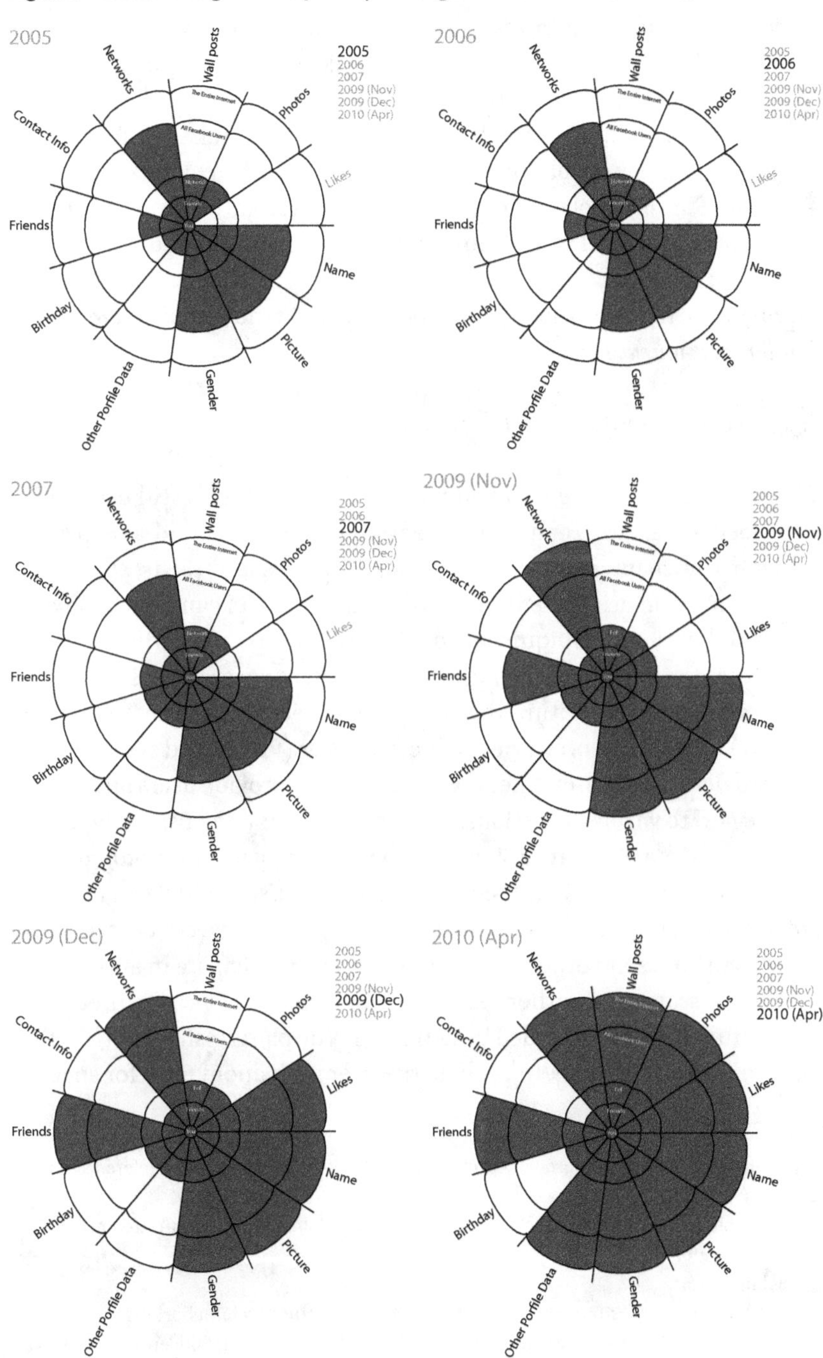

© Matt McKeon, http://mattmckeon.com/facebook-privacy/

Valk; Fidler (2015); and Petersen (2018). Readers interested in mass surveillance issues in Internet infrastructure are referred to a two-part study on mass surveillance published in 2015 by the Science and Technology Options Assessment (STOA) panel of the European Parliament:

- 'Mass Surveillance – Part 1: Risks and opportunities raised by the current generation of network services and applications',[73] 12 January 2015.
- 'Mass Surveillance – Part 2: Technology foresight, options for longer term security and privacy improvements',[74] 13 January 2015.

(In 2017, an article was subsequently published[75] in the *Computer Standards & Interfaces journal.*)

5.5.3 Privacy-enhancing technologies (PETs)

Whereas the worldwide web and most Internet protocols still used today were not designed with end-user privacy requirements in mind, as explained earlier in this chapter, Privacy-Enhancing Technologies[76] (PETs) such as Tor Browser provide privacy-enhanced alternatives. PETs can help overcome undesired effects of ubiquitous identifiability. Privacy while browsing websites is the strongest when using Tor Browser to access websites hosted as an onion service[77] within the Tor network, recognizable by the '.onion' top level domain. Also, a variety of peer-to-peer (P2P) based systems exist that provide closed-circuit networks designed to provide users anonymity with regard to various functions. One example is I2P (https://geti2p.net/), which provides a platform design from the bottom up using cryptography to achieve specific privacy and security properties. Users of I2P can host and browse .i2p sites (referred to as 'Eep sites'; not unlike the concept of onion services in Tor), send other I2P users email, and participate in anonymous instant messaging. Another example is GNUnet[78] (https://gnunet.org/). Similar functions are provided by RetroShare (http://retroshare.net/). Other platforms exist that seek to provide the user with anonymity for specific

73 Available at http://www.europarl.europa.eu/thinktank/en/document.html?reference=EPRS_STU(2015)527409.

74 Available at http://www.europarl.europa.eu/thinktank/nl/document.html?reference=EPRS_STU(2015)527410.

75 Schuster 2017.

76 The term 'Privacy-Enhancing Technology' (PET) was coined by John Borking.

77 'Onion service' refers to the concept that was formerly referred to as 'hidden service'. Maybe move this note to page 24, see comment 8.

78 Grothoff 2017.

functions, such as MUTE (http://mute-net.sourceforge.net/) for file sharing. Furthermore, research on mechanisms for anonymous authorization, that allow users to make use of a service without the service provider having to know the user's identity or even a pseudonym, remains relevant; as well as research on (vulnerabilities in) protocols and implementations of software that claims[79] to provide secure and private communication, such as Silent Circle, Signal, and Telegram.

Another relevant development is the emergence of self-hosted storage and communication platforms, such as the free and open source software ownCloud and Nextcloud. These platforms allow individuals and organizations to run their own 'cloud' and keep in full control over their data. Such platforms can incorporate PETs: for instance, end-to-end encryption for file sharing is scheduled to be part of Nextcloud 13. Self-hosted also means that the user or organization, rather than a provider, is responsible for security. And due to the amount of functionality and hence complexity, vulnerabilities are bound to be found in the platforms and in the underlying software; the discovery, disclosure, and timely patching of vulnerabilities will be a challenge, as is true for any complex system. It can be argued that cloud providers and self-hosted platforms protect against different threat models, have different user groups, and different usage scenarios; and hence complement rather than compete with each other.

PETs are developed within and outside academic contexts. Academic research on PETs is encouraged, for instance to improve their robustness, privacy, and security. A key dilemma is if, and how, PETs can and should be designed in a way that still caters to reasonable and legitimate interests of law enforcement agencies and intelligence and security services. Although criminals, too, make use of Tor (and other privacy-enhancing technologies and platforms), it is important to keep in mind that Tor is also widely in use to protect legitimate interests, such as whistleblower protection. The Dutch Publeaks[80] Foundation and Italian Anti-Corruption Authority[81] (ANAC) rely on onion services; and so do news media that use the SecureDrop[82]

79 Insofar the protocols and code of such systems are not openly published, a healthy level of skepticism is recommended. Academics can play an important role in discovering weaknesses, which may be caused by accidental bugs or be intentional backdoors.

80 The Publeaks Foundation is a joint initiative by various Dutch media. The Publeaks website was established in September 2013 and is available at https://publeaks.nl/.

81 'Italian Anti-Corruption Authority (ANAC) Adopts Onion Services', 13 February 2018, https://blog.torproject.org/italian-anti-corruption-authority-anac-adopts-onion-services. The ANAC website is available at http://www.anticorruzione.it/portal/public/classic/.

82 The SecureDrop project website is available at https://securedrop.org/.

software, which include The Intercept, The New York Times, The Guardian, The Washington Post, and Bloomberg News.

Research and development of PETs and of novel cryptologic building blocks for new, yet to be invented categories of PETs, is key to the future of privacy. The work on Privacy Patterns[83] is highly recommended for readers interested in privacy by design. Publications on applications of privacy by design can also be found in journals or at conferences in disciplines that are not focused on computer security and privacy. To give one example, work on privacy by design in the context of intelligent transportation systems has been published[84] in the domain-specific journal published in the *Journal of Transportation Planning and Technology*.

5.5.4 Digital vulnerabilities in current and emerging technology

Research into digital security and vulnerabilities has proven successful in improving the security of digital communication: the discovery of design flaws and bugs in implementations of older versions of SSL/TLS, novel cryptanalytic attacks against cryptographic methods supported by those older mechanisms, and so on, have led to TLSv1.2 (standardized in RFC 5246) and TLSv1.3 (still a draft at the time of writing). These newer protocols are significantly more robust in delivering security and privacy. Similarly, over the past decades, research into software vulnerabilities has led to the discovery – and subsequent patching – of a plethora of vulnerabilities in operating systems and end-user applications such as browsers. As mentioned earlier, security is a systems property, and failure of a component can mean failure of the system as a whole. Two recent examples of this are the Meltdown and Spectre[85] vulnerabilities that affect Intel processors in a way that essentially compromises the security of systems as a whole.

The digital 'threat landscape' is vast, and research into security and vulnerabilities will for the foreseeable future remain a crucial pillar in improving trustworthiness of digital systems. Readers interested in these matters are referred to the following publications by the European Network & Information Security Agency (ENISA), that lay out threat landscapes for big data, hardware, and the Internet:

83 Colesky 2015; Colesky 2016; Colesky 2018. Privacy Patterns website available at https://privacypatterns.eu/. For privacy by design, see Cavoukian (2009); Hoepman (2014).

84 Lederman 2016.

85 Kocher 2018. Also see https://meltdownattack.com/.

- 'Big Data Threat Landscape' (January 2016) https://www.enisa.europa.eu/publications/bigdata-threat-landscape
- 'Hardware Threat Landscape' (December 2016) https://www.enisa.europa.eu/publications/hardware-threat-landscape
- 'Cyber Threat Landscape' (November 2017) https://www.enisa.europa.eu/news/enisa-news/enisa-report-the-2017-cyber-threat-landscape

For security challenges related to the Internet of Things, readers are referred to the NIST Interagency Report 8200[86] and to the informational text[87] produced by the Internet Research Task Force (IRTF) Thing-to-Thing Research Group, both of which are still drafts at the time of this writing. Section 5.7 of latter document reinforces the need for testing IoT devices to discover (and patch) vulnerabilities:

> 5.7. Testing: bug hunting and vulnerabilities
>
> Given that IoT devices often have inadvertent vulnerabilities, both users and developers would want to perform extensive testing on their IoT devices, networks, and systems. Nonetheless, since the devices are resource-constrained and manufactured by multiple vendors, some of them very small, devices might be shipped with very limited testing, so that bugs can remain and can be exploited at a later stage. This leads to two main types of challenges:
>
> 1. It remains to be seen how the software testing and quality assurance mechanisms used from the desktop and mobile world will be applied to IoT devices to give end users the confidence that the purchased devices are robust.
>
> 2. It is also an open question how the combination of devices from multiple vendors might actually lead to dangerous network configurations, for example, if combination of specific devices can trigger unexpected behavior.

86 NIST Interagency Report (NISTIR) 8200 on the Status of International Cybersecurity Standardization for the Internet of Things (IoT), draft of February 2018. Available at https://csrc.nist.gov/publications/detail/nistir/8200.

87 IETF draft-irtf-t2trg-iot-seccons: 'State-of-the-Art and Challenges for IoT Security'. Available at https://datatracker.ietf.org/doc/draft-irtf-t2trg-iot-seccons/. At the time of writing, the current draft version is number ten, released in February 2018. This draft expires on 16 August 2018, after which a new draft or final release is expected.

Similar challenges exist in other current and emerging technology, such as virtualization,[88]software-defined networking,[89] speech recognition,[90] robotics, e-health technology, and so on.

Dilemmas can exist in computer vulnerability research where privacy interests collide with interests protected by law enforcement and intelligence. For instance, implemented and used correctly end-to-end encryption makes communication inaccessible to *anyone* but the sender and receiver, including to government agencies tasked with investigating crime and threats to national security. When no other viable means are available to carry out their legal tasks, these agencies resort to the exploitation of computer vulnerabilities to compromise devices (laptops, smartphones, etc.), for instance to locate and identify suspects or to eavesdrop on communication before it gets encrypted on devices ('pre-encryption'). This has led to the emergence of a market for 0-days, and knowledge about vulnerabilities is now often sold rather than publicly disclosed.[91] Many technology vendors and service providers have 'bug bounty' programmes, offering money to anyone who discovers a serious vulnerability and reports it to them in accordance with their guidelines, encouraging bug hunters to allow them to assess and patch the vulnerability before it is publicly disclosed. Bug bounties can be high, depending on the impact of a vulnerability and how difficult it is: for instance, Intel in February 2018 started a programme offering up to USD 250,000 for side-channel vulnerabilities. The 0-day market can be lucrative as well: in 2015, 0-day acquisition firm Zerodium, which was established by the French digital spyware company Vupen, offered[92] a million USD for a full iOS 9 jailbreak:

> Zerodium will pay out one million U.S. dollars ($1,000,000.00) to each individual or team who creates and submits to Zerodium an exclusive, browser-based, and untethered jailbreak for the latest Apple iOS 9 operating system and devices.

88 For instance: 'guest-to-host escapes', Rowhammer attacks, and other attacks that compromise the isolation of guests in shared virtualized environments.

89 For instance: unauthorized rerouting or mirroring of traffic.

90 Carlini 2018.

91 Allodi 2017. Also see Coriens Prins. (2014). 'Handel in geheime digitale lekken', *Nederlands Juristenblad* 89(17), 865-865.

92 See https://www.zerodium.com/ios9.html.

Zerodium reported[93] that 'only one team' received that bounty. An overview[94] is available of current bounties for 0-day exploits for various software, both desktop/server software and mobile software, ranging from USD 5,000 to USD 1,500,000 per submission. Pricing is based on the difficulty of finding exploitable vulnerabilities in a particular piece of software and market demand for a capability of exploiting that software. In a bug bounty programme seeking exploits against Tor Browser, an 0-day (or series of 0-days) that yield root/system access on Tor Browser users running Tails (based on GNU/Linux) or Windows 10 operating systems and have the Tor Browser security setting set to 'HIGH' (the default setting of Tor Browser, which blocks JavaScript) was awarded with bounties in the order of USD 200,000 to USD 250,000.

As long as systems remain vulnerable, hacking capabilities provide some redress for the challenges that strong end-to-end encryption pose to governments. Governments also seek alternative methods, for instance by imposing mandatory key escrow or pursuing 'kleptographic' methods, i.e. cryptographic methods that are designed to still allow access under certain conditions.[95] The 'crypto problem' remains an open problem to governments, policymakers and technologists. Interested readers are referred to two publications[96] released in February 2018.

5.6 Conclusion

An intuition of informatics and privacy has been provided, and it was argued that the relation between informatics and privacy can be viewed from two perspectives: ICT poses privacy challenges, and privacy poses ICT challenges. Selected topics relating to both perspectives have been discussed. From a

93 Ibid.

94 See https://www.zerodium.com/program.html.

95 For instance depending on some secret knowledge or cryptanalytic capabilities; that hopefully do not become available to criminals or hostile states.

96 1) 'The Risks of "Responsible Encryption"', February 2018. White paper by Stanford cryptologist Riana Pfefferkorn discussing risks of pursuing a requirement 'that vendors must retain the ability to decrypt for law enforcement the devices they manufacture or communications their services transmit'. Available at https://assets.documentcloud.org/documents/4374283/2018-02-05-Technical-Response-to-Rosenstein-Wray.pdf. 2) 'Decrypting the Encryption Debate – A Framework for Decision Makers', released 15 February 2018. Consensus Study Report of a study chaired by Fred Cate, with input from, among many others, noted Stanford cryptographer Dan Boneh. Available at https://www.nap.edu/catalog/25010/decrypting-the-encryption-debate-a-framework-for-decision-makers.

technical perspective, cryptography, PETs, and access controls are building blocks for privacy and data protection. Readers interested in privacy from a technological perspective are suggested to look at the resources listed below.

Finally, it can be noted that privacy-related publications exist in branches of informatics that directly deal with identified or identifiable personal data, for instance bioinformatics (e.g. processing genetic data), health informatics (e.g. processing electronic medication or health records), urban informatics (technologies for use in cities and urban environments), security informatics (e.g. identifying potential terrorists, spies, and criminals; and depending on regime, dissidents), and certain areas of robotics research. Due to length restrictions, these were not discussed here.

Further reading

Academic conferences

- PET Symposium (PETS) (https://petsymposium.org/). Recent proceedings:
  - Journal Proceedings on Privacy Enhancing Technologies (https://www.degruyter.com/view/j/popets)
- Computers, Privacy, and Data Protection (CPDP) (http://www.cpdpconferences.org/). Recent proceedings:
  - CPDP 2017: 'Data Protection and Privacy: The Age of Intelligent Machines', Ronald Leenes, Rosamunde van Brakel, Serge Gutwirth, and Paul De Hert (eds.), Oxford: Hart Publishing, 2017.
  - CPDP 2016: 'Computers, Privacy and Data Protection: Invisibilities & Infrastructures', Ronald Leenes, Rosamunde van Brakel, Serge Gutwirth, and Paul De Hert (eds.), Dordrecht: Springer, 2017.
  - CPDP 2015: 'Data Protection on the Move', Serge Gutwirth, Ronald Leenes, and Paul De Hert (eds.), Dordrecht: Springer, 2016.
  - CPDP 2014: 'Reforming European Data Protection Law', Serge Gutwirth, Ronald Leenes, and Paul De Hert (eds.), Dordrecht: Springer, 2015.
- IFIP International Information Security and Privacy Conference (IFIP SEC) (https://www.ifipsec.org/).
- Amsterdam Privacy Conference (APC), organized bi/tri-annually by the Institute of Information Law (IViR) of the University of Amsterdam. See e.g.: https://apc2018.com/
- ACM SIGSAC (https://www.sigsac.org/) conferences, including:
  - ACM Conference on Computer and Communications Security (CCS) (https://www.sigsac.org/ccs.html)
  - WiSec: ACM Conference on Security and Privacy in Wireless and Mobile Networks (https://www.sigsac.org/wisec/)
  - CODASPY: ACM Conference on Data and Application Security and Privacy (http://www.codaspy.org/)
- IEEE Symposium on Security & Privacy (S&P) (https://www.ieee-security.org/)
  - Co-hosted: IEEE International Workshop on Privacy Engineering (IWPE) (http://iwpe.info/)

- USENIX Security (https://www.usenix.org/) and co-hosted workshops, such as:
    - Workshop on Offensive Technologies (WOOT) (e.g. WOOT'18: https://www.usenix.org/conference/woot18)
    - Symposium on Usable Privacy and Security (SOUPS) (e.g. SOUPS'18: https://www.usenix.org/conference/soups2018)
- Network and Distributed System Security Symposium (NDSS) (https://www.ndss-symposium.org/)
- Events sponsored by the International Association for Cryptologic Research (IACR), a non-profit scientific organization. Including:
    - Crypto (https://www.iacr.org/meetings/crypto/)
    - Eurocrypt (https://www.iacr.org/meetings/eurocrypt/)
    - Asiacrypt (https://www.iacr.org/meetings/asiacrypt/),
    - Cryptographic Hardware and Embedded Systems (CHES) (https://ches.iacr.org/)
    - Real World Cryptography (RWC) (https://rwc.iacr.org/)
- Financial Cryptography and Data Security, organized by the International Financial Cryptography Association (IFCA) (https://ifca.ai)
- The International Symposium on Research in Attacks, Intrusions, and Defenses (RAID) (http://www.raid-symposium.org/)

Hacker conferences

Novel and high-quality work on privacy in relation to technology, both in defence (e.g. new PETs and security mechanisms) and offence (e.g. new vulnerabilities and attacks) is not only presented at academic conferences, but often also first, or even only, at hacker conferences. Large(r)-scale hacker conferences include:
- Chaos Communication Congress. See: https://ccc.de/en/
- DEF CON. See: https://www.defcon.org/
- Black Hat. See: https://www.blackhat.com/
- Hack in the Box. See: http://www.hitb.org/
- Four-yearly hacker conference organized in the Netherlands, new name for each event. Most recent event: Still Hacking Anyway (SHA) 2017. See https://sha2017.org/.

Small(er)-scale hacker conferences include, inter alia:
- INFILTRATE
- PHDays
- PH-Neutral (a speakers-only event)
- t2.fi

It is recommended to browse through conference materials (papers, slides, videos, code) of past conferences, which are usually publicly available and archived on the web.

Books

Academic works are included in the bibliography at the end of this chapter. These non-academic publications are further recommended:
- *Privacy in Technology*, J.C. Cannon (ed.), International Association of Privacy Professionals (IAPP), 2014.
- *Introduction to IT Privacy – A Handbook for Technologists*, Travis Breaux (ed.), International Association of Privacy Professionals (IAPP), 2014.

Miscellaneous resources

- *Anonymity Bibliography*, Freehaven. Selected papers and bibliography on anonymity, 1977-present. See https://www.freehaven.net/anonbib/
- *Dcypher*. Dutch platform for scientific research on information security. See https://www.dcypher.nl.

Bibliography

Acar, Gunes, Christian Eubank, Steven Englehardt, Marc Juarez, Arvind Narayanan, and Claudia Diaz. (2014). 'The Web Never Forgets: Persistent Tracking Mechanisms in the Wild' in *Proceedings of the 2014 ACM Conference on Computer and Communications Security* (*CCS 2014*). New York: ACM.

Allodi, Luca. (2017). 'Economic Factors of Vulnerability Trade and Exploitation' in *Proceedings of the 2017 ACM SIGSAC Conference on Computer and Communications Security* (*CCS '17*). New York: ACM, 1483-1499. doi: 10.1145/3133956.3133960.

Altman, Irwin. (1975). *The Environment and Social Behaviour: Privacy, Personal Space, Territory, Crowding*. Monterey: Brooks/Cole Pub. Co.

Asghari, Hadi, Michel J.G. van Eeten, Axel M. Arnbak, and Nico A.N.M. van Eijk (2012). 'Security Economics in the HTTPS Value Chain', presented at *TPRC* 2012: the *R*esearch Conference on Communication, *I*nformation and *I*nternet *P*olicy.

Martin Bangemann et al. (1994). *Europe and the Global Information Society* ('Bangemann report'). Recommendations of the High-level Group on the Information Society to the Corfu European Council. *Bulletin of the European Union*, Supplement No. 2/94.

Bernstein, Daniel J., Nadia Heninger, Paul Lou, and Luke Valent. (year). *Post-quantum RSA*, Cryptology ePrint Archive: Report 2017/351.

Borking, John. (2010). *Privacyrecht is code. Over het gebruik van Privacy Enhancing Technologies*. Deventer: Kluwer.

Brundage, Miles, Shahar Avin, Jack Clark, Helen Toner, Peter Eckersley, Ben Garfinkel, Allan Dafoe, Paul Scharre, Thomas Zeitzoff, Bobby Filar, Hyrum Anderson, Heather Roff, Gregory C. Allen, Jacob Steinhardt, Carrick Flynn, Seán Ó hÉigeartaigh, Simon Beard, Haydn Belfield, Sebastian Farquhar, Clare Lyle, Rebecca Crootof, Owain Evans, Michael Page, Joanna Bryson, Roman Yampolskiy, and Dario Amodei (year). 'The Malicious Use of Artificial Intelligence: Forecasting, Prevention, and Mitigation', arXiv:1802.07228 [cs.AI].

danah boyd. (2008). *Taken Out of Context: American Teen Sociality in Networked Publics*, PhD thesis University of Berkeley, California.

Stephen Budiansky. (2010). 'What's the Use of Cryptologic History?' in *Intelligence and National Security*, 25(6), 767-777. doi: 10.1080/02684527.2010.537875.

Randy Burkett. (2013). 'An Alternative Framework for Agent Recruitment: From MICE to RASCLS'. *Studies in Intelligence* 57(1).

Carlini, Nicholas and David Wagner(2018). 'Audio Adversarial Examples: Targeted Attacks on Speech-to-Text', arXiv:1801.01944 [cs.LG], submitted 5 January 2018.

Cavoukian, Ann (2009 [revised 2011]). *Privacy by Design: The 7 Foundational Principles*. Available at https://www.ipc.on.ca/wp-content/uploads/Resources/7foundationalprinciples.pdf.

Clayton, Richard. (2008). 'The Phorm Webwise System', technical analysis, University of Cambridge. Available at: http://www.cl.cam.ac.uk/~rnc1/080518-phorm.pdf.

Colesky, Michael, Jaap-Henk Hoepman, Christoph Boesch, Frank Kargl, Henning Kopp, Patrick Mosby, Daniel Daniel Le Métayer, Olha Drozd, José M. del Álamo, Yod-Samuel Martín, Mohit Gupta, and Nick Doty. (2015). *Privacy Patterns* (website), Contribution by EU FP7 project 'PRIPARE'. Available at https://privacypatterns.org.

Colesky, Michael, Jaap-Henk Hoepman, and Christiaan Hillen. (2016). 'A Critical Analysis of Privacy Design Strategies' in *Security and Privacy Workshops (SPW)*, IEEE, 33-40.

Colesky, Michael, Julio C. Caiza, José M. del Álamo, Jaap-Henk Hoepman, and Yod-Samuel Martín. (2018). 'A System or Privacy Patterns for User Control' in *Proceedings of SAC 2018: Symposium on Applied Computing, Pau, France, April 9-13, 2018 (SAC2018)*.

Dingledine, Roger, Nick Mathewson, and Paul Syverson. (2004). 'Tor: the Second-generation Onion Router' in *Proceedings of the 13th conference on USENIX Security Symposium* (SSYM'04). Berkeley: USENIX,.

Dujmovic, Nicholas. (2018). 'Tech Stars on the Wall: The Human Cost of Intelligence Technology' in *International Journal of Intelligence and CounterIntelligence* 31(1), 126-138. doi: 10.1080/08850607.2017.1337447.

Dulek, Yfke, Christian Shaffner, and Florian Speelman. (2016). 'Quantum Homomorphic Encryption for Polynomial-sized Circuits' in *Advances in Cryptology – CRYPTO 2016*, Matthew Robshaw and Jonathan Katz (eds.), Lecture Notes in Computer Science 9816. Berlin/Heidelberg: Springer.

Durumeric, Zakir, James Kasten, Michael Bailey, and J. Alex Halderman. (2013). 'Analysis of the HTTPS Certificate Ecosystem' in *Proceedings of the 2013 Internet Measurement Conference* (IMC 2013). New York: ACM.

Peter Eckersley. (2010). 'How Unique Is Your Web Browser?' in M.J. Atallah and Nick J. Hopper (eds.). *Proceedings of the 10th International Conference on Privacy-enhancing Technologies.* Lecture Notes in Computer Science 6205. Berlin/Heidelberg: Springer. Lecture Notes in Computer Science 6205. doi: 10.1007/978-3-642-14527-8_1.

Fidler, David (ed.). (2015). *The Snowden Reader.* City: Indiana University Press.

Forer, Louis. (1989). *A Chilling Effect: The Mounting Threat of Libel and Invasion of Privacy Actions to the First Amendment.* City: Norton.

Gentry, Craig. (2009). *A Fully Homomorphic Encryption Scheme*, PhD thesis Stanford University. Available at https://crypto.stanford.edu/craig.

Ryan M. Gerdes, Ryan M., Mani Mina, Steve F. Russell, and Thomas E. Daniels. (2012). 'Physical-Layer Identification of Wired Ethernet Devices' in *IEEE Transactions on Information Forensics and Security* 7, 1339-1353.

Grothoff, Christian. (2017). *The GNUnet System*, PhD thesis University of Rennes 1. Available at https://grothoff.org/christian/habil.pdf.

Hellman, Martin and Whitfield Diffie. (1976). 'New Directions in Cryptography' in *IEEE Transactions on Information Theory* 22(6), 644–654. doi: 10.1109/TIT.1976.1055638.

Hoepman, Jaap-Henk. (2014). 'Privacy design strategies' in *ICT Systems Security and Privacy Protection*, 446-459.

Hou, Weikun, Xianbin Wang, Jean-Yves Chouinard, and Ahmed Refaey. (2014). 'Physical Layer Authentication for Mobile Systems with Time-Varying Carrier Frequency Offsets' in *IEEE Transactions on Communications* 62, 1658-1667.

Holvast, Jan. (1986). *Op weg naar een risicoloze maatschappij? De vrijheid van de mens in de informatiesamenleving.* Leiden: Publisher.

Horn, Gayle. (2005). 'Online Searches and Offline Challenges: the Chilling Effect, Anonymity and the New FBI Guidelines' in *New York University Annual Survey of American Law* 60, 735.

Jansen, J.R.P. and C.E.W. Hesselman. (2016). 'Ervaringen met privacybeheer voor DNS-"big data"-toepassingen' in *Privacy & Informatie* 4.

Kocher, Paul, Daniel Genkin, Daniel Gruss, Werner Haas, Mike Hamburg, Moritz Lipp, Stefan Mangard, Thomas Prescher, Michael Schwarz, and Yuval Yarom. (2018). 'Spectre Attacks: Exploiting Speculative Execution', arXiv:1801.01203 [cs.CR], submitted 3 January 2018. Available at https://arxiv.org/abs/1801.01203.

Koot, Matthijs R. (2012). *Measuring and Predicting Anonymity*, PhD thesis University of Amsterdam.

Lederman, Jaimee, Brian D. Taylor, and Mark Garrett. (2016). 'A Private Matter: the Implications of Privacy Regulations for Intelligent Transportation Systems' in *Transportation Planning and Technology* 39(2). doi: 10.1080/03081060.2015.1127537.

Leeuw, Karl de. (2015). 'The Institution of Modern Cryptology in the Netherlands and in the Netherlands East Indies, 1914–1935' in *Intelligence and National Security* 30(1), 26-46. doi: 10.1080/02684527.2013.867223.

Lessig, Lawrence. (1999). *Code and Other Laws of Cyberspace*. New York: Basic Books.

Macrakis, Kristie. (2010). 'Confessing Secrets: Secret Communication and the Origins of Modern Science' in *Intelligence and National Security* 25(2), 183-197. doi: 10.1080/02684527.2010.489275.

McKeon, Matt. (year).'The Evolution of Privacy on Facebook', http://mattmckeon.com/facebook-privacy/. Graphics used with permission.

Martin, Jeremy, Travis Mayberry, Collin Donahue, Lucas Foppe, Lamont Brown, Chadwick Riggins, Erik C. Rye, and Dane Brown. (2017). 'A Study of MAC Address Randomization in Mobile Devices and When it Fails' in *Proceedings on Privacy Enhancing Technologies*, 4, 365-383. doi: 10.1515/popets-2017-0054.

Matte, Célestin, Mathieu Cunche, Franck Rousseau, and Mathy Vanhoef. (2016). 'Defeating MAC Address Randomization Through Timing Attacks' in *Proceedings of the 9th ACM Conference on Security & Privacy in Wireless and Mobile Networks* (WiSec 2016). New York: ACM, 15-20. doi: 10.1145/2939918.2939930.

Matte, Célestin. (2017). *Wi-Fi Tracking: Fingerprinting Attacks and Counter-Measures*, PhD thesis INSA Lyon. Available at https://hal.archives-ouvertes.fr/tel-01659783.

Ralph C. Merkle, Ralph C. (1978). 'Secure Communication over an Insecure Channel' in *Communications of the ACM* 21(4), 294-299. doi: 10.1145/359460.359473.

Milan, Stefania and Niels ten Oever. (2017). 'Coding and Encoding Rights in Internet Infrastructure' in *Internet Policy Review* 6(1). doi: 10.14763/2017.1.442.

Mowery, Keaton and Hovav Shacham. (2012). 'Pixel Perfect: Fingerprinting Canvas in HTML5' in *Proceedings of W2SP 2012*, IEEE Computer Society.

Murdoch, Steven, Mike Bond, and Ross Anderson. (2012). 'How Certification Systems Fail: Lessons from the Ware Report' in *IEEE Security & Privacy* 10(6), 40–44. doi: 10.1109/MSP.2012.89.

Nissenbaum, Helen. (2010). *Privacy in Context: Technology, Policy, and the Integrity of Social Life*. Stanford: Stanford Law Books.

Pariser, Eli. (2011). *The Filter Bubble: What the Internet Is Hiding from You*. New York: Penguin Press.

Lund Petersen, Karen and Vibeke Schou Tjalve. (2018). 'Intelligence Expertise in the age of information sharing: public–private "collection" and its Challenges to Democratic Control and Accountability' in *Journal of Intelligence and National Security* 33(1) 21-35. doi: 10.1080/02684527.2017.1316956.

Pfitzmann, Andreas and Marit Hansen. (2010). 'A Terminology for Talking about Privacy by Data Minimization: Anonymity, Unlinkability, Undetectability, Unobservability, Pseudonymity, and Identity Management', version 0.34, 10 August. Available at https://dud.inf.tu-dresden.de/Anon_Terminology.shtml.

Roosendaal, Arnold. (2013). *Digital Personae and Profiles in Law*.City: Wolf Legal Publishers.

Ross, Ron, Michael McEvilley, and Janet Oren. (2016). 'Systems Security Engineering – Considerations for a Multidisciplinary Approach in the Engineering of Trustworthy Secure Systems', NIST Special Publication 800-160 (updated January 2018). Available at https://csrc.nist.gov/publications/detail/sp/800-160/final.

Schuster, Stefan, Melle van den Berg, Xabier Larrucea, Ton Slewe, and Peter Ide-Kostic. (2017). 'Mass Surveillance and Technological Policy Options: Improving Security of Private Communications' in *Computer Standards & Interfaces* 50, 76-82. doi: 10.1016/j.csi.2016.09.011.

Shi, Yan and Micheal A. Jensen. (2011). 'Improved Radiometric Identification of Wireless Devices Using MIMO Transmission' in *IEEE Transactions on Information Forensics and Security* 6(4), 1346-1354. doi: 10.1109/TIFS.2011.2162949.

Shor, Peter. (1997). 'Polynomial-time Algorithms for Prime Factorization and Discrete Logarithms on a Quantum Computer' in *SIAM Journal of Computing* 26, 1484-1509.

Shostack, Adam. (2014). *Threat Modeling: Designing for Security*. Indianapolis: Wiley.

Thierer, Adam. (2013). 'Technopanics, Threat Inflation, and the Danger of an Information Technology Precautionary Principle' in *Minn. J.L. Sci. & Tech.* 14(1). Available at https://scholarship.law.umn.edu/mjlst/vol14/iss1/8.

Wang, Wenhao, Zhi Sun, Kui Ren, and Bocheng Zhu. (2016). 'Increasing User Capacity of Wireless Physical-Layer Identification in Internet of Things' in *Global Communications Conference (GLOBECOM)*, 1-6. doi: 10.1109/GLOCOM.2016.7841894.

Samuel D. Warren, Samuel D. and Louis D. Brandeis. (1890).'The Right to Privacy' in *Harvard Law Review* 4(5).

Webb, Diana. (2007). *Privacy and Solitude in the Middle Ages*. London: Hambledon Continuum.

Wullink, Maarten, Giovane C.M. Moura, Moritz Müller, and Cristian Hesselman. (2016). 'ENTRADA: a High-Performance Network Traffic Data Streaming Warehouse' in *IEEE/IFIP Network Operations and Management Symposium (NOMS'16)*. doi: 10.1109/NOMS.2016.7502925.

Political Science and Privacy

Charles Raab

Privacy-related and surveillance issues are salient in public and political consciousness. The monitoring of human behaviour, and the collection, processing, use, and communication of personal information, are well-established practices in the history of states. The exercise, legitimacy, and organization of power, and the processes of politics and policy-making, have crucial information and communication dimensions. Edward Snowden's revelations in 2013 of surveillance activities by states highlighted the political importance of comprehending these processes and policies, and of responding to them, as many activists have done. Less dramatic contexts, such as in business activity, have existed for a very long time and, with the flourishing of the Internet, online commerce, and social media, are now matters of major social and political concern. The nature and means of contemporary database accumulation, intensive data analysis, data sharing, and other surveillance processes – usually seen as a huge benefit for public policy, law enforcement, security, and democratic governance – increasingly pose dilemmas for the protection of citizens' rights, including privacy, and for the nature of citizenship itself. The Facebook/Cambridge Analytica debacle erupted into public attention and showed the dangers to democratic elections that data disclosures may pose.

Questions about data protection – which relates to privacy and surveillance – have also been to the fore, not least with regard to the development of the European Union's General Data Protection Regulation (GDPR) and its potentially profound effects on global as well as local flows of personal data in commerce and the public sector. Lawyers have dissected the legal provisions and novel concepts regarding rights and obligations, and technical specialists have pondered the feasibility of new requirements for the transparency and accountability of opaque information systems and processes such as data analytics. Data protection can potentially leverage changes in the distribution of power between citizens and the state or companies.

These, and many more, are all subjects of great political importance and there has been no shortage of commentary on them by many critics and other observers. However, subjects like these are under-researched by political scientists using the frameworks, tools, and concepts of that academic discipline. For the most part, this discipline has not taken advantage of the opportunity to revisit some core concepts and theories in the light of these issues and of surveillance capacities that could bring profound changes in the relationship of states and citizens as well as major shifts in the processes and power structures of states and interstate regimes. The study of information privacy regulation also seems to be largely outside the attention span of the academic study of politics. Political science's findings and perspectives could make an important contribution to understanding, and to policy and public debate about, the institutions, processes, and behaviour involved in the 'surveillance society'. Some practicing political scientists and closely related specialists (e.g. in public administration, policy studies, and security studies) as well as some with political science degrees but working elsewhere in academia, do conduct research on topics in which privacy and surveillance considerations are implicated (e.g. official secrecy, open government, 'e-government', law enforcement, national security, online commerce, information technologies, and freedom of information). However, there is little sustained effort within political science to analyse surveillance and related practices affecting privacy and other rights and freedoms against the background of political theories about power, democracy, fairness, and the liberal state, or using the theories and methods of empirical political science – and its cognate, international relations – to examine organizational and policy processes in and around privacy and surveillance. In addition, the politics of information makes only rare appearances as a discrete course of study in the formal curricula of political science.

Studying privacy and surveillance has been left largely to scholars who define themselves or who are identified as working within theories, approaches, concepts, and methods of other disciplines in the social sciences and humanities or in legal or technological studies. A critical mass of identifiable political scientists and closely allied specialists is slow to emerge, contributing work to academic journals, editing books, or writing monographs on privacy and related issues and topics. Of course, what constitutes a 'field' or a 'discipline' can be debated, and disciplinary boundaries are blurred. There have indeed been some noteworthy writings produced by those trained in the study of politics or working academically in this field, alongside works by specialists located elsewhere in academia. But exceptions prove the rule: much of the work of shedding light on politically charged developments in surveillance, information technologies and information rights-implicating practices and policies has been driven more by a frequently ahistorical concern over contemporary issues and social problems, or by a fascination with events and new technologies, than by a less dramatic attempt to apply the theories and academic training of political science and thereby to bring these privacy-related subject matters into focus as an object of political-scientific study.

One consequence is that privacy research from the perspective of the study of politics tends to be non-comparative in the sense of systematic, theory-driven investigation that either cuts across jurisdictions (e.g. countries) or across policy areas (e.g. health, education, transport, the environment), or across both. Regulatory policy studies of privacy tend to be 'sectorized' or 'ghettoized'; yet they may offer empirical material for comparative study of topics that are germane to many such areas – such as the nature of risk assessment, the deployment of policy instruments, the configuration of policy networks and institutions operating at different levels, the dynamic processes of technological change and innovation, and the role of the media and public opinion – but they do not engage in comparative analysis themselves. Arguably, the focus of attention is blurred by the way in which privacy, surveillance, and their regulation cut across more conventional subjects or policy areas, pervading many of these as well as featuring strongly in the everyday life of people. Within political science, privacy is not normally comprehended as both a distinctive, definable social or individual phenomenon or as a distinctive, definable area of politics and policy. This makes it an intriguing subject for academic study but presents opportunity and career costs for researchers that make its incorporation into the mainstream of political science a less inviting prospect.

Political science also has only a small and underpowered presence in the emergent multidiscipline of 'surveillance studies'. It is scarce within the many European Union and other funded projects featuring information policy-related subjects that document policies and describe practices in different fields of application, or that compare contexts and national settings of practice, governance, and regulation, or that survey public attitudes and knowledge. However, some projects do seek to reformulate terms and discourses around privacy, surveillance, resilience, resistance, security, liberty, rights, identity, freedom of information, and other concepts, and devise analytic frameworks for empirical research. The formal study of politics has long engaged with these empirical and theoretical matters as part of an interest in political theory, the state, political and governmental institutions, and political behaviour. But in order to cast light on – and indeed to be critical of – what we see happening all around us in an informatized and digitized world – and given a critical mass of research and researchers – these approaches and concepts could be fruitfully leveraged and applied in order to enable us to see that which we do not yet see, and that upon which other specialisms do not focus, without remaining in the grooves of discursive analysis and critique. This, in turn, also has a payoff in terms of research on the freedom-of-information side of the story of information politics and policy, in which a parallel story could perhaps be constructed of the limited impact made hitherto by political science.

In one sense, it does not matter that there is no obvious 'political science of privacy'. So what, if contemporary and historical developments are illuminated by works, including journalism and films, that are eclectic and non-specific in their provenance and that focus attention on privacy and surveillance in everyday life, organisational behaviour or political decision-making; or that expose power imbalances, democratic deficiencies, regulatory inadequacy, digital divides, and the like? So what, if there happens to be a rich literature created by scholars working in and across other disciplines, or by non-academic commentators, that illuminates political phenomena, problems, and issues? There is certainly a growing issue-and-problem-driven 'politics of privacy', broadly construed, in the literature. But is there a distinctive and lower-temperature 'political science' of the study of privacy, surveillance, or data protection? Perhaps not, and many might argue that this gap is of little consequence. Take power, for example: power is a central concept that serves to define this discipline; the relationship of information to power is a classic theme of political studies, and so, too, is the study of the exercise, legitimacy, and organization of power. The privacy and surveillance implications of power and its (mal)distribution, exercise, and organization are already comprehended by theories and empirical research conducted within other disciplines, including sociology, psychology, behavioural economics, history, media and cultural studies, public policy and administration, law, and several more. In any case, a 'discipline' may not provide the best lenses for understanding, and affecting through practical and political action, the sources and effects of power in and across societies with regard to privacy and its cognate values. Moreover, the political science 'canon' of studies, hypothesizing, and theory about power has itself been shaped and enriched by a wide variety of academic specialisms. Why, therefore, worry? But there are gaps in the existing avenues of research that political scientists, more than other types of academic, might fill. What might these be?[1]

Privacy and related subjects and issues could be enhanced by a variety of political science perspectives and approaches drawn, and applied systematically, from (e.g.) comparative politics and public policy, international relations, policy studies of governance and regulation, (multilevel) governance, theories about institutions and their development, political economy, survey design and analysis, and normative political theory about power and democracy. This enhancement would help to shape multidisciplinary research agendas as well as inform public and policy debate on these issues. It would bring to light new

1 A fuller discussion of research topics, concepts, and approaches, with citations of literature across disciplines, is found in Charles Raab (2008), 'Beyond Activism: Research Perspectives on Privacy', Tilburg University Legal Studies Working Paper No. 004/2008 (TILT Law & Technology Working Paper Series No. 007/2008). Available at SSRN: http://ssrn.com/abstract=1096562; accessed 15 February 2018.

information and new ways of looking at it that fall outside the provinces of other avenues of academic analysis. In turn, political scientists can create knowledge about surveillance and data protection- or privacy-related practices to enrich many conventional subjects that do not have privacy or surveillance as their specific focus. This two-way general highway of subject-matter includes:

- the involvement of interest and pressure groups, and the media, in policy-making processes and in public communication;
- the role of legislation and implementation in rapidly changing technological environments;
- the relationship of national to international arenas for regulation;
- the implications of practices and policies for our understanding of security, human rights, law-enforcement and public-service delivery;
- institutional accountability and transparency, and the politics of trust and trustworthiness;
- the influence of economically important industries upon policy and practice;
- societal and political resilience and resistance in the face of commercial, governmental and international pressures upon everyday life;

and many others topics germane to subfields.

The collection, processing, use, and communication of personal information, or on the other hand the openness and transparency of official information, are old phenomena. Studying the politics of surveillance and privacy more intensively and systematically would enhance the possibility for political scientists to compare findings and approaches across different subject matters that are more frequently studied within the framework of the discipline, such as economic transactions, healthcare provision, public transport, international relations of various kinds, among others. It would enable rigorous comparative analysis of privacy or surveillance policy with other areas of policy-making and state or private activity.

This chapter has argued that contributions to the study of information privacy issues can be grounded in empirical research, theories, concepts, and analytical approaches derived from the discipline of political science. But information privacy is not the only privacy show in town. Research and commentary on other dimensions of privacy besides the informational one serve to broaden the field and constructively blur the boundary that has developed between information privacy and other domains of privacy: e.g. the body, public and private space, thoughts and movement. Governance and regulatory regimes (including the law) and policy activity for these other objects of study could also be investigated as part of the analysis. Here too, as with information privacy, the academic study of privacy touches base with the world of privacy advocacy, with science and technology studies, and with a wide range of social sciences and humanistic disciplines.

In highlighting the potential contribution that political science might make to understanding current, past, and future phenomena in the broad area of privacy and surveillance, the aim is not to dismiss the many insights and the lively, engaged stimulation that have been offered by the many books, edited collections, articles, reports, and other media that have contributed to our understanding of, and concern about, the economic, social, and political processes that are fraught with implications for privacy, even without waving the flag of any academic discipline, least of all political science. Nor is it to supplant other avenues to knowledge, but to invite political scientists to join with colleagues who have already been travelling this road for some time, and to support their endeavours in this pursuit.

6. Privacy from an Intelligence Perspective

Willemijn Aerdts & Giliam de Valk

6.1 Introduction

In 1974, the Rolling Stones released their song Fingerprint File. In this song Mick Jagger sings about the file the FBI created about him[1] and states 'these days it's all secrecy; no privacy'.[2] Is this statement true? What is the relationship between privacy and the work of intelligence- and security services? Has this relationship changed over the years? The first section of this chapter will give an explanation about intelligence, intelligence gathering and the role of intelligence- and security services in the democratic legal order.[3] The second section deals with the meaning and function of privacy in relation to intelligence. The third section focuses on the classic texts and authors and current dilemma's (for example accountability and oversight and big data and surveillance). Future challenges will be discussed in the fifth section. Because there will always be new threats, new modus operandi of the agencies as well as their opponents, and new ways of intelligence gathering. The sixth section offers some conclusions and an extensive list of literature. First, we will discuss some basics of intelligence studies.

6.1.1 What is intelligence?

There is an extensive debate within the discipline of intelligence studies about its exact definition.[4] In the field of science, several authors differ of opinion whether intelligence only encompasses secret information or if it

1 The possible existence of this file has never been proven, but in 2006 it became public that the FBI kept an extensive file on John Lennon after his concert for the release of the manager of the MC5 John Sinclair (New York Times, 'While Nixon Campaigned, the F.B.I. Watched John Lennon', Adam Cohen, 21 September 2006.).

2 The Rolling Stones, Fingerprint File, album: It's Only Rock 'n Roll, 1974. This song wasn't the only protest song made against government surveillance and interrogation in 1974. David Bowie released his song '1984'.

3 This chapter will deal with the intelligence- and security services in the light of democratic legal orders (in contrast to the work of these services in for example dictatorial regimes whose first aim is surveillance, in order to control its citizens.

4 For example: Warner 2002; Breakspear 2013, 678-693.

also contains information that is publicly available; is covert action also part of intelligence and should black propaganda[5] or double-cross operations also be considered part of the definition?[6]

Furthermore, the term and concept of intelligence is used in (academic) literature in a broad range of contexts. Sherman Kent made the distinction between: knowledge, organization type, and activities in 1949.[7] Academic literature usually refers to the concept of intelligence within the context of the intelligence gathering process, occasionally it refers to the intelligence product itself or the intelligence organization.[8]

Without losing ourselves in this debate, we could state that intelligence consists of the organized collection of both specific public and secret information, with the overall intention of supporting the executive branch in matters of national security.[9] Intelligence commonly focuses on intentions and capabilities of adversaries that could potentially harm or disadvantage state and democracy.[10]

In a number of countries, a clear distinction is made between intelligence and prosecution (e.g. the Netherlands and Germany). In other countries, like the United States and France, a number of these tasks are integrated into the same organization.

Supplementary to the production of intelligence for intelligence consumers, services also have the opportunity to act themselves to deter and/or mitigate certain threats. A known technique is called 'obtrusive approach'.[11] For example, a service can make itself known to one of its targets, to let the target know that he or she is actively being surveilled to discourage this individual from carrying out an attack, and to sow distrust within a group.[12]

5 One of the examples of soft covert action, propaganda that appears to be spread from one side, but actually comes from the other side (of for example the conflict, see Herman 1996, 55; H. Becker. (1949). 'The Nature and Consequences of Black Propaganda' in *American Sociological Review* 14(2), 221.

6 For example: de Graaff 2012, 11-12; Hijzen 2015, 14.

7 Kent 1949, ix.

8 Lowenthal 2017, 1-2; De Valk 2005, 8-9; Scott and Jackson 2004, 141-43; De Graaff 2012, 11-14.

9 The intelligence cycle is an instrument that can be used to distinguish the different stages of intelligence gathering. The cycle consists of the following five stages: 1. Planning and direction, 2. Collection, 3. Processing, 4. Analysis and 5. Dissemination. See for example Gill and Phytian 2006, 3. It's important to note that the intelligence process in not always linear, and that intelligence is more than only the gathering of information. See for example Gill and Phytian 2006, 3-7

10 Hijzen and Aerdts 2017., 521-554. Compare Herman 1996, 49-53; Scott and Jackson 2004, 154; Gill and Phythian 2006, 6-7.

11 Valk 2005, 57.

12 To give an example, the Dutch Domestic Security Service (Binnenlandse Veiligheidsdienst) used this instrument in the 1970s for the Dutch Red Youth (Rode Jeugd) and in regard to the

6.1.2 Intelligence gathering

Civil and military intelligence services are tasked with the collection of (offensive) intelligence abroad; security services are primarily focused on defensive intelligence gathering within their national territory. One of their primary tasks (in Western democracies) is to defend the democratic legal order and state security.[13]

Lowenthal states that intelligence- and security services exist for the following reasons: to avoid strategic surprise (early warning), to warn in advance about severe threats to the national security (for example in relation to the prevention of terrorist attacks) and to provide long-term expertise and policy support to government organizations such as the police, ministerial departments, and other services as well as the provision of adequate and timely delivery of intelligence.[14]

In order to do so, services go through different stages in the intelligence process.[15] These different phases show the different activities of the services in the light of intelligence. However, it should be stated that this is a theoretical model; in practice one can see that some stages can occur simultaneously.[16]

In the first stage the services make an inventory of the needs of their consumers. These needs can be laid down in annual plans but can also be the result of acute crisis situations.[17] During the second stage, the actual intelligence is being gathered. In intelligence literature, a distinction is made between different means of intelligence gathering. Human intelligence (humint), signals intelligence (sigint), imagery intelligence (imint), geospatial intelligence (geoint), and open source intelligence (osint) are the most mentioned techniques.[18] Nowadays, social media intelligence (SOCMINT) is being described as a separate means of intelligence gathering.[19] During the third phase, the gathered information is prepared for analysis.

groups of south-Moluccan youth, see Engelen 2007, 158.

13 De Valk 2005, 8-9.

14 Lowenthal 2017, 2-4.

15 These phases are described in for example the intelligence cycle and the intelligence matrix. The intelligence matrix shows the possible concurrence of the different stages in the process, but because the intelligence cycle effectively shows the different kinds of activities of the services more clearly, we decided to show the cycle in this chapter to illustrate these different activities.

16 Hijzen and Aerdts 2017, 529.

17 Hijzen and Aerdts 2017, 530.

18 See for example Lowenthal Intelligence Secret's to policy 2017, Gill and Phytian 2006, 63-76 and Herman 1996, 61-82.

19 Omand, Bartlett, and Miller 2012, 801-823.

Fig. 6.1: the Intelligence Cycle (source: CIA Intelligence Agency. (1993). *A Consumers Handbook to Intelligence.* Langley: Central Intelligence Agency).

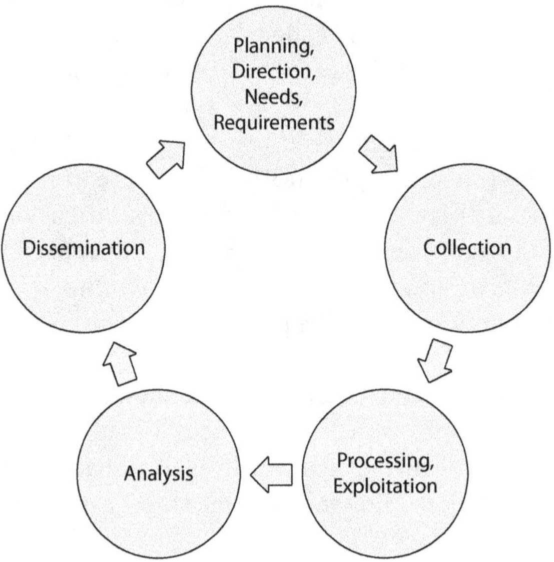

One can think of the deciphering of codes and the visualization of binary materials.[20] During the fourth stage of the analysis, analysts try to combine all the information into an intelligence analysis.[21] One of the important elements of this stage is the assessment of the sources: the determination of the accuracy of the information and the trustworthiness of the sources itself. During the last stage, the information is being disseminated to the different intelligence consumers. This can be done in written products, but also by word of mouth.[22]

One of the dilemmas of intelligence- and security agencies is the fact that they are notable not to disclose their modus operandi, their current state of knowledge and their targets because this could potentially inform adversaries. It is therefore difficult to be held accountable by the general public, which remains one of the most important reasons why institutionalized oversight is of utmost importance.

20 Hijzen and Aerdts 2017, 540.

21 De Valk 2005, 13.

22 Hijzen and Aerdts 2017, 542-543.

6.1.3 Role of intelligence- and security services in the democratic legal order

One of the tasks of a state is providing safety and security for its population.[23] One could state that in a democratic legal order, intelligence and security services are in place to protect the rule of law. In order to be able to do so, one could model the construct of a 'social contract' in which the general public states that in order for intelligence and security services to protect them, they provide them with a mandate and matching special powers to do so. These special powers can, however, infringe individual rights of civilians.

The relation between the intelligence community and privacy is, in the view of the public, dominated by the fear that agencies should be considered a threat to the individual privacy of citizens.[24] This viewpoint is also echoed in the academic community. Gill and Pythian, for example, state that surveillance is central to contemporary governance.[25] The intelligence community fulfils in this respect an imperative role. Not all authors, though, take the infringement on privacy as a necessary starting point. Mary DeRosa pleas for a new model to protect individual privacy – one that relies less on prohibiting the collection and dissemination of private information and more on effective oversight and control of government activity.[26]

This leads to the question of a more fundamental relationship between an intelligence community and the democratic legal order. Without aiming at a definite answer, this relationship is defined by two main dimensions. The first concerns the integrity of the state, including the *Trias Politica*, and the second concerns its individual citizens, including fundamental freedoms and rights.

To start with the state, there are three main pillars for intelligence and security agencies. First of all, there is the integrity of the state. This encompasses the foundation of the state, as laid down by Max Weber. This relates to the state's monopoly of the legitimate use of physical force,[27] including its ability to recover taxation. This is the foremost task of security

23 Eijkman and Weggemans 2013, 285.

24 See for example the most recent discussion in the Netherlands on the referendum on 21 March 2018 on the new Intelligence- and Security Act.

25 Gill and Phythian 2008, 29 ff., 149.

26 Mary DeRosa, 'Privacy in the Age of Terror,' *The Washington Quarterly*, Volume 26, 2003, issue 3, 27-41.

27 Weber 1919. Weber has used it first as a 'Rechtsbegriff', although the concept itself can be dated back as far as the 16th and 17th century (resp. Jean Bodin and Thomas Hobbes).

agencies and police forces. They should thus not only counter domestic insurgency groups, but also outlaw motor gangs and organized crime, that control areas, including demanding protection money from businesses. Secondly, there is the integrity of state borders. This encompasses foremost the intelligence agencies. A new dimension is the digital world, in which the division of tasks between security and intelligence agencies tends to get more blurred. And thirdly, there is – in a democratic legal order – Montesquieu's distribution of power, or the *Trias Politica* principle, with its division in a legislature, an executive, and a judiciary branch.[28] Infringements by ruling political parties to put aside the independency of the judiciary, as for example happened in Venezuela in 2017 would then be part of the task of its domestic security agency – in order to protect the integrity of the democratic legal order.

The second relationship concerns citizenship in a democratic legal order. The debate on the role and the position of citizens in the state has a long history. Already with Aristotle conceptual thinking started, to be followed by many others, among others Baruch Spinoza (*Tractatus theologico-politicus*) and Hannah Arendt. This conceptual debate is added by, since 1215, a tradition of chartas. In the context of a democratic legal order, agencies are crucial for the protection of fundamental civil rights and freedoms, including freedom of speech and the private sphere – in which an individual enjoys a degree of authority, free of interference by governmental, institutions, and groupings. Agencies in a democratic legal order are then tasked to protect such fundamental rights against, for example, groups that attempt to block certain legal political views or violate the private sphere for political aims. This may include, for example, the protection of mink breeders against animal rights extremists, or personnel of abortion clinics against religion-motivated assassinations. Such violations are not only the task of a security agency but can also belong to the tasks of an intelligence agency when it concerns infringements of the private sphere by foreign states in an attempt to control and manipulate its diaspora.

So, security agencies, operating in the context of a democratic legal order, have by definition the obligation to protect and defend fundamental civil rights and freedoms, and by that the privacy of their citizens. As this may, and often will, incur the existence of conflicting values, their activities have to meet the requirements that they stay within the existing legal framework. In order to do so, services have to adhere to the principles of

28 Montesquieu 1748.

proportionality and subsidiarity. Extra tensions may arise when a security service is an integral part of a country's police organization and thus has executive police powers.

As put by former security service official Peter Keller, a non-police security service may have a wide mandate of action but prefers to use its (special) powers as sparingly as possible: thus, limiting the chances of compromising its actions and to be able to continue to operate for an indefinite period of time, as many operations play out itself over many years. A police organization mostly works on a far shorter timeline and ultimately will have to present the results of its investigation in an (open) court. Therefore, its information must meet the judicial standards of admissible, valid, and convincing proof. The court will decide whether this has been the case, including the proper use of police powers to obtain the information. Police and security service activities thus function in a different paradigm.[29] This debate is relevant as it is directly related to potential intensity of infringements on the private sphere.

In short, the intelligence community is not just there to protect the state against perceived threats, or to surveil its citizens. This view would merely apply to the role of agencies in the context of totalitarian regimes. Agencies in the context of a democratic legal order have two main objectives. First of all, the integrity of the state, and second, to protect the fundamental rights and freedoms of its citizens. In this, agencies are neither liberal nor conservative – they follow basic principles of necessity, proportionality, and subsidiarity. And as their acting implies also the protection of the privacy of its civilians – being part of the fundamental right on a private sphere – their acting not only needs a legal foundation, but also to meet the aforementioned principles. In a democratic legal order, it is, up to a certain extent, the population itself, through its elections for parliament, that grant special powers to the services to execute their tasks. The efforts of the intelligence- and security services should be able to be held accountable by processes of superintendence, so-called oversight.[30]

29 Mail Peter Keller to Giliam de Valk and Willemijn Aerdts, 13 December 2017. For example, the principles of proportionality and subsidiarity are laid down in the Dutch Law on the Intelligence and Security Agencies of 2002. In debates between, for example, the head of the Dutch security agency (AIVD) and a privacy watchdog, they aim at the same goal – a free democratic society. The head of AIVD stressed that he did not feel he was visiting the lion's den when visiting a privacy watchdog (consulted in August 2017).

30 Gill and Phytian 2006, 151.

6.2 Meaning and function of privacy

After this theoretical exploration of the relationship between agencies and privacy in light of the democratic legal order, this paragraph continues with a description of the meaning of privacy in regard to intelligence.

There is a strained relationship between privacy on one side and the special powers granted by law to intelligence and security services on the other. Some of these special powers infringe the private life of citizens. For example, the wiretapping of telephone communications, hacking computers, the observation and following of persons will interfere with the right to privacy.[31] Like Bellaby states: 'Intelligence communities face a tension created by, on the one hand, the duty to protect the political community and, on the other hand, the reality that intelligence collection may entail activities that negatively affect individuals'.[32]

If one would look at the European Charter of Human Rights (ECHR), human rights are protected and can be classified into three categories: absolute, limited, and qualified. Privacy is one of the rights that might be affected by intelligence and falls in the categories of the qualified rights. Where absolute rights (e.g. right to life, article 2 ECHR) are non-derogable (cannot be restricted in any circumstances) and limited rights can be derogable 'in times of war or other public emergency threatening the life of the nation'[33] situations.[34] Qualified rights are stated in a positive form and can be limited in the 'circumstances in which the general public interest (including national security, public safety, prevention of disorder, protection of the rights and freedoms of others) can be taken into account'.[35] Another

31 Not only the right to privacy can be a point of concern, also for example the principle of non-discrimination can be concerned, see e.g. Eijkman 2011, 90-101.

32 Bellaby 2012, 95.

33 Article 15 EHRM states: Derogation in time of emergency: 1. In time of war or other public emergency threatening the life of the nation any High Contracting Party may take measures derogating from its obligations under this Convention to the extent strictly required by the exigencies of the situation, provided that such measures are not inconsistent with its other obligations under international law; 2. No derogation from Article 2, except in respect of deaths resulting from lawful acts of war, or from Articles 3, 4 (paragraph 1) and 7 shall be made under this provision; 3. Any High Contracting Party availing itself of this right of derogation shall keep the Secretary General of the Council of Europe fully informed of the measures which it has taken and the reasons therefor. It shall also inform the Secretary General of the Council of Europe when such measures have ceased to operate and the provisions of the Convention are again being fully executed.

34 Gill 2009, 86.

35 Gill 2009, 86.

provision dealing with the right of privacy is article 17 of the International Covenant on Civil and Political Rights.[36]

In countries that are party to the ECHR, the use of these special powers by intelligence- and security services is restricted by law.[37] In the Netherlands, there needs to be (written) approval by the concerned minister or the head of the service before deploying special powers.[38] Furthermore, the use of special powers is only allowed in case the requirements of necessity (necessary in a democratic society in the interests of national security, see article 8 of the ECHR[39] and paragraph 2.3 of the chapter 'Privacy from a legal perspective' of this book), proportionally,[40] subsidiary and finality (information gathered by services should be used only for the purpose for which it was gathered)[41] are fulfilled. Also, the use of special powers is only allowed for as long as deemed necessary.[42]

As discussed earlier, intelligence- and security services cannot disclose their modus operandi, the actual level of information and their list of (potential) targets to the larger public. One can imagine that not only the interested public would like to know more about the modus operandi of a service, but also targets would be very interested in this information. The same goes for the level of information and the list of targets. In case

36 Article 17 ICCPR states: 1. No one shall be subjected to arbitrary or unlawful interference with his privacy, family, home or correspondence, nor to unlawful attacks on his honour and reputation; 2. Everyone has the right to the protection of the law against such interference or attacks.

37 See article 8 EHRM.

38 Art. 19 lid 1 en lid 2 van de Wet op de Inlichtingen- en Veiligheidsdiensten 2002.

39 Article 8, sub 2 of the European Convention of Human Rights states: 2. There shall be no interference by a public authority with the exercise of this right except such as is in accordance with the law and is necessary in a democratic society in the interests of national security, public safety or the economic well-being of the country, for the prevention of disorder or crime, for the protection of health or morals, or for the protection of the rights and freedoms of others.

40 Some practical guidance on the concept of proportionality in the light of intelligence was given by the McDonald Commission (Commission of Inquiry Concerning Certain Activities of theRoyal Canadian Mounted Police). In its report, the Commissions proposed that the special powers of the services should be used only in cases proportionate to the threat under investigation. This should be weighed against the possible results (damage to civil liberties and the democratic legal order. This also in light of the principle of subsidiarity (less intrusive alternative specials power should be used whenever possible). (Report Commission of Inquiry Concerning Certain Activities of the Royal Canadian Mounted Police 1981, 513.)

41 Gill 2009, 92.

42 Lid 3 van artikel 19 van de Wet op de Inlichtingen- en Veiligheidsdiensten 2002 states: 'de toestemming wordt, voor zover bij of krachtens de wet niet anders is bepaald, verleend voor een periode van ten hoogste drie maanden en kan telkens op een daartoe strekkend verzoek worden verlengd voor eenzelfde periode'.

the name of a specific target is known, not only the target knows to take precautions, but also the people around this person can be aware of the fact that they are being watched.

On the other hand, the ways intelligence gathering changes over time will be kept secret (see the previous paragraph). With the advancement of technology, people become more dependent on technology that enables spatial and temporal data access and collection. In the last years, it became clear that intelligence and security services monitor social media. SOCMINT is for a large part in the realm of open sources. Agencies improved the useful exploitation of technology and open source data by intelligence- and security agencies.[43]

This emphasises the need for checks and balances. All phases in the intelligence process are vulnerable to the abuse of rights. Relatively small decisions may result in great distress. As an example, Gill mentions the use of intelligence in regard to the Operation Iraqi Freedom in 2003, in which case intelligence was used for political purposes.[44] And, as Lowenthal states: 'For most citizens, the trade-off between ethics and increased security is acceptable, provided that the intelligence community operates with rules, oversight, and accountability'.[45]

Some authors state that there is a trade-off between security and human rights. Others, like Gill and Phythian, state that this is a false contradiction because 'rights and security cannot simply be traded off against each other'.[46] This is discussed in more detail in the chapter on ethics by Marijn Sax. One of the main objectives of democratic legal orders is to protect rights and freedoms of their civilians. In the long term, states can only achieve long-term legitimacy in case they respect these rights and freedoms. They state that rules and (ethical) regulations 'will contribute to the effectiveness of security as much as to propriety'.[47]

6.3 Classic texts and authors

The discipline of intelligence studies is relatively new. In contrast to for example the discipline of law, where certain texts can be seen as the

43 Pulver & Medina 2017, 241-256.
44 Gill 2009, 91.
45 Lowenthal 2017, 460.
46 Gill and Phytian 2006, 155-156.
47 Gill and Phytian 2006, 155-156.

foundation in regard to the debate on privacy and law, in intelligence studies the traditional debates and dominant school cannot be seen separate from the relatively new 'classic' texts.

Some of the authorities in the field of intelligence are Michael Herman, Peter Gill, and Mark Phytian. In their textbooks and articles on Intelligence, they all touch upon intelligence and privacy, surveillance, and other special powers that are used by intelligence- and security agencies.

The British Michael Herman wrote an article on ethics and intelligence after the 9/11 attacks. In this article he states that a new paradigm is needed in regard to intelligence and ethics. He states that although he does not consider this the biggest problem of the international society, because of the increasing importance of the use of intelligence, 'intelligence has to fit into the ethics of an increasingly co-operative system of states, perhaps with bigger changes in thinking than previously seemed possible'.[48] He does not specify these changes in this article but hopes that academics as well as practitioners will elaborate on this subject together.[49]

Gill and Phytian adopt surveillance as one of the core concepts of intelligence, in their eyes it is part of modern governance. They state that one of the core goals of governance (or Foulcault's governmentality) entails control and understanding. General surveillance can be seen as an intrinsic part of the global developments that occurred after the attacks of 9/11.[50]

Others, like DeRosa and the authors of this article, argue that the prohibition of the collection of intelligence is outdated, and that the focus should be on oversight and control of the intelligence gathering activities as well as governments and private companies.[51]

Toni Erskine wrote an article 'As Rays of Light to the Human Soul? Moral Agents and Intelligence Gathering' where she explains three approaches to the justification of the use of intelligence by nation states. Hobbesian realism states that it is the moral duty of the state to protect the population. This results in the argument that the activities of intelligence- and security services are justified if they serve the collective well-being of nations. The second justification focuses on the actions by looking at the results and consequences they have outside the direct national policies. In this approach, the activities of intelligence- and security agencies will be accepted in case they 'they maximize the good through balancing the benefits of increased

48 Herman 1996, 355-356.

49 Herman 1996, 356.

50 Gill and Phytian 2006, 29.

51 DeRosa 2003, 36.

knowledge against the costs of how it might have been acquired, in a way similar to that in which the principles of "double effect" operated in the field of just war theory'.[52]

The difficulty of the use of this just war theory, resides in the highly complex computations of goods and harms required in order to draw up Michael Herman's 'ethical balance sheet'. The phrase 'if they maximize the good through balancing the benefits of increased knowledge against the costs', tends to utilitarian ethics, in which the utility is the sum of all benefits for national security, minus the loss of privacy of anyone involved in the action. This may cause tensions with a legal order, which is based on rule. This tension is also present in the first Hobbesian approach.

This is contrary to the last and third approach. This is called the deontological approach, and it is based on the work of Emmanuel Kant. Rules simply prohibit certain actions. And like Erskine states, many of the methods used by intelligence- and security services fail to meet these deontological standards.[53] This last approach seems, as a way of organizing society, to cause the least tensions with the concept of a democratic legal order. It is most in line with the role of intelligence and security services in a legal order as described in the first section of this chapter. Although the deontological approach seems the logical choice with respect to the legal order, doubts are cast on its feasibility.

Gill and Phythian state that the final resolution of ethical dilemmas in the light of intelligence cannot be found in legislation or human rights treaties.[54] They state that 'since intelligence cannot be disinvented, and current practice is denominated by realist ethics, perhaps the most we can strive for is harm minimization; we need to regulate the "second oldest profession" in such a way as to minimize the harm it does to producers, consumers and citizens'.[55] Ross Bellaby states in his article 'What's the Harm? The Ethics of Intelligence Collection' of 2012 that academic experts as well as intelligence professionals ask for an ethical framework in regard to the use of the means of intelligence gathering.[56] Authors struggle in getting the harm that is being done and the protection that is needed in one model. An exception seems to be the already mentioned DeRosa. She

52 Erskine 2004, 359-381; Gill and Phytian 2006, 154.

53 Erskine 2004, 359-381.

54 Gill and Phytian 2006, 155.

55 Gill and Phytian 2006, 155.

56 Bellaby 2012, 93.

proposes to rely less on prohibiting the collection and dissemination and more on effective oversight and control.

What is lacking, is a wider debate of the function of intelligence and security services within a democratic order. Should they only be there for the protection of the state, or to protect the liberties of its citizens? In the first section of this chapter, some elements for discussion are presented. However, this is a debate that is still in its infancy. As a result, this gives authors as Gill and Phytian, by referring to for example Foucault, the opportunity to adopt surveillance as the core concept of intelligence. However, other concepts, as those of Arendt, could be used for a rather opposite point of view – services in the context a legal order should protect the public sphere and the versatile debate. As a result, we are left with Erskine's general observations, without much direction of what the role of services in a legal order ought to be. Although the political theoretical debate is still open for refinement, in the more judicial oriented literature, the general impression is that infringements to the right of privacy can be justified, but they should be embedded in law and be balanced by a form of accountability (see below).[57]

6.4 Traditional debates and dominant schools

As stated before, the academic field of intelligence studies is relatively young. Methodology is still developing. In the Netherlands, there are different initiatives in regard to these developments. These initiatives take place on both the academic level (for example within the research group Intelligence & Security of the Institute of Security and Global Affairs of Leiden University, as within the intelligence- and security agencies, including the Defensie Inlichtingen en Veiligheidsinsituut (DIVI, Defence Intelligence and Security Institute). This research on methodology and techniques does not only contribute to the improvement of the analytical capabilities of intelligence and security services, but also to the broader and more fundamental issue of how not to miss relevant relationships in the domain of security and threats.

57 Also (supra)national government bodies state the necessity of reviewing the powers of the services, the need for accountability and a code of intelligence ethics. See for example the House of Lords Selects Committee on the Constitution report 'Surveillance: Citizens and the State 2nd Report of Session 2008-2009 (February 2009) and Parliamentary Assembly of the Council of Europe, Democratic Oversight of the Security Sector in Member States, Recommendation 1713 (Strasbourg, 23 June 2005).

6.4.1 Accountability & Oversight

Lowenthal states that 'for most citizens, the trade-off between ethics and increased security is acceptable, provided that the intelligence community operates with rules, oversight, and accountability'.[58] The aspects of oversight, proportionality, the prevention of the abuse of power, trust, big data, surveillance, and the privatization of intelligence will be discussed below.

6.4.1.1 Oversight

One of the roles of oversight in a democratic legal order is to ensure that intelligence (and its instruments) are conducted proportionately, not to feed the false contradiction or mythical balance between human rights and security. Oversight is also important in regard to maintaining national security because abuse of power by intelligence- and security agencies might also increase the threat.[59] Furthermore, oversight plays a crucial role in creating trust in the services and to increase the legitimacy of the intelligence- and security services.[60]

6.4.1.2 Proportionality

In a democracy, oversights instruments play an important role in the protection of the rights of civilians. One of the key tasks for intelligence oversight organizations is to ensure that the services operate within the official mandate and special powers.[61] Intelligence oversight can be considered a relatively new institutionalized activity in democratic legal orders. In the last 25-30 years, progress has been made in the democratic control of intelligence activities as part of their checks and balances system in regard to the executive powers.[62] As Gill mentions: 'The objective of intelligence oversight is to increase both its efficacy and propriety'.[63]

Intelligence oversight refers to the process of superintendence of the intelligence- and security agencies to ensure that policy and legal mandate of the agencies are consistent. Oversight should take place at different levels, internal as well as external.[64]

58 Lowenthal 2017, 460.
59 Gill 2012, 217-218.
60 Wegge 2017, 692.
61 Wegge 2017, 687.
62 See for example Gill 2012 217-218; Wegge 2017, 687.
63 Gill 2012, 217.
64 See for example Gill and Phytian 2006,. 151;Gill 2009, 86; and Gil and Phytian 2006, 148-71.

6.4.1.3 *Prevention of abuse of power and trust (of society)*

In the last decades, national legislation made the external oversight procedures intelligible, but it remains 'devilishly difficult'[65] in practice.[66] One of the elements that deserved attention in this light is the role of (members of) oversight committees. Wegge states that they need to be aware of their role in the democratic legal order. They should ensure that all activities of the intelligence agencies are executed within the legal frameworks.[67] This in contrast to policymakers who mostly don't handle the practical implementation, but have the authority to make legislation in regard to the work of the intelligence community.

Another element is the fact that 'oversight might enhance the risk of leaks and unauthorized disclosure of state secrets'.[68] Other possible problems that may arise in the light of transparency are that some secrets are legitimately worth protecting and that information can quite easily be misused or misinterpreted.[69]

No oversight committee has the resources or capacity to keep track of all activities of intelligence- and security agencies. Therefore, focus on oversight is very important, but attention also must be paid at all times to recruitment, the training of the personnel (also in the light of ethics). And like Gill states: 'If left simply to "insiders", the issues may be dealt with from the mindset of law and rights as minimal standard for practice or, worst, as minimal standards for reporting on practice'.[70]

One last remark in the light of oversight: although the oversight efforts of the intelligence community have been intrinsically intensified over the last thirty years, Gill and Phythian state that this legal reform is party symbolic. Even with these new governmental architectures of legality and accountability, it is still possible to construct reports in such a way that the work of intelligence and security agencies only appear to comply with the adopted legal standards.[71] And to add to that, Aldrich states in his article 'Beyond the vigilant state: globalisation and intelligence', we also know

65 Gill 2012 218.

66 Gill 2012, 218.

67 Wegge 2017, 691.

68 Wegge 2017, 691. Other risks that are mentioned by Wegge are the risk of weakening 'the aggressiveness and willingness to push limits, fostering an unhealthy risk-averse culture within the services' (p. 691). 'The fear of wrongdoing could overshadow the reasonability to avert threat and collect information in ways that sometimes need to be aggressive and forward-leaning', according to Wegge (p. 691).

69 Florini 1998, 60-61.

70 Gill 2009, 101.

71 Gill and Phytian 2006, 152.

much more than we could have ever imagined about intelligence in the past ten years. However, 'little of this information has come to us through the formal channels of oversight and enquiry'.[72] Most has been revealed by, for example, investigation journalism, leaks, or whistle-blowers.

Institutionalized oversight could also potentially influence international intelligence cooperation. It influences the willingness of certain states to actively engage in intelligence sharing, and national overseers lack authority over their national borders.[73] This issue will be addressed in depth in the fifth section.

6.4.2 Big data & surveillance

In this paragraph, the impact of big data[74] on state surveillance and privacy will be discussed.

6.4.2.1 Surveillance and big data

In this chapter, the following definition of big data is used: gathering of masses of data of an undefined number of people without a pre-established purpose. These data are being processed on a group or aggregated level through the use of algorithms.[75] Gill and Phythian describe the concept of surveillance on the basis of two components: gathering and storage of information and supervision by the state of people's behaviour.[76]

One can further make a distinction between two forms of surveillance: data surveillance and personal surveillance. Data surveillance is the systematic (mass) monitoring of persons' communication through the use of technology. This form of surveillance is used to identify possible targets (Van Buuren also calls it 'suspicion generator' because one of the purposes of mass surveillance is to find the specific persons, on the basis of certain indicators, that will possibly pose a threat and therefore can be allocated personal surveillance).[77] Personal surveillance can be defined as surveillance with a focus on a specific person.[78]

72 Aldrich refers to leaks that yield information on services. Aldrich 2008, 902.

73 Wegge 2017, 692; Gill 2012, 218. For example, this has to do with the different policies in regard to public disclosure of materials, methods of working etc.

74 In chapter 6 of this book, technical aspects with regard to big data will be explained.

75 Van der Sloot 2016, 3.

76 Gill and Phytian 2006, 29.

77 Van Buuren 2017, 229-248.

78 See Clarke 1988, 498; Van Buuren 2017, 238.

6.4.2.2 *Surveillance and intelligence*

Some authors see state surveillance as a key concept of contemporary governance.[79] Yet, these views remain uncontested for a lack of a more general political theoretical reflection of what the position of services in a legal order *ought* to be – surveillance, or the protection of the civil liberties of its citizens. But also, more detailed, concerning the data itself, there are major issues to be dealt with. Nowadays, it's almost impossible to see surveillance activities of intelligence and security services apart from the surveillance activities developed by other actors like the corporate community. The borderline between the agencies in regard to the collection of data, but surveillance techniques is rather thin.[80]

6.4.2.3 *Big data, surveillance and privacy*

Van Buuren states in his article 'Ethical challenges of data surveillance' that it seems odd to challenge the concept of privacy in the light of intelligence surveillance when people voluntarily give away all their personal information for communication services free of charge, discounts, and online convenience.[81] He even states that privacy is not the issue, and that the real problem is the shift of power from specific categories of civilians to a 'state-corporate complex'[82] that composes risk categories based on 'algorithms covered in black boxes of socio-technical assemblages'.[83] He states that the underlying ethical question has to do with the shift of power and its consequences.[84]

Others state that the notion of harm in regard to privacy is difficult in the light of big data gathering, because it is difficult to substantiate the specific harm to a specify person[85] and current state surveillance practices are not in line with the right to privacy (as laid down in for example article 8 ECHR and article 17 ICCPR).[86]

One of the dilemmas in regard to big data and surveillance has to with storage and (re)distribution of the gathered data. Stored data is constantly

79 Gill and Phytian 2006, 149; Gill 2006, 28.

80 Buuren 2017, 240. For example, Van Buuren states, that data collected by public as well as private actors for a different purpose (e.g. social security, criminal investigations, online purchases etc. are permanently resold, combined, analysed etc. by several different actors.

81 Van Buuren 2017, 238.

82 Van Buuren 2017, 238.

83 Van Buuren 2017, 239.

84 Van Buuren 2017, 242.

85 Van der Sloot 2016, 4.

86 For example, Georgieva 2015, 127.

combined, exchanged, refined, and redistributed by a whole range of actors.[87] And as mentioned before, all this data can be used to created risk profiles (most of the times without the subjects being aware of this happening). And again, also Eijkman and Schuurmans argue that it is not giving away the information per se, but the disconnection of the data from its context that people are concerned about.[88]

There have been several court cases of the European Court of Human Rights that dealt with the use of mass surveillance and the article 8 ECHR.[89] One of the difficulties in regard to these cases is the that the Court focuses on natural persons and individual harm (victim-requirement), and that this is difficult to prove in regard to big data and surveillance.[90] However, this is solved by the Court by stretching the focus on individual harm '1. When there is a reasonable chance that the applicant has been harmed, 2/ when it is likely that the applicant will be affected by the practice in the future and 3) when the mere existence of a law or policy as such lead to violation'.[91] The Court adopted this approach because of the fact that most citizens are not aware they are being followed to this extent. 'Mostly, the issue is simply the presumed abuse of power by national authorities'.[92] In regard to this large-scale data gathering, there doesn't seem to be a *relative* interest at stake (that can be 'balanced' against other stakes), but *absolute* interests.[93]

6.4.3 Privatization of intelligence

This paragraph deals with the intelligence gathering and processing by private parties. This is not a new phenomenon, but how does it influence the intelligence process, and privacy related matters nowadays?

6.4.3.1 Privatization

Exchange of intelligence between state and non-state actors is not a new phenomenon. For example, already in the 1950s a public-private partnership in intelligence exchange was created by a former MI5 employee, the International Diamond Security Organization. It is said that this private intelligence network eliminated illicit diamond trade when being wound

87 Van Buuren 2017, 240.

88 Eijkman and Weggemans 2013, 291-292.

89 See also section 3.2.1.3 in the chapter 'Privacy from a legal perspective'.

90 Van der Sloot 2016, 9.

91 Van der Sloot 2016, 9.

92 Van der Sloot 2016, 20.

93 Van der Sloot 2016, 21.

up after two years. Other examples are the public-private cooperation during the war in North Yemen from 1962-1970 and the cooperation between government and private companies in countries with incoming Marxist governments.[94] Notable is the provision of training by former government personnel to private security companies. Steele states 'that the private sector could support government agencies through more efficient provision of specific services such as commercial imagery, foreign language assistance, market research and media monitoring, all categorised as open source intelligence'.[95] The attacks of 9/11 led to new legislation, and therefore a significant rise of the outsourcing of intelligence activities.[96]

In regard to the privatization of intelligence, one could make a distinction[97] between the processing of data obtained by the government from private parties, and the gathering (and processing) of information by these private parties themselves (think for example of the financial and telecom sectors).[98]

Privatization of intelligence is a discussion within the intelligence community itself.[99] For example, Michael Herman states that the population itself has become an important provider of information, but he rejects the assumption that this would also imply that the state has lost ground. Only the collection of information is being outsourced, the analysis is still in the hands of the government. Uri Bar-Joseph and Gregory Treverton even go further when stating that the collection of intelligence is no longer considered the core business of services by the intelligence community. However, Herbert as well as Treverton come to the conclusion that information is never only 'raw data', they state that collection and analysis can be considered parallel and sometimes even integral processes (see also the explanation on the intelligence cycle in the first section of this chapter).[100]

94 Delaforce 2013, 23-25.

95 Steel 2013, 22.

96 See for example Voelz 2009, 568-613; Chesterman 2008, 1055-1074.

97 There is one other category of privatization of intelligence, the so-called 'butts in the seat', where contractors, literally, sit next to the governmental counterparts performing the same tasks, see Abbot 2017; Delaforce 2013, 23, where the authors speaks about 'future loss of distinction between public-private sector employment for aspiring intelligence professionals'. However, this category will not be discussed further in this chapter.

98 Delaforce 2013, 21-39, 28.

99 'There are disputes over everything from cost to quality', according to Sebastian Abbot (News21) https://news21.com/story/2006/07/28/the_outsourcing_of_u_s_intelligence, last accessed 19 December 2017.

100 Petersen and Tjalve 2017, 26.

6.4.3.2 *Possible disadvantages of privatization in the light of privacy*

There are potential problems concerning this privatization. One could think of the relationship between the private parties and state dominance[101] in the exchange of intelligence, the fragmentation of the different elements of the intelligence cycle, transparency,[102] and connected to this, in regard to privacy probably the most problematic area, the accountability gap. The private sector is bound by less legal restrictions than the government organizations where the collection and distribution of intelligence is concerned. Private collectors of intelligence are not bound by the public rules in regard to for example retention, storage or (later) public access to the information.[103]

6.4.3.3 *Future of privatization of intelligence*

Delaforce states that it is unlikely that the intensity of public-private intelligence exchange will decline. She even states that because of the advance of digital technology not only the volume of gathered information will increase but also the type of information that can be collected will change.[104] This will have consequences for how states, the general public and the intelligence community think about accountability and oversight. This also relates to the proportionality and prevention of abuse of powers (see above).

6.4.4 Conclusion

This fourth paragraph of this chapter dealt firstly with the impact of accountability and oversight. It showed the importance of oversight for intelligence. Not only in the light of trust and the prevention of the abuse of power, but also in light of proportionality. The second issue discussed in this paragraph was the impact of big data on surveillance and privacy. Issues that were discussed had to do with the impact of the gathering of big data and privacy and for example the storage of information. To conclude, the issue of the privatization of intelligence was discussed. This is not a new development, but the impact (amongst other things on intelligence) became more significant. And this has implications, for example, for the issue of oversight.

101 And/or vulnerability to state pressure (Michaels year, 901-966).

102 'The potential consequences from gaps in the historical records, related to archival limitations on privately collected intelligence', Delaforce, 2013, 22.

103 See Delaforce, 2013, 23; Michaels 2008, 902, 926-928; and Eijkman and Weggemans 2013, 294.

104 Delaforce, 2013, 30.

6.5 New challenges and topical discussions

This chapter will discuss some new challenges and topical discussions in the light of privacy and intelligence. How will possible new means of intelligence gathering influence the right to privacy? How will cyber developments and the use of big data influence the way we look at intelligence and privacy? And how is international intelligence cooperation influenced by the right to privacy?

6.5.1 New means of intelligence gathering

Since the development of the discipline of intelligence, there always have been new ways of intelligence collection. For example, SIGINT started with the interception of telegraph cables, and moved from satellite interception to the current interception of Internet cables. IMINT started with hot air balloons and moved to reconnaissance aircraft and satellites. Nowadays GEOINT is a new discipline that evolved in the mid-1990s from IMINT into a full-blown analytical mapping discipline, fit for asymmetrical warfare.[105]

And there will always be new developments. The field of intelligence can be seen as a cat-and-mouse game in regard to the services and their state and non-state adversaries. These new developments will come with new dilemmas in regard to privacy. For example, the rise of the Internet provided new possibilities for intelligence gathering, but also new challenges regarding privacy. The same goes for the introduction of social media.

To deal with these continuous developments, some authors state that new (intelligence) legislation should be technology neutral.[106] Arguments in favour of this position are: consistency (the need to avoid technologies being treated differently when they should be treated alike), the speed of technological changes, and recognition of the shortcomings of institutions (do regulators really understand the ins and outs of all the new technologies?).[107] Opponents of this opinion state that especially surveillance laws should be targeted at specific technologies. They state that some differences deserve to be treated differently[108] (One of the examples in the article of Ohm is a laptop. Should it be treated the same as a notepad or be considered more

105 Warfare between powers/entities whose (military) powers differ significantly, for example, a guerilla organization versus a professional national army.

106 To give an example, American Congress adopted the Protect America Act in 2007 to amend the Foreign Intelligence Surveillance Act (FISA) in a more technology-neutral manner.

107 Ohm 2010, 1685.-1713.

108 Ohm 2010, 1695-1969; Lessig 2006, 77-79.

like a house because of all the information it contains?) and that in case the technology changes, the executive branch has to consult with the legislator to review the legislation and form an opinion about the new technology. Another argument made by the opponents of technology-neutral legislation is that the more specific surveillance legislation is, the better oversight bodies will able to function.[109]

In any case, at least the accountability and oversight options should be taken care of, also in the light of new collection possibilities. One of the possible solutions in this regard could be to also include people with a technical background in the oversight committees, and people that are familiar with the methodology that is used by services. A good step is being taken in the Netherlands, with the introduction of an ICT expert with experience from within the service in the new review board for the exercise of investigatory powers (Toetsingscommissie Inzet Bevoegdheden, TIB) that will have to authorize the use of special powers before they can be used by the Dutch services.[110]

6.5.2 Cyber developments and the use of big data

On 11 July 2017, the renewed Intelligence and Security Services Act (Wet op de Inlichtingen- en Veiligheidsdiensten) in the Netherlands was adopted by the senate. The Act should have entered into force on 1 January 2018.[111] One of the new elements in the law has to do with the gathering of bulk information with the use of cable interception.

In his weblog Electrospaces.net, Peter Koop illustrated the practical implications of bulk collection under the new law. It is interesting the see, that the proposed oversight is not designed along the lines of the working processes. The approval moments 2a and 2b, for example, are not in line with the direct stage in the working processes. The act appears to be designed from the perspective of oversight. To have adequate oversight, it is of utmost importance to make sure that oversight has good insights into the working processes. This could cause problems in the execution of oversight in this regard.

109 Ohm 2010, 1713.

110 Van den Dool and Versteegh, 'AIVD straks gecontroleerd door oud-AIVD'er en topadvocaat, NRC Handelsblad, 23 January 2018, last accessed 8 February 2018. See article 36 of the new Wet op de Inlichtingen- en Veiligheidsdiensten (Intelligence and Security Services Act).

111 However, because of the fact that 350,000 signatures were assembled, a non-binding referendum on the adopted law will be organized on 21 March 2018. The entry into force is postponed until 1 May 2018.

Fig. 6.1

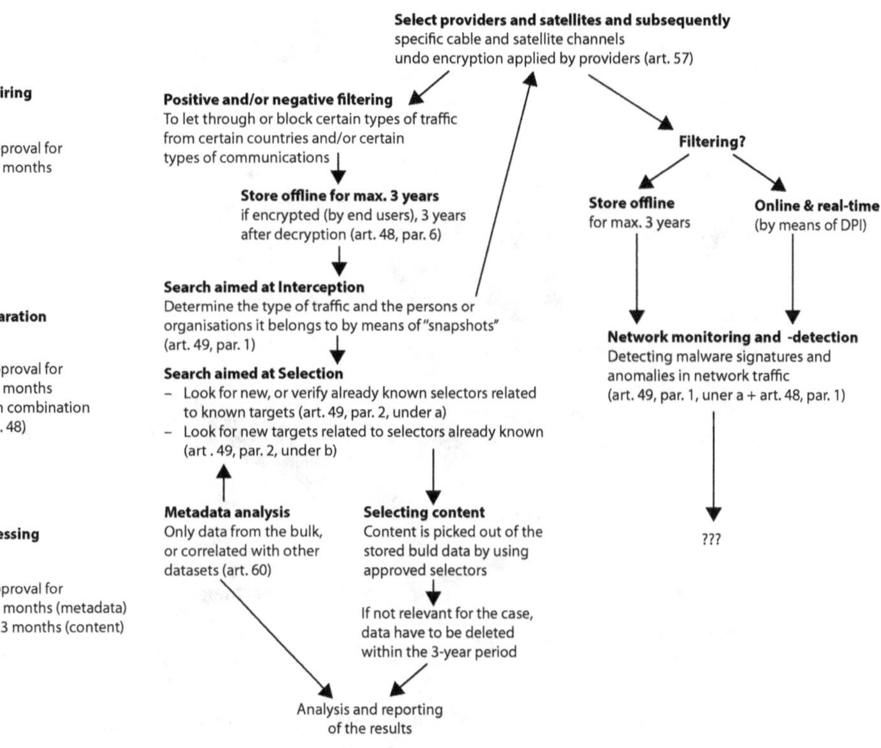

© Peter Koop/Electrospaces.net

6.5.3 (International) intelligence cooperation

International cooperation in regard to intelligence is not a new phenomenon. Already in the 19th century diplomats acted as liaison between different countries, exchanging intelligence related information.

6.5.3.1 International intelligence cooperation

International cooperation intensified during the twentieth century. Herman states that international cooperation during peacetime, is something that did not take place on a regular basis until after the institutionalization of

Fig. 6.2

Bulk Collection unde the new Dutch Secret Services Act
with the 3 stages for which external approval is required

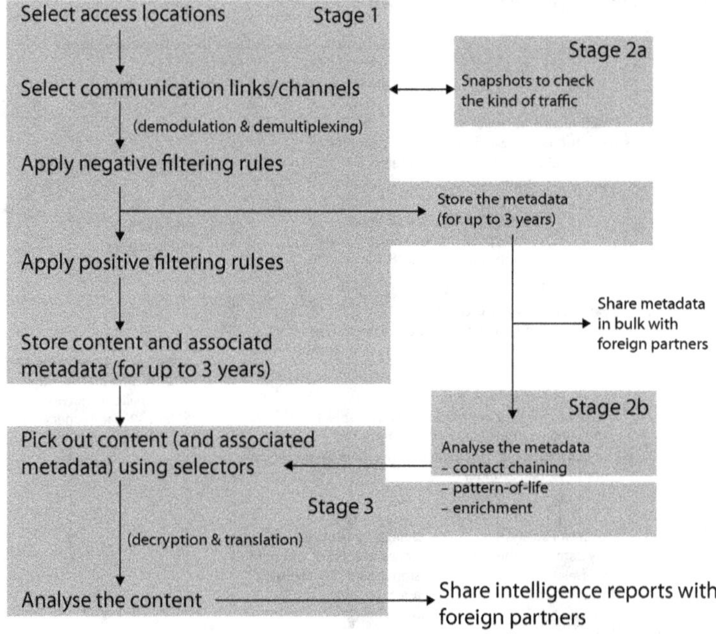

© Peter Koop/Electrospaces.net

intelligence in the nineteenth century.[112] This cooperation between services has implication for privacy of civilians. And because of the changing nature of this cooperation, this topic is discussed here.

Services of different countries have several reasons to want to work together. Important reasons to work together on the technical level are pooling and burden-sharing (two agencies know more than one). Most intelligence- and security services have their own specialty when it comes to the collection of intelligence or a specific geographical location. These forms of international cooperation are hardly formally written down, but one can assume that most output of Western intelligence services is shared with other (national and international) parties. This also has to do with the 'quid pro quo' principle. This 'something for something' in the light of intelligence cooperation means that in order to get information from another service, the requesting service needs to offer something in exchange. This does

112 Herman 1996, 200.

not have to be a simultaneous exchange, as long as it happens eventually. Cooperation on strategic levels can also serve political objectives.[113]

6.5.3.2 *Possible disadvantages of international intelligence cooperation in regard to privacy*

Possible disadvantages or restraints in intelligence cooperation also exist. Many services consider themselves the best or most reliable, however, there are considerable risks for sources in the light of intelligence cooperation. How do they make sure that others will take care of your sources in the same way that you do? Do you want to partly disclose your modus operandi to these other countries? And remember, you have no permanent friends or permanent enemies in regard to international intelligence cooperation, only common interests. And one can imagine that multilateral cooperation (as opposed to bilateral cooperation) even magnifies these risks.[114] Furthermore, and more important in the light of the theme of this chapter, are the possible implications of cooperation for human rights protection.[115] Because you want to make sure that other (international) parties handle the privacy of civilians and target in the same way, within the same legal framework as you do.

6.5.3.3 *Future of international intelligence cooperation*

How to deal with this cooperation in the future? Again, also in this regard, oversight and accountability play an important role (see the fourth section). However, in this light of oversight, Wegge makes a remark when he states that oversight might also have a restraining effect on the willingness of services to cooperate with international partners because representatives of the foreign services might be more distrustful of the members of oversight committees than of their international counterparts.[116]

Also, with regard to oversight, this international cooperation poses a challenge. Oversight procedures have been adopted within the national intelligence systems and national oversight authorities lack the authority outside their own countries.[117] Gill states that even within the European context (for example Europol, Sitcen, and Frontex) the contribution of these supranational bodies is limited because of the bilateral nature of most intelligence sharing.[118]

113 Herman 1996, 204-218.
114 Herman 1996, 207-208.
115 Gill year, 96.
116 Wegge year, 692.
117 Gill 2012, 218. Gill states in his article two examples where the CIA deliberately kept the bilateral arrangements out of the mechanism of oversight (p. 219).
118 Gill 2012, 220.

To conclude this paragraph, one of the most important elements within international intelligence cooperation and privacy, has to do with the bulk collection of information. As may be clear, this is still a developing topic. Under what circumstance are services allowed to share this bulk data with foreign partners and how is this exchange of information controlled by oversight committees? We argue, in line with DeRosa, that oversight needs to be reformed to be able to deal with these new matters effectively.

6.5.4 Conclusion

This paragraph of this chapter on Privacy and Intelligence dealt with future challenges in regard to this topic. It is stated that new ways of intelligence gathering, will come with new dilemmas for the future. This is not new, but it is good to realize that this is an ever-evolving discipline. Furthermore, dilemmas in regard to cyber developments and oversight were discussed. The final topic that was discussed in this paragraph was (international intelligence cooperation. Also, this issue of not new, but the dilemmas that comes with this cooperation are evolving as well, for example in the light of human rights.

6.6 Conclusion

To conclude this chapter on intelligence and privacy, we would like to come back to the article of DeRosa. She, and others with her,[119] opt for a new model to protect individual privacy in times of big data and dataveillance.[120] This new model could be congruent with the position of the intelligence and security services as described in the first section. Deontology seems to be the most appropriate ethical approach to the justification of the use of intelligence in a democratic legal order.

This is not only a technical debate about the degree in which intelligence- and security services are allowed to invade personal space and infringe upon the right of privacy, but also a debate how services are to actively protect civilians and their personal rights.

Data mining is an important instrument of intelligence- and security services. Being able to collect, process, and analyse big data and the search for suspicious correlations seems to be indispensable to avert threats.

119 For example, Van der Sloot 2014, Chapter 2, paragraph 1; Van der Sloot 2016.
120 DeRosa 2003, 27-41.

Henceforth, adequate oversight is of utmost importance. As shown in this chapter, this relates to the position of services and their special powers in the democratic legal order (proportionality), the confidence and trust society has in the services, and the prevention of the abuse of these special powers.

Is the current model of institutionalized oversight capable of coping with these (new) challenges? If we would take, for example, the Dutch agencies as an illustration, the model of DeRosa would imply a twofold revision of the new Dutch intelligence legislation. First, in addition to oversight with regard to the legality (statutory framework) of the intelligence services (are the services allowed to collect the data and are they gathered in the correct manner?), an assessment of the effectiveness of its actions is necessary. Furthermore, some oversight committees consist of individuals solely with a legal background. Oversight committees should also be composed of people with a more technical background and (insider) knowledge of intelligence research, analysis, and methodology.

On the other hand, we must not forget that oversight is not solely accomplished by institutionalized oversight committees. Also, the informal coalitions of jurists, researchers, investigative journalists, and persons working for civil society organizations may play an important role in this regard.[121]

And although the information revolution and the social media changed intelligence collection, and the private sector has completely changed the playing field in the last 25 years, already before that the change was characterized: 'these days it's all secrecy; no privacy'.

Further reading

General literature

J. Waldo, H.S. Lin, and L.I. Millet, L.I. (eds.), *Engaging Privacy and Information Technology in a Digital Age*. Washington: The National Academies Press.

Coffey, A. (2015). 'Fusion Centers as Buffers: Re-Examining the Role of State and Local Organizations in Securing the Homeland and Safeguarding Civil Liberties'. *Sine Fine*, 24 September www.adminsociety.com.

Donohue, L.K. (2016). *Future of Foreign Intelligence: Privacy and Surveillance in a Digital Age*. New York: Oxford University Press.

Verdonck Klooster and Associates. (2017). 'Privacy & Business Intelligence: water en vuur of Yin en Yang?' Journal? 30 March 2017, www.vka.nl.

Hijzen, C.W. and W.J.M. Aerdts. (2017). 'Voor de aanslag: terrorismebestrijding voor inlichtingen- en veiligheidsdiensten' in E. Bakker, E.R. Muller, U. Rosenthal, and R. De Wijk (eds.), *Terrorisme*. Deventer: Kluwer, 521-554.

121 Gill 2012, 221.

Lever, R. (2017). 'Privacy Fears over Artificial Intelligence as Crimestopper'. Journal? 12 November www.phys.org.

Stanley, J. and Steinhardt, B. (2007). 'Even Bigger, Even Weaker – The Emerging Surveillance Society: Where Are We Now? An updte to the ACLU Report "Bigger Monster, Weaker Chains"'. City: ACLU Technology and Liberty Program.

EDPS. (2016). *Artificial Intelligence, Robotics, Privacy and Data Protection. Room document for the 38th International Conference of Data Protection and Privacy Commissioners.* City? Date?

CSIS. (2014). *Balancing Security and Civil Liberties. Liberties for Rebuilding Trust in Intelligence Activities.* City: Publisher.

Regan, P.M., Monahan, T., and Craven, K. (2015). Constructing the Suspicious: Data Production, Circulation, and Interpretation by DHS Fusion Centers. *Administration and Society* 47(6), 740-762.

Munnichs, G., Kouw, M., and Kool, L. (2017). *Een nooit gelopen race. Over cyberdreigingen en versterking van weerbaarheid.* The Hague: Rathenau Instituut.

Eijkman, Q. and Weggemans, D. (2013). 'Open source intelligence and privacy dilemmas: Is it time to reassess state accountability?' *Security and Human Rights* 4, 285-296.

Special Committee to Review the Freedom of Information and Protection of Privacy Act (2004). *Enhancing the Province's Public Sector Access and Privacy Law.* Victoria: The Legislative Assembly of British Columbia.

Bergkamp, L. (2002). 'EU Data Protection Policy. The Privacy Fallacy: Adverse Effects of Europe's Data Protection Policy in an Information-Driven Economy'. *Computer Law & Security Report* 18(1), 31-47.

Brey, Ph. (2006). 'Freedom and Privacy in Ambient Intelligence'. *Ethics and Information Technology* 7, 157-166. 2005 or 2006?

Loof et al. (2015). *Het mensenrechtenkader voor het Nederlandse stelsel van toezicht op de inlichtingen- en veiligheidsdiensten.* Leiden: Afdeling staats- en bestuursrecht, Universiteit Leiden.

RØnn, K.V. (2016). 'Intelligence Ethics: A Critical Review and Future Perspectives'. *International Journal of Intelligence and CounterIntelligence* 29(4), 760-784.

Van der Sloot, B. (2014). 'Privacy in the Post-NSA Era: Time for a Fundamental Revision?' JIPITEC 5(2), paragraph 1.

Gendron, A. (2005). 'Just War, Just Intelligence: An Ethical Framework for Foreign Espionage'. *International Journal of Intelligence and CounterIntelligence* 18(3), 398-434.

Van der Sloot, B. (2012). 'Langs lijnen van gelijkheid: een jurisprudentieanalyse van artikel 15 EVRM'. *NTM NJCM-Bull* 37(2), 208-229.

Lewis, J.A. (2014). *Underestimating Risk in the Surveillance Debate.* Washington: CSIS.

Solove, D.J. (2011). 'Introduction' in *Nothing to Hide: The False Tradeoff between Privacy and Security.* New Haven/London: Yale University Press.

Hribar, G., I. Podbregar, and T. Ivanusa. (2014). 'OSINT: A Grey "Zone"?' *International Journal of intelligence and CounterIntelligence* 27(3), 529-549.

Millet, L.I., H.S. Lin, and J. Waldo. (2010). 'Engaging Privacy and Information Technology in a Digital Age: Executive Summary'. *Journal of Privacy and Confidentiality* 2(1), 5-18.

Calo, R. (2010). 'Peeping HALs: Making Sense of Artificial Intelligence and Privacy'. *European Journal of Legal Studies* 2(3), pages.

Koops et al. (2016). *Privacy Impact Assessment Wet op de inlichtingen- en veiligheidsdiensten 20xx.* Delft: TNO, Privacy & Identity Lab.

Bargh et al. (2014). *Privacy Protection in Data Sharing: Towards Feedback Based Solutions.* Guimares: Association for Computing Machinery.

Friedewald et al. (2005). *Privacy, Identity and Security in Ambient Intelligence: A Scenario Analysis.* City: Elseview Ltd.

Directorate-General for External Policies (2015). *Study: Surveillance and censorship: The impact of technologies on human rights.* Brussels: European Parliament, Policy Department.

Lachmeyer, K. and Witzleb, N. (2014). 'The Challenge to Privacy from Ever Increasing State Surveillance: A Comparitive Perpective'. *UNSW Law Journal* 37(2), 748-783.

Georgieva, I. (2015). 'The Right to Privacy under Fire – Foreign Surveillance under the NSA and the GCHQ and Its Compatibility with Art. 17 ICCPR and Art. 8 ECHR'. *Utrecht Journal of International and European Law* 31(80), 104-130.

Committee on Technical and Privacy Dimensions of Information for Terrorism Prevention and Other National Goals, National Research Council (2010). 'The Science and Technology of Privacy Protection: Appendix L of "Protecting Individual Privacy in the Struggle Against Terrorists"'. *Journal of Privacy and Confidentiality* 2(1), 57-71.

Lord, J. (2015). 'Undercover Under Threat: Cover Identity, Clandestine Activity, and Covert Action in the Digital Age'. *International Journal of Intelligence and CounterIntelligence* 28(4), 666-691.

German, M. and J. Stanley. (2007). *What's Wrong With Fusion Centers?* Washington: ACLU.

Accountability and Oversight

Rapport (2009). UK House of Lord Surveillance, Citizens and the State ?

Bos-Ollermann, H. (2016). *New Surveillance Legislation & Intelligence Oversight Challenges: the Dutch Experience.* Presentation at the International Intelligence Oversight Forum, Bucharest, 11-12 October.

Wegge, N. (2017). 'Intelligence Oversight and the Security of the State', *International Journal of Intelligence and CounterIntelligence* 30(4), 687-700.

Commissie van Toezicht op de Inlichtingen- en Veiligheidsdiensten (2015). *Reactie CTIVD op het concept-wetsvoorstel Wet op de inlichtingen- en veiligheidsdiensten 20xx*, 26 August.

H. Brouwer. (2014). *A Call for More Transparency: A Dutch Perspective on Large-scale Intelligence Gathering and International Cooperation.* Speech at the International Intelligence Review Agencies Conference, London, 8 July 2014.

Big data and Surveillance

D. Anderson Q.C. (2016). *Report of the Bulk Powers Review.* London: The National Archives.

Lim, K. (2016). 'Big Data and Strategic Intelligence'. *Intelligence and National Security* 31(4), 619-635.

William J. Lahneman, William J. (2016). 'IC Data Mining in the Post-Snowden Era'. *International Journal of Intelligence and CounterIntelligence* 29(4), 700-723.

B. van der Sloot, B. (2016). 'Is the Human Rights Framework Still Fit for the Big Data Era? A Discussion of the ECHR's Case Law on Privacy Violations Arising from Surveillance Activities' in S. Gutwirth et al. (eds.), *Data Protection on the Move*, Law, Governance and Technology Series 24. Dordrecht: Springer Science + Business Media.

European Union Agency for Network and Information Security (ENISA) (2015). *Privacy by Design in Big Data. An Overview of Privacy Enhancing Technologies in the Era of Big Data Analytics.* City: Publisher.

Wetenschappelijke Raad voor het Regeringsbeleid (WRR) (2016). *Big Data in een vrije en veilige samenleving.* Amsterdam: Amsterdam University Press.

The Netherlands Scientific Council for Government Policy (WRR) (2016). *Exploring the Boundaries of Big Data.* Amsterdam: Amsterdam University Press.

Wetenschappelijke Raad voor het Regeringsbeleid (WRR) (2016). *Het Gebruik van Big Data door de MIVD en AIVD. Working Paper 18*. Den Haag:

The Netherlands Scientific Council for Government Policy (2017). *WRR-Policy Brief 6, Big Data and Security Policies: Serving Security, Protecting Freedom*. The Hague: WRR.

B. van der Sloot, B. (2015). 'Privacy as Personality Right: Why the ECtHR's Focus on Ulterior Interests Might Prove Indispensable in the Age of "Big Data"'. *Utrecht Journal of International and European Law* 31(80), 25-50.

Privatization of intelligence

Storrock, T. (2016). 'Five Corporations Now Dominate Our Privatized Intelligence Industry'. *The Nation*, 8 September.

Scahill, J. (2008). 'Blackwater's Private Spies'. *The Nation*, 5 June.

Voelz, G.J. (2009). 'Contractors and Intelligence: The Private Sector in the Intelligence Community'. *International Journal of Intelligence and Counter Intelligence* 22(4), 586-613.

DemMars, W.E. (2010). 'Hazardous Partnership: NGOs and United States Intelligence in Small Wars'. *International Journal of Intelligence and Counter Intelligence* 14(2), 193-222.

Hansen, M. (2012). 'Intelligence Contracting: On the Motivations, Interests, and Capabilities of Core Personnel Contractors in the US Intelligence Community'. *Intelligence and National Security* 29(1), 58-81.

Petersen, K.L. and V.S. Tjalve. (2017). 'Intelligence Expertise in the Age of Information Sharing: Public-Private "Collection" and Its Challenges to Democratic Control and Accountability'. *Intelligence and National Security* 33(1), 21-35.

Bruneau, T.C. (2014). 'Patriots for Profit: Contractors and the Military in US National Security'. *Intelligence and National Security* 30(1), 178-179.

Delaforce, R. (2013). 'Public and Private Intelligence: Historical and Contemporary Perspectives, research article'. *Salus Journal* 1(2), 21-39.

Abbot, S. (2006). 'The Outsourcing of U.S. Intelligence Analysis'. *News 21*, 28 July.

Delaforce, R. (2017). *The Privatisation of Intelligence: Challenges for the Mediterranean and Balkan Regions*. City: Research Institute for European and American Studies, www.rieas.gr.

Van Puyvelde, D. (2014). *The US Intelligence Community and the Private Sector: How Rational Are Privatization Rationales*? El Paso: National Security Studies Institute, University of Texas.

Gentry, J.A. (2015). 'Toward a Theory of Non-State Actors' Intelligence'. *Intelligence and National Security* 31(4), 465-489.

Chesterman, S. (2008). '"We Can't Spy ... If We Can't Buy!": The Privatization of Intelligence and the Limits of Outsourcing "Inherently Governmental Functions"'. *The European Journal of International Law* 19(5), 1055-1074.

References

S. Abbot. (2017). 'The Outsourcing of US Intelligence Analysis', web page, last accessed 19 December 2017 https://news21.com/story/2006/07/28/the_outsourcing_of_u_s_intelligence.

R.J. Aldrich. (2008). 'Beyond the Vigilant State: Globalization and Intelligence'. *Review of International Studies* 35, 889-902. doi: 10.10171S0260210509990337.

H. Becker. (1949). 'The Nature and Consequences of Black Propaganda'. *American Sociological Review* 14(2), 221-235.

R. Bellaby. (2012). 'What's the Harm? The Ethics of Intelligence Collections'. *Intelligence and National Security* 27(1), 93-117.

G.M. van Buuren. (2017). 'Ethical Challenges of Data Surveillance' in *Ethics of Counter Terrorism*. Amsterdam: Boom, 229-248.

A. Breakspear. (2013). 'A new definition of intelligence' in *Intelligence and National Security* 28(5), 678-693.

S. Chesterman. (2008). 'We Can't Spy.. If We Can't Buy!: The Privatization of Intelligence and the Limits of Outsourcing Inherently Governmental Functions'. *The European Journal of International Law* 19(5), 1055-1074.

R.A. Clarke. (1988). 'Information Technology and Dataveillance'. *Communications of the ACM* 31(5), 498-512.

R. Delaforce. (2013). 'Public and Private Intelligence: Historical and Contemporary perspectives'. *Salus Journal* 1(2), 21-39, p. 32-33.

Q. Eijkman. (2011). 'Preventive Counter-terrorism and Non-discrimination Assessment in the European Union'. *Security and Human Rights* 2, 90-101.

Q. Eijkman and D. Weggemans. (2013). Open source intelligence and privacy dilemma's: is it time to reassess state accountability? *Security and Human Rights* 23(4), 285-296

T. Erskine. (2004). '"As Rays of Light to the Human Soul?" Moral Agents and Intelligence Gathering'. *Intelligence and National Security* 19(2), 359-381, doi: 10.1080/0268452042000302047.

A. Florini. (1998) 'The End of Secrecy'. *Foreign Policy* 111, 50-63.

I. Georgieva. (2015). 'The Right to Privacy under Fire- Foreign Surveillance under the NSA and the GCHQ and Its Compatibility with Art. 17 ICCPR and Art. 8 ECHR'. *Utrecht Journal of International and European Law* 104, 104-128.

P. Gill (2009). 'Security Intelligence and Human Rights: Illuminating the 'Heart of Darkness', *Intelligence and National Security* 24(1), 78-102.

P. Gill. (2012). 'Intelligence, Threat, Risk and the Challenge of Oversight'. *Intelligence and National Security* 27(2), 206-222.

P. Gill and M. Phythian. (2006). *Intelligence in an Insecure World*. Polity Press, Malden.

B.G.J de Graaff (2012). '*De ontbrekende dimensie: Intelligence binnen de studie van internationale betrekkingen*' (inaurgural lecture, 2 March 2012, Utrecht: Universiteit Utrecht.

B. Hayes. (2010). 'Spying in a See-through World: The 'Open Source' Intelligence Industry'. *Statewatch Bulletin* 1, 1-10.

Report Commission of Inquiry Concerning Certain Activities of the Royal Canadian Mounted Police (1981)

M. Herman. (1996). *Intelligence Power in Peace and War*. Cambridge: Cambridge University Press 1996.

M. Herman. (2010). 'Ethics and Intelligence after September 2001'. *Intelligence and National Security* 19(2), 342-358.

C.W. Hijzen. (2015). 'De vijand en zijn geheimen: Over de inlichtingengeschiedenis als vakgebied' Leidschrift 30(3), 7-24.

S. Kent, '*Strategic intelligence for American world policy*, New Jersey: Princeton 1949.

M.M. Lowenthal. (2017). *Intelligence: From Secrets to Policy*. Los Angeles: CG Press.

J.D. Michaels. (2008). 'All the President's Spies: Private-Public Intelligence Partnerships in the War on Terror'. *California Law Review* 96(4), 901-966.

Omand, Bartlett and Miller (2012). 'Introducing Social Media Intelligence'. *Intelligence & National Security* 27(6), 801-823.

Petersen, K.L. and V.S. Tjalve. (2017). Intelligence Expertise in the Age of Information Sharing: Public-Private "collection" and Its Challenges to Democratic Control and Accountability'. *Intelligence and National Security* 33(1), 21-35.

A. Pulver and R.M. Medina. (2017). 'A Review of Security and Privacy Concerns in Digital Intelligence Collections'. *Intelligence and National Security* 33(2), 241-256.

M. DeRosa. (2003). 'Privacy in the Age of Terror'. *The Washington Quarterly* 26(3), 27-41.

Montesquieu. (1748). *De l'esprit des lois*. Genève: Barrillot et Fils.

L. Scott & P. Jackson, The study of intelligence in theory of practice', *Intelligence and National Security* 19(2), 139-169.

Van der Sloot, B. (2016). 'Is the Human Rights Framework Still Fit for the Big Data Era? A Discussion of the ECHR's Case Law on Privacy Violations Arising from Surveillance Activities' in S. Gutwirth et al. (eds.), *Data Protection on the Move*, Law, Governance and Technology Series 24. Dordrecht: Springer Science + Business Media, 3.

Valk, G.G de, (2005). *Dutch Intelligence: Towards a Qualitative Framework*. Groningen: Eleven International Publishing.

G.J. Voelz. (2009). 'Contractors and Intelligence: The Private Sector in the Intelligence Community'. *International Journal of Intelligence and Counter Intelligence* 22(4), 568-613. doi: 10.1080/08850600903143106.

M. Warner. (2002). 'Wanted: A definition of intelligence', *Studies in Intelligence* 2002 46(3)

M. Weber. (1919). *Politik als Beruf.* City: Publisher.

N. Wegge. (2017). 'Intelligence Oversight and the Security of the State'. *International Journal of Intelligence and CounterIntelligence* 30(4), 687-700.

A Privacy Doctrine for the Cyber Age

Amitai Etzioni

A privacy doctrine built for the cyber age must address a radical change in the type and scale of violations that the nation – and the world – face, namely that the greatest threats to privacy come not at the point that personal information is collected, but rather from the secondary uses of such information. Often-cited court cases, such as *Katz, Berger, Smith, Karo, Knotts, Kyllo* – and most recently *Jones* – concern whether or not the initial collection of information was legal. They do not address the fact that personal information that was legally obtained may nevertheless be used later to violate privacy. That the ways such information is stored, collated with other pieces of information, analysed, and distributed or accessed – often entails very significant violations of privacy.[1] While a considerable number of laws and court cases cover these secondary usages of information, they do not come together to make a coherent doctrine of privacy – and most assuredly not one of them addresses the unique challenges of the cyber age.[2]

Here I attempt to show that in order to maintain privacy in the cyber age, boundaries on information that may be used by the government should be considered along three major dimensions: The level of sensitivity of the information, the volume of information collected, and the extent of cybernation. These considerations guide one to find the lowest level of intrusiveness holding constant the level of common good. A society ought to tolerate more intrusiveness if there are valid reasons to hold that the threat to the public has significantly increased (e.g. there is a pandemic), and reassert a lower level of intrusiveness when such a threat has subsided.

Sensitivity

One dimension is the *level of sensitivity* of the information. For instance, data about a person's medical condition is considered highly sensitive, as are one's political beliefs and conduct (e.g. voting) and personal thoughts. Financial information is ranked as less sensitive than medical information, while publicly

1 Etzioni 2012. For more details, see Etzioni 2015.

2 Swire 2002, 912. ('The increasing storage of telephone calls is part of the much broader expansion since 1967 of stored records in the hands of third parties. Although there are no Supreme Court cases on most of these categories of stored records, the *Miller* and *Smith* line of cases make it quite possible that the government can take all of these records without navigating Fourth Amendment protections').

presented information (e.g. licence plates) and routine consumer choices much less so.

These rankings are not based on 'expectations of privacy' or on what this or that judge divines as societal expectations.[3] Rather, they reflect shared social values and are the product of politics in the good sense of the term, of liberal democratic processes, and moral dialogues.[4] Different nations may rank differently what they consider sensitive. For example, France strongly restricts the collection of information by the government about race, ethnicity, and religion (although its rationale is not the protection of privacy but rather a strong assimilationist policy and separation of state and church). For those who analyse the law in terms of the law and economics paradigm, disclosure of sensitive data causes more harm to the person by objective standards than data that are not sensitive. Thus, disclosure of one's medical condition may lead to losing one's job or not to be hired, to be unable to obtain a loan, or incur higher insurance costs, among other harms. In contrast, disclosure of the kinds of bread, cheese, or sheets one buys may affect mainly the kind and amount of spam one receives.

Volume

The second dimension on which a cyber-age privacy doctrine should draw is the volume of information collected. Volume refers to the total amount of information collected about the same person *holding constant* the level of sensitivity. Volume reflects the extent of time surveillance is applied (the issue raised in *Jones*), the amount of information collected at each point in time (e.g. just emails sent to a specific person or all emails stored on a hard drive?), and the bandwidth of information collected at any one point in time (e.g. only the addresses of email sent or also their content?). A single piece of low-sensitivity data deserves the least protection, and a high volume of sensitive information should receive the most protection.

Under such a cyber-age privacy doctrine, different surveillance and search technologies differ in their intrusiveness. Least intrusive are those that collect only discreet pieces of information of the least sensitive kind. These include speed detection cameras, tollbooths, and screening gates, because they all

3 Shaun Spencer raises concerns around legislating privacy protections. See Spencer 2002, 860. ('Given the powerful influence of various lobbies opposed to strong privacy protection, that role may best be described as a sine qua non. That is, unless the public has a strong desire for privacy in a particular area, attempts to pass legislation establishing that area as a private sphere are doomed to fail (...) To the extent that legislatures base privacy legislation on social values and norms, they necessarily rely on the same changing expectations as the judicial conception of privacy').

4 Etzioni 2004, 67-71.

reveal, basically, one piece of information of relatively low sensitivity. Radiation detectors, heat reading devices and bomb and drug-sniffing dogs belong to this category, not only because of the kind of information (i.e. low or not sensitive) they collect, but also because the bandwidth of the information they collect is very low (i.e. just one facet, indeed a very narrow one, and for a short duration).

Typical closed-circuit televisions (CCTVs) – privately owned, mounted on one's business, parking lot, or residential lobby – belong to the middle range because they pick up several facets (e.g. location, physical appearance, who one associates with), but do so for a brief period of time only and in one locality. The opposite holds for Microsoft's Domain Awareness System, first tested in New York City in 2012. The programme collates thousands of pieces of information about the same person from public sources – such as that from the city's numerous CCTV cameras, arrest records, 911 calls, licence plate readers, and radiation detectors – and makes them easily and instantly accessible to the police. While the system does not yet utilize facial recognition, it could be readily expanded to include such technology.

Phone tapping – especially if not minimized and continued for extended periods of time – and computer searches, collect more volume. (This should not be conflated with considerations that come under the third dimension: Whether these facts are stored, collated, analysed, and distributed i.e. the elements of cybernation). Drones are particularly intrusive because they involve much greater bandwidth and have the potential to engage in very prolonged surveillance at relatively low costs compared to, say, a stake-out. These volume rankings must be adapted as technologies change. The extent to which combining technologies is intrusive depends on the volume (duration and bandwidth, holding sensitivity constant) of information collected. High-volume searches should be much more circumscribed than low-volume ones.

Cybernation: Storing, analysis, and access

The third dimension is the one that is increasing in importance and regarding which law and legal theory have the most catching up to do. Historically, much attention was paid to the question whether the government can legally collect certain kinds of information under specific conditions. This was reasonable because most violations of privacy occurred through search and surveillance that implicated this first-level collection of spot information. True, some significant violations also occurred as a result of collating information, storing it, analysing it and distributing it. However, to reiterate, as long as records were paper bound, which practically all were, these secondary violations of privacy were inherently limited when compared to those enabled by the digitization of data and the use of computers (i.e. by cybernation).

To illustrate the scope and effects of cybernation, a comparison follows: In one state, a car passes through a tollbooth, a picture of its licence plate (but not the driver or others in the front seat) is taken – and then this information is immediately deleted from the computer if the proper payment has been made. In another state, the same information, augmented with a photo of the passengers, is automatically transmitted to a central data bank. There, it is combined with many thousands of other pieces of information about the same person, from locations s/he has visited (e.g. based on cell tower triangulation) to his magazine subscriptions and recent purchases and so on. The information is regularly analysed by artificial intelligence systems to determine if people are engaged in any unusual behaviour, what places of worship they frequent (e.g. flagging mosques), which political events they attend (e.g. flagging those who are often involved in protests), and if they stop at gun shows and so on and on. The findings are widely distributed to local police and the intelligence community, and can be gained by the press and divorce lawyers.

Both systems are based on spot information, that is, pieces of information pertaining to a very limited, specific event or point in time – as is the case in the first state. However, if such information is combined with other information, analysed, and distributed, as depicted in the second scenario, it provides a very comprehensive and revealing profile of one's personal life. In short, the most serious violations of privacy are often perpetuated not by surveillance or information collection *per se*, but by combination, manipulation, and data sharing – by cybernation. The more information is cybernated, the more intrusive it becomes.

References

Etzioni, Amitai. 2004. *From Empire to Community: A New Approach to International Relations*. New York: Palgrave Macmillan.

Etzioni, Amitai. 2012. 'The Privacy Merchants: What Is To Be Done?', *University of Pennsylvania Journal of Constitutional Law* 14(4): 929.

Etzioni, Amitai. 2015. *Privacy in a Cyber Age: Policy and Practice*. New York: Palgrave Macmillan.

Spencer, Shaun B. 2002. 'Reasonable Expectations and the Erosion of Privacy', *San Diego Law Review* 39(3): 843.

Swire, Peter P. 2002. '*Katz* is Dead. Long Live *Katz*', *Michigan Law Review* 102(5): 904.

7. Privacy from an Archival Perspective

Tjeerd Schiphof

7.1 Introduction

Archival science is an academic and professional discipline. As such it is concerned with the theory and methodology, as well as the practice, of the creation, use and preservation of records and archives: coherent collections of records. Archives are kept for several reasons: democratic states feel the need to be transparent; archives can be used to make political accountability possible; archives are essential in assisting the protection of citizens' rights, and they are necessary to answer a societal need for safeguarding collective memory and cultural heritage.

It is not uncommon to distinguish between records management and archives management. The former is concerned with records in general, yet more in particular with records that are relatively young and still used in their functional context. A fraction of those records will be selected for long-term preservation and transferred to a repository. This chapter will deal with the privacy aspects of long-term preservation. Information specialists that work in this field are generally called 'archivists'. The term 'archive' is used both for an archival institution and for a collection of records.

Records, in the sense of pieces of evidence or information, might well contain personal data. Seeing that records will be collected and arranged, and can be accessed, altered, or deleted, it becomes clear that privacy and data protection issues must be taken into consideration.

Is there sufficient reason to discuss the privacy aspects of long-term preservation of information? Elena S. Danielson, author of the book The Ethical Archivist, summarizes it neatly: violation of privacy is part of the archival process. 'The real question is how it can be meliorated.' (Danielson 2010, 9) Much in this chapter will deal with this field of tension between long-term archiving of personal data and its objectives, and the protection of privacy and personal data.

Broadly speaking there are three ways in which information becomes eligible for long-term preservation. Firstly, there is government-generated information, which might be transferred, after a selection procedure, into a government-linked repository (archival institution). A birth certificate would follow this route. Secondly, a government-linked archival organization can acquire materials at will. A national archive could for instance take up the personal archive of an author or photographer. Thirdly, there are private archival institutions.

In the context of government-related records management (the first possibility), a record would initially serve its primary purpose. At a certain point in time, if it is not already deleted as a result of archival retention rules, a decision has to be made whether it deserves to be transferred to a repository for future historical or research purposes. The archivist has an advisory role in the process of transfer, and, if there are identifiable personal data involved, privacy considerations play a role in deciding whether or not certain materials are eligible for long-term retention. In these cases, it will have to be decided whether access restrictions will be necessary. In order to accomplish these tasks, the archivist needs to have knowledge of relevant archival and privacy regulations. If these cannot offer sufficient guidance it might be necessary to consider ethical aspects as well. The ethical aspects will be further explained in this chapter.

In these same government-related, archival surroundings it is common to take care of materials from other sources as well, mostly private collections (the second possibility). The archivist then has the role of reaching agreements with donors. Again, these collections might contain personal data, which makes that the receiving archivist must be aware of privacy regulations when negotiating the conditions for long-term preservation.

Alongside the government-linked archives holding government-related materials as well as private collections, there are private organizations collecting a wide variety of material. This is the third possibility. These three different contexts are regulated differently, depending on the relevant jurisdiction. In the United States private archives are relatively free from regulation; in the European Union this is not the case. Data protection rules apply here to the government sector as well as to the private sector, and certain exemptions are only available for 'archiving in the public interest' and historical research. Private collections in public archives have a middle position: they are kept in a public context, but privacy-based access limitations depend very much on the contractual terms under which these were accepted.

The structure of this chapter is as follows. Section 2 serves two purposes. It examines more in detail where issues of privacy and data protection play a role in the archival processes. Furthermore, the role and function of privacy in the United States and in the European Union are put in the perspective of this chapter. In section 3, the protection of personal data in an archival context is connected with the EU General Data Protection Regulation and its underlying principles. This is supplemented with a description of a semi-legal British code of conduct and some professional ethical codes, addressing also the function and value of these codes in the archival field. Section 4 introduces current discussions in archival science; section 5 discusses relevant trends and dilemmas. Finally, conclusions and suggestions for further reading are provided in section 6.

7.2 Role and function of privacy

This section will go deeper into the relation between privacy and personal data protection on the one hand, and the archival field on the other.

7.2.1 Privacy and the archival process

In this chapter it is assumed that the archivist is concerned with the long-term retention of information. In fact this is a rather traditional conception: in the archival field it is broadly accepted nowadays that an archivist should not 'wait' until records arrive at the gate of his long-term repository, but that he should actively be concerned with records and information in their earlier phases as well. Doing this he would enter what is generally called the professional field of records management. It would be outside the scope of a book chapter like this to deal with the privacy aspects of records management as a whole, reason why this contribution limits itself to the traditional view of an archivist's tasks.

The archivist has a professional duty to guard a number of interests, one of which is the protection of privacy of persons that are somehow involved in the archival processes. There will be people whose data are in the archive, with their permission but more often unwittingly; but also users. Where and when will she be confronted with privacy issues, and what are her resources in dealing with these? What are the competing interests? The following inventory can be made, with reference to the three types, as already set out in the preceding subsection: archiving in a government-linked setting; acquisitions made by government-linked archival institutions; and long-term archiving in a private setting. In all these circumstances the privacy of the user of archival materials is also relevant. Normally there will be use records, which might be consulted in case materials appear to have been stolen or damaged. Use records exist also elsewhere (in libraries, for instance), and for this reason the connected privacy aspects will not be discussed in this chapter.

1. Government-related activities generate an enormous amount of information, which will be documented in records. Only a fraction of these records will end up in archival institutions for the long term. Many files containing personal data will be destroyed as a result of legal or

institutional retention rules; state schools for instance might destroy files of pupils some time after they have left school. Generally government records of a certain age have to be transferred to a state archive, for which selection criteria and procedures will be in place. At this same time access restrictions can be formulated: a record could for instance be declared closed for a given period or only accessible for academic research. Archivists will be involved in developing and applying mentioned criteria and restrictions. Privacy considerations play a role in these processes, and have to be balanced against others interests, like retention costs and future access to the records. The issues can be complex: how to formulate selection criteria? Which materials will be needed in the future? Is it sensible to keep materials with sensitive personal data, knowing they will hardly be used due to restrictions? The law gives guidance, as do institutional rules and professional ethics (see section 3).

2. State archives are the designated institutions to keep government records for the long term. Generally they see it as part of their mission to acquire other materials as well, in order to give a fuller picture of national history, to contribute to safeguarding cultural heritage and to be able to serve the public better (for ...?). An example would be the acquisition of the private archive of a prominent politician. The acquisition of such privately owned records might involve the privacy of the donor himself, or the privacy of third parties (for instance the senders of received letters). This means that the archivist has to be aware of possible future harm to individuals. Archivist and donor might disagree on the question which restrictions (if any) have to be imposed. It happens that donors ask for unreasonable restrictions, which can leave the archivist with the dilemma whether to acquire materials with restricted access, or to refuse the donation and have nothing.
3. Private archival institutions will in general be operating under different regulations than state archives. Depending on national legislation this can work out either way: the regulation of private archives can be less strict, yet on the other hand it might also be the case that legislation grants public archiving certain privileges. In case of acquisitions the archivist must assess the materials and negotiate access restrictions with the donor, much like we saw in type two.

The next subsection will reflect on some relevant differences and similarities between the US and the EU. Normally, archival institutions strive to give the widest possible access to their holdings. Freedom of information regulation might support this stance. From this perspective access restrictions must be minimized. On the other hand, there are legal obligations and professional ethical codes that aim to protect the privacy of all parties involved. The archivist has to balance these interests.

7.2.2 Legal aspects: some differences between the European Union and the United States

It would be outside the scope of this contribution to investigate privacy-related regulations around the world. This chapter will focus on the US and the EU; it is important to stress that there are major differences between the two with regard to privacy regulation. That has its effects on the role and function of privacy in the archival field.

It has been estimated that the US has some 700 state and federal privacy laws. Danielson speaks of 'a bewildering thicket of legislation' (2010, 195). A useful introduction to the current topic, but not up to date anymore, is the book *Navigating Legal Issues in Archives* (2008) by Menzi L. Behrnd-Klodt. Useful is also the reader, edited by Behrnd-Klodt and Peter Wosh: *Privacy and Confidentiality Perspectives: Archivists and Archival Records* (2005). The US privacy regulation 'model' is to enact privacy laws for specific areas. There is little privacy regulation for the private sector. Danielson argues that Americans, no matter their political conviction, are more suspicious of privacy intrusions by the government than of privacy threats coming from the market or from research (2010, 201). Lord Neuberger, speaking of the differences between the US and the EU regarding privacy-related regulation, suggests the following aspects. In Europe there is more faith in formal regulations, where market-based solutions are favoured in the US. He stresses the fact that the economical interests in the US are greater in the EU. Europe has its recent history with totalitarian governments, which has stimulated privacy protection. Moreover, there is a contrast with the US and its commitment to the First Amendment (Neuberger 2015, no. 20). Danielson also notes the perceived high value of open access in the US. American archivists and historians associate open access with a vivid democracy. In dealing with conflicts, this has its effect on the balancing of privacy and access. 'Strategies are put in place to attempt to open as much data as legally possible', she says (Danielson 2010, 201).

The European legal model with respect to data protection differs from the situation in the US. There is uniform, extensive, and detailed data protection regulation, which applies both to the private sector as to government-related processing of personal data. This is embodied now in the General Data Protection Regulation (GDPR). This regulation can function as a good frame of reference for researching the relations between its underlying principles on the one hand, and the principles of long-time data preservation on the other hand. Yet it is important to realize that the law and the underlying principles will not always offer sufficient guidance while balancing interests. In case the law does not provide sufficient guidance a code of professional ethics or a branch code might assist (section 3).

7.2.3 Conclusion

Summing up, it can be said that collecting and arranging historical records and providing access to them is the primary mission in the archival field. Often these records will contain personal data, so privacy issues abound. Privacy as a principle can be said to legitimize curbing otherwise unrestrained processing of personal data. More in particular tensions between access and the privacy of those involved (donors and third parties) have to be solved. On balance, in the United States access appears to be more valued than privacy: privacy functions as a correction on the right to access. In Europe privacy has more footing as a basic right in itself, with extensive data protection rules as its corollary.

7.3 Basic texts and authors

This section discusses a number of texts that shed light on aspects of processing personal data in an archival context. Ketelaar distinguishes five what he calls 'layers' of privacy protection with regard to public archives:

1. Legislation. This might concern privacy rules or archival legislation, and includes access rules. Legislation also defines the role(s) of archivists and the archive as institution. Access rules will also deal with the possibility of requests for dispensation. At this same level of legislation are freedom of information rules. For the purpose of this chapter this layer includes 'soft law': rules that are voluntarily adhered to and so cannot be enforced, but have a certain authority nonetheless.
2. The conditions of transfer to a repository, which specify access restrictions.
3. Researcher's undertakings: researchers sign for their use of sensitive data. This is often required by the conditions of transfer.
4. Physical and practical regulations of the archives. For example with regard to storage, reading room rules, and lending procedures. Occasionally this might include a prohibition to make pictures of materials.
5. Professional ethics. (Ketelaar 1995)

There is reason to add an extra layer:

6. Codes of conduct: codes, agreed upon by relevant sectors in society in cooperation with the data protection authority.

Three influential texts will be discussed to provide the reader with a basic understanding of privacy issues occurring in the field of archival science and archival practice. Successively in the category 'soft law': a Recommendation of the Council of Europe (section 3.1); a piece of legislation: the GDPR (subsection 3.2); and a Sector Code (a code of conduct): the United Kingdom Code of practice for archivists and records managers (subsection 3.3). The subsection closes with the presentation of some relevant ethical codes (subsection 3.4).

7.3.1 Recommendation of the Committee of Ministers to member states on a European policy on access to archives

As the title (*Recommendation*) already indicates this document has no binding force.('Council of Europe: Recommendation No. R (2000) 13 on a European Policy on Access to Archives' n.d.) It is issued on behalf of the Council of Europe, which should not be confused with the European Union. It addresses the Council of Europe member states and it does not specify citizens' rights or duties. Nevertheless, this is an interesting document. It is 'soft law' but it serves well the current aim to explore the tension between two competing interests, access, and protection of privacy in an archival setting. The preamble points out that historians wish to study civil society in order to better understand the complexity of the historical process in general, and of that of the twentieth century in particular. Moreover, this understanding of recent European history could contribute to the prevention of conflicts. In this light, Council of Europe member states are called upon to: 'adopt legislation on access to archives inspired by the principles outlined in this recommendation, or to bring existing legislation into line with the same principles'.

These are the principles that are relevant for the protection of personal data ('archives' are records in this context):

- 'Access to public archives is a right. In a political system which respects democratic values, this right should apply to all users regardless of their nationality, status or function'.
- 'The legislation should provide for:
  - either the opening of public archives without particular restriction; or

  - a general closure period'.
  - 'Exceptions to this general rule (...) can, if the case arises, be provided to ensure the protection of (...) private individuals against the release of information concerning their private lives'.
- 'Finding aids should cover the totality of the archives (...)'.
- 'The applicable rules should allow for the possibility of seeking special permission from the competent authority for access to documents that are not openly available'.
- 'If the requested archive is not openly accessible (...) special permission may be given for access to extracts or with partial blanking'.
- 'Any refusal of access or of special permission for access shall be communicated in writing, and the person making the request shall have the opportunity to appeal against a negative decision, and in the last resort to a court of law'.
- 'Wherever possible, *mutatis mutandis*, attempts should be made to bring arrangements for access to private archives in line with those for public archives'('Council of Europe : Recommendation No. R (2000) 13 on a European Policy on Access to Archives' n.d.) (Appendix 5-12).

In other words, Council of Europe member states should have or adopt access rules that ideally contain these provisions in order to deal with access versus privacy tensions. Privacy-based restrictions are allowed. It should always be possible to ask for access to restricted documents, and there should be a possibility to ask for extracts or for 'censored' (partly blanked) documents. If access is denied, the applicant has a right to have this in writing and to bring it to court. The background of the rule that finding aids should cover the totality of records is that sometimes documents are not described or indexed, for instance if they contain personal data that are thought to be too embarrassing. Causing them to be unfindable defies access and amounts to a kind of censorship. The last recommendation, on the position of private archives, refers to the circumstance that donors are often tempted to ask for more privacy-related restrictions than reasonably necessary. This might be difficult to redress with legislation, and there is a task for receiving state archivists to negotiate for reasonable restrictions (subsection 3.4).

7.3.2 General Data Protection Regulation (GDPR)

At EU level there is this recent privacy-related regulation, the General Data Protection Regulation (GDPR). It contains rules and principles that address member states, as well as individuals (for instance data subjects) and organizations, among which archival institutions. This subsection sets out to consider to what extent there is something we could call a 'privileged' position of processing personal data for 'archiving in the public interest' and for 'historical research' in this regulation. Broadly speaking, government-linked archives can make use of exemptions for 'archiving in the public interest'. References to relevant exemptions can be found at several places in the GDPR. It appears that these have been dealt with in several ways:

1. in terms of an exemption on some data protection principles (purpose limitation, storage limitation, stricter rules for special categories of personal data; subsection 3.2.1);
2. as exemption in relation to other GDPR rules (duties corresponding with information rights of the data subject, and the right to be forgotten; subsection 3.2.2); and lastly
3. to open the possibility for Union law or member state law to regulate certain specific areas, e.g. with regard to rectification of data – implying national complementary regulation (subsection 3.2.3).

'Archiving in the public interest' will generally be associated with what is done by government-linked institutions, although the text seems to allow for the possibility that non-public sector archives will be covered. French law, for instance, used the term 'non-profit archival activities' in relation to the exemption for scientific, historical, and statistical research. This was under the regime of the Data Protection Directive, the precursor of the GDPR (Iacovino and Todd 2007, 117). In general, though, any privileged position of non-public sector archival organizations, like for instance private archives or corporate archives, will be determined by provisions concerning historical research.

As to other relevant GDPR terminology: 'data subject' is the natural, living person to whom certain data refer. This means that EU member states are free to regulate what we might call 'post mortem' data protection, and that there will be no harmonization of laws regarding this issue. Certain official documents (like a birth certificate) might not be accessible in one member state during the actual lifetime, in other countries for a fixed period of time (e.g. 100 or 110 years after birth).

'Processing personal data' does not concern unstructured personal data; data in digital format are always considered 'structured' data. 'Unstructured' means in this context that scattered personal data in for instance a collection of paper letters cannot be found via an index or other systematic entry. Digitization of archival documents will turn unstructured personal data (not governed by the GDPR) into structured personal data that do fall under the GDPR. Section 5 will comment on this aspect.

'Controller' is the person or body that determines the purposes and the means of the processing of personal data. In the framework of this chapter it will often be an archival institution or a government agency that is responsible (and thus a controller) in GDPR terms.

7.3.2.1 Exemptions regarding data protection principles

A central principle of data protection is 'purpose limitation': the demand that personal data that has been collected for a certain purpose, cannot be used for other, incompatible purposes. The purpose limitation principle is also known as purpose *specification* principle. Archiving for the long term serves, almost by definition, other goals than the original purpose, the new goal being for instance availability for scientific and historical research. The GDPR provides an exception to this requirement for the processing of personal data for archiving in the public interest and for historical research (GDPR article 5.1.b).

A next principle of data protection is 'storage limitation'. Personal data should only be kept in a form that refers to an identifiable individual as long as this is necessary for the original processing. Archives store documents that contain personal data for an indefinite period of time. Anonymization is generally not an option as it would destroy the historical value of the information, which needs context and authenticity, among other things. Acts relating to archiving in the public interest and for historical research are therefore exempted (GDPR article 5.1.e).

Processing sensitive data ('special categories of personal data')

The GDPR prohibits in article 9.1 the processing of sensitive data. This prohibition can be a major setback for archives, as holdings might well contain data that concern, for instance, health, sexual orientation, or religion. Yet this prohibition does not apply if processing is necessary for archiving purposes in the public interest, scientific, or historical research purposes. For this exemption to work it is necessary that it is embedded in member state law and that there are measures to safeguard the fundamental rights and the interests of the data subject (GDPR article 9.2.j).

Other principles of data protection, as mentioned in article 5 of the GDPR, are not formulated with any exemption for archiving in the public interest: lawfulness, fairness, transparency, data minimization, accuracy, integrity, confidentiality, and accountability. It seems reasonable that all principles must be considered while assessing ethical aspects of processing personal data for archiving purposes (subsections 3.5 and 3.6).

All exemptions are conditional, in the sense that there must be appropriate safeguards (GDPR article 89[1]; more on this in subsections 3.4.2 and 3.4.3).

7.3.2.2 *Exemptions from controllers' duties as formulated in the GDPR*

If personal data have been obtained from another source than the data subject himself, the controller normally has the duty to inform the data subject on quite a few aspects of the situation. This relates to the principle of transparency. After establishing that this duty exists with regard to a given data subject, the controller must for example inform him about the legal basis for the processing; about the period for which the data will be stored; and about the source from which the personal data originate. This duty to provide information does, however, not apply if 'the provision of such information proves impossible or would involve a disproportionate effort, in particular for processing for archiving purposes in the public interest, scientific or historical research purposes or statistical purposes subject to the conditions and safeguards referred to in Article 89(1)'.

The GDPR contains a much discussed if not controversial 'right to be forgotten' (GDPR article 17). In certain circumstances a data controller has the duty to erase data if a data subject requests this. Some of these circumstances are, for instance, that the personal data are no longer necessary in relation to the purposes for which they were collected or otherwise processed; or because the data subject withdraws his consent to process the data. Clearly, a right to erasure might result in harming the integrity of archives. Comparable to the case of the information duties discussed earlier, the position of archives has been taken into account: the right to be forgotten does not apply in case of 'archiving purposes in the public interest, scientific or historical research purposes or statistical purposes (...) in so far as the right (...) is likely to render impossible or seriously impair the achievement of the objectives of that processing' (GDPR article 17.3.d).

7.3.2.3 *National complementary rules*

Thus far most of the named rules of the GDPR are harmonized and have direct effect: they create concrete rights and duties equally for citizens and EU member states. There is a number of subjects, however, that the GDPR leaves for the EU member states to fill in, be it within given parameters. What follows is an inventory of relevant issues EU member states can regulate themselves. Whether member states actually make use of the possibilities they have is outside the scope of this chapter. Iacovino and Todd remarked that under the predecessor of the GDPR, the Data Protection Directive, 'many member states could have made broader use of the Directive's derogation powers or could have made them more specific in their application to archives or archival processes' (Iacovino and Todd 2007, 117). One might assume that member states also will not use to the full their possibilities to facilitate archiving in the public interest under the GDPR. The political intent may be lacking, and member states have had little time for the legislative process. These are not the ideal circumstances to bring about substantial changes facilitating the work of archival institutions.

Freedom of information and privacy

Protection of personal data can collide with the right of access to (government) information. This subject matter was addressed in subsection 3.1 (the Council of Europe recommendation). The GDPR of the EU has no specific provision to reconcile these two rights. It seems reasonable to assume that this area of tension is governed by GDPR article 6.1.e, which says that processing personal data will be lawful if this processing is necessary for the performance of a task carried out in the public interest or in the exercise of official authority vested in the controller. As a result national law applies.

7.3.2.4 *Possible derogations in national law*

Article 89(1) of the GDPR makes it possible that member states formulate rules for archiving purposes in the public interest and for scientific or historical research purposes. Such a national rule then prevails over the GDPR rule. But this is conditional: these national deviating rules must 'be subject to appropriate safeguards, in accordance with this Regulation, for the rights and freedoms of the data subject. Those safeguards shall ensure that technical and organizational measures are in place in particular in order to ensure respect for the principle of data minimization'. It concerns the duties a controller normally has on the basis of articles 15, 16, 18, 19, 20, and 21 of the GDPR.

In what follows these possible derogations will be looked at more closely. What can the EU member states do to regulate (and to facilitate) their archives with regard to GDPR data protection rules? All derogations below are available for archiving in the public interest; all except the numbers 4 and 5 can be used by member states to facilitate historical research:

1. On the basis of article 15 GDPR the data subject has *the right to obtain information* from the controller. This 'right of access' includes the right to know whether personal data concerning himself are being processed, to know the source of the data, and to obtain a copy of the data (section 3). Complying with article 15 might prove to be very cumbersome to archives.

2. Article 16 GDPR contains *a right to rectification*: 'The data subject shall have the right to obtain from the controller without undue delay the rectification of inaccurate personal data concerning him or her. Taking into account the purposes of the processing, the data subject shall have the right to have incomplete personal data completed, including by means of providing a supplementary statement'. Given the nature and mission of archives, there will be much 'inaccurate' and 'incomplete' personal data in their holdings. It is their task to collect and store information for future use, often (not always) irrespective of correctness or completeness. In other words, archives will need some sort of exemption or restriction.
3. Article 18 GDPR formulates *a right to restriction of processing*. It enables a data subject to stop a controller from processing personal data in a number of cases, for instance in case of inaccuracy of data, unlawful processing, or the processing of superfluous data.
4. Article 19 builds on the articles 16, 17, and 18 and creates a duty for a controller to inform recipients to whom personal data have been disclosed of any rectification or erasure of data, and of any restriction of processing. Notification is not necessary if it proves impossible or if it involves disproportionate effort. Nevertheless, a national lawmaker in the EU is allowed to accommodate archives in the public interest by deviating from this rule.
5. Article 20 contains *a right to data portability*.
6. Article 21, lastly, expresses *a right to object*. It allows a data subject to object to processing his personal data on grounds of his or her particular situation, even if the processing in itself is lawful. The controller (for the purposes of this text: the archival organization) can continue processing the data if it 'demonstrates compelling legitimate grounds for the processing which override the interest, rights and freedoms of the data subject'. Member states might consider that it is not very attractive to expose their own archival services to discussions of this kind, and introduce an exemption or limitation with regard to article 21.

Summed up, it is fair to say that the GDPR is very important for archives that process personal data. The precise impact in given circumstances can only be assessed by also taking into account national complementary law. On the highest level the GDPR lays down guiding principles which might serve to remind us of the tensions between personal data protection and the preservation of data for the long term.

7.3.3 Code of Practice

An interesting document in the field of archival science is the code of practice for archivists and records managers under Section 51(4) of the Data Protection Act 1998 (UK) (The National Archives, The Society of Archivists, the Records Management Society and the National Association for Information Management, 2007). This code is not a law in itself, but linked to the UK Data Protection Act 1998, which in its turn served as the implementation of the EU Data Protection Directive. The GDPR encourages representative organizations to draw up codes of conduct, as did its predecessor, the DPD. (DPD: recital 61; GDPR: recital 98). Yet this type of codes was not very common under the DPD. Iacovino and Todd could only find this (draft) code for the archives sector (2007, 119).

The Code can be envisaged as being at a level between legislation (subsection 3.2) and professional ethical codes (subsection 3.4). As it does not qualify as a law, decisions that are inspired by it might not hold in court. On the other hand, compared to professional ethical codes it provides more concrete assistance in solving a range of smaller and bigger problems with regard to data protection and long-term retention of personal data. The code has a chapter called 'responsibilities of archivists', from which the following issues have been selected.

- The code warns for too strict an interpretation of the law, as this can lead to the loss of material for posterity. This is the problem of what might be perceived as the danger of the 'empty archives' (Henttonen 2017, 290). An over-cautious interpretation of the UK Data Protection Act 1998 (or indeed: any data protection regulation) may have the effect of anonymizing, weeding, or the destruction of files containing personal data that would otherwise have ended up in the archives repository (Code 4.3.2).
- On the other hand, it is advised to consider in an early stage whether it is, on balance, useful to collect and retain materials that might well be closed for a long period. It might be foreseeable that documents, especially when these contain sensitive information, might only be used for research because of tight rules. Archivists should weigh up whether preservation of these data is in the public interest (Code 4.3.3.).
- The Code touches upon an interesting matter where it states that with regard to the principle of accuracy (GDPR article 5.d) it should be clear that personal data in archives are not kept up to date in the way one might expect. Archives are concerned with historical integrity. Archivists (and data subjects) should rely on the use of supplementary

statements or certificates in order to rectify without damaging archival integrity (Code 4.7.1, see also subsection 3.4.3 on GDPR article 16).
- The processing of information should be 'fair'. This is a general data protection principle that is also found in the GDPR (5.1.a). Within the context of access/disclosure the Code adds to this that the guiding principle is: 'when in doubt, withhold the data'. Whether or not to disclose involves an assessment of possible substantial damage or substantial distress. This assessment should take into account factors like the nature of the information, the age of the information, the credibility and possible consequences of disclosure (Code 4.9.4).
- The 'age of the information' as mentioned in the foregoing sentence is also known as the 'lapse of time factor'. It is relevant according to the Code because it might affect the extent of distress. It also touches upon the level of security measures.

7.3.4 Ethical Codes

There are professional ethical codes that address archivists. Relevant parts of three of these are quoted here, in relation to the following questions: What is an archivist? What does he do? What are his responsibilities with regard to access to the collections, and to the protection of privacy? Whose privacy is referred to?

7.3.4.1 Code of Ethics of the International Council on Archives 1996 (ICA)

The ICA is an international, non-governmental organization that aims at promoting international cooperation in the archival field. Membership is open to organizations, groups, and individuals. Under its auspices an ethical code has been drafted (ICA 1996). In 2008 a questionnaire was distributed in order to assess how the Code had been used. It appeared that the major ethical issue respondents had faced was 'access'. This covered some different issues, one of them being access versus privacy. Asked where their primary loyalty lay a majority answered 'the law'. One of the recommendations following this survey was that the ICA Code did not need major revision, as the ethical principles it provides remained relevant (ICA 2008).

The code describes 'archivist' as anyone concerned with the control, care, custody, preservation, and administration of archives. It is stressed that she must keep in mind the – possibly conflicting – rights and interests of employers, owners, data subjects, and users. Archivists should promote access to archival material. A next rule says that there is a duty to 'respect both access and privacy', especially the privacy of those who had had no say in the processing of the records (article 7). It is remarkable that the ICA Code acknowledges corporate privacy (ICA 1996).

7.3.4.2 *Code of Ethics and Professional Conduct, Association of Canadian Archivists (2017)*

Finalized in 2017, the Code of Ethics of the Association of Canadian Archivists (ACA) is the most recent of the three codes discussed here. The code is not an ethical code in the strict sense, rather than an aspirational text: what the archivist should be striving for rather than what he must do – in terms of the code 'what we do' and 'what we are'. The code formulates nine principles, one of which is 'access'. Access is making records available to the widest possible audience, while respecting the privacy of individuals that created the records or are the subjects of records. There is special attention for those who were not involved in processing these records. The code advises to draw guidance from privacy law as well as data protection law (ACA 2017).

7.3.4.3 *Code of Ethics for Archivists, The Society of American Archivists (2005)*

The Society of American Archivists (SAA) is a very large association with more than 5000 institutional and individual members. The association is very active and has made publications on ethical and legal aspects possible, a number of which has been used for this chapter. The SAA also issued a code, the Code of Ethics for Archivists. The text is rather concise, stating nine principles. It addresses in its preamble 'all those concerned with the selection, control, care, preservation, and administration of historical and documentary records of enduring value'. It is the responsibility of archivists to further the use of records, this being a fundamental goal of keeping archives. Yet the use of records must be in accordance with legal requirements. Restrictions on access with an eye to privacy or confidentiality are allowed. Archivists are held also to protect the privacy rights of those who are the subject of records, individuals, or groups, and the privacy rights of donors (SAA 2005).

Taken together, the following remarks can be made about these three texts. The ICA code does not refer explicitly to privacy rights of users, as do the other two texts. Privacy is primarily a personal right; this text however recognizes a corporate privacy right. The SAA code recognizes privacy rights of groups. All three underline the privacy rights of those who were not involved in the disposition of their personal data. In the light of the fact that there is tension between the principle of access and the duty to respect privacy, and dilemmas to be solved, none of the texts provides much guidance. The question whether this renders them without value for the problem area under discussion will be taken up in section 4.

7.3.5 Conclusion

This section started out by distinguishing six layers of privacy protection with regard to public archives. In principle the same division is valid for non-public archives. Examples of three of these layers have been presented: law and soft law; a code of conduct, and ethical codes. We have seen that the GDPR is an important yet complex piece of legislation that regulates the processing of personal data in general, covering activities of public and private archival institutions as well. The picture is broad and the more complicated because EU member states have been given the opportunity to complement the regulations on public archives and historical research with regard to a number of issues. It appears that discussions in the archival world concerning privacy for an important part revolve around access versus privacy. The Council of Europe recommendation addresses this issue, as do professional ethical codes.

7.4 Traditional debates and dominant schools

This section then explores this central issue for archivists: how to balance access to archival holdings and protecting the privacy rights of individuals whose personal data are held in the records. After dealing in the preceding section with a situation in which there is legislation (notably the GDPR in the EU), the assumption in this section is that the law provides limited guidance. In such circumstances the discussion as to what professional ethics could, and should mean for balancing competing interests gains much more weight. The individuals whose interests need protection fall into three categories: persons whose data are in government records; donors of materials that have been acquired by archival institutions; and third parties or 'blind donors'. 'Access' can also be differentiated: it might be unconditional, meaning available to the general public, or possible for (some) research purposes only. Access restrictions can take the form of a limited period or the necessity to obtain permission.

7.4.1 Ethical codes and balancing access and privacy

Ethical codes, some of which have been mentioned in the preceding section, summon to promote both access and to protect privacy, without specifying how to resolve dilemmas. The history of the ethical code of the Society of American Archivists (SAA) illustrates the problematic character of the issue.

Within the SAA, there has been a debate about finding the balance between protecting the individual privacy of individuals and providing open access. In the early 1970s, the Watergate era, open access appeared to be the more important interest. The SAA code of ethics, in its 1980 and 1992 versions, reflected some confusion on the most desirable archivist's role. On the one hand the code seemed to urge the archivist to increase restrictions in some instances of donations, because donors did not always understand the sensitive nature of the materials they intended to transfer. Elsewhere the code gave the instruction to 'discourage the imposition of restrictions by donors'. Archivists were thus left with ambiguous advice (Behrnd-Klodt and Wosh 2005, 61-62).

This leads to the preliminary question whether ethical codes for archivists are relevant in the first place. And, if they do not contribute substantially to a fruitful discussion on the solving of dilemmas, how else could professional ethics be operationalized?

Elena S. Danielson is very critical of ethical codes as a tool to solve the dilemma under discussion. 'It is disingenuous to write ethical guidelines saying that archivist should protect the privacy rights of data subjects. Violation is part of the process. The real question is how it can be meliorated' (Danielson 2010, 9). Balancing privacy and open access has always been an issue for archivists, she says, adding that it is a growing dilemma in the light of information technology (2010, 184).

Like Danielson, Marybeth Gaudette has no favourable vision on ethical codes as instruments to help solve access-privacy dilemmas. In her article 'Playing fair with the right to privacy' she deals with the balance between the needs of researchers (access) and the privacy rights of individuals. She focuses on private collections (not controlled by the government). More in particular she is concerned with 'blind-donors': those people whose data have entered the collection without their knowledge. Archivists have, she argues, refused 'to take a definite profession-wide stance on the issue of allowing access to the writings, correspondence, and other creations of blind-donors' (Gaudette 2003, 23). The profession as a whole has not come any further than 'perfunctory nods' in the form of various ethical codes. Gaudette brings forward four arguments. Firstly, ethical codes have little meaning in the legal sense. Secondly, they are indicative of unethical behaviour rather than a means to correct this. Thirdly, instead of showing the highest standards, codes represent 'the lowest common denominator of professional agreement'. And fourthly, ethical codes are 'frustratingly vague' (Gaudette 2003, 23). The thought imposes itself that ethical codes are not of much help in balancing access and privacy.

7.4.2 Access and privacy in public sector archives

In her book *Without Consent. The Ethics of Disclosing Personal Information in Public Archives* (1992) Heather MacNeil explores ethical dilemmas that the archivist might face. In the public sector, she says, statutes, regulations, or agency policies will in general have determined access restrictions. This means that the basic decisions as to the sensitivity of the records will already have been made. However, important questions may be left open. She focuses on the field of tension between privacy interests of the individual on the one hand, and research interests on the other. Her discussion concerns the administration of access to personal information in the US and Canada. She sees a growing demand in socio-historical research for records containing personal information as well as administrative and legal difficulties archivists encounter while facilitating access to these records. Furthermore, she researches the ethics of providing access to personal information for research purposes, and the principles that archivists should apply while making access decisions. Lastly, she discusses policies and practices that would contribute to balancing the competing interests of individual privacy of the one, and the right to know of the other (MacNeil 1992, 6-7).

MacNeil finally formulates a number of obligations for the archivist:

> Among them are the obligations to strengthen and systematize policies and procedures for the administration of access to records containing personal information within their institutions; to learn about and debate privacy issues as a profession and to participate, through professional organizations, in improving the archival status quo in matters relating to privacy; and, lastly, to contribute, again as a profession, to the larger public debates concerning the protection of information privacy in an increasingly public society. (MacNeil 1992, 182)

The archivist deals not only with questions about general access but also about access to accommodate research. What is needed, says MacNeil, is a framework for reviewing requests for access to facilitate research and statistical purposes. Especially in cases where obtaining consent of data subjects is often impossible or impracticable, there is an important role for archivists. They must act on behalf of record subjects, who are unaware that information concerning them can be used for research. This calls for a review process and an ethical review board. (MacNeil 1992, 183-186).

This review process, to be put in place to determine whether requests to access restricted records could be granted, would involve an ethical review board. This board, to be established under the head of the archival institution, is envisaged as a standing body dealing with research proposals. It would screen them, specify access conditions, monitor the research process, and even impose penalties in case access conditions would be violated. A request to access restricted records for research purposes would entail a number of elements, for instance an explanation why the research could not be done without personal data, and what safeguards were foreseen to prevent unauthorized further processing of the data (MacNeil 1992, 186-187). On the whole, MacNeil advocates coherent policies and clear procedures to deal with access to restricted public records (1992, 192).

7.4.3 Access and privacy with regard to non-government-controlled collections

This subsection deals with situations in which the law and agency procedures play a subordinate role. For this type of collections Marybeth Gaudette distinguishes three 'camps' with regard to the question at what point the rights of blind-donors should be superseded by research interests of historians. Proponents of the first group, such as Heather MacNeil, argue that archivists should preferably work with mandatory closed periods. At the other end of the spectrum one finds the camp that is in favour of unhindered access to everything that has not been explicitly closed by the terms of the donor agreement. Thirdly, there is a group that wants to determine access requests on the basis of the sensitivity of the records. Each case will then be decided upon merits (Gaudette 2003, 24).

A proponent of the second approach is Mark Greene, who argues that the archivist could not, and should not, accept the burden of deciding the type of privacy issues under discussion. Why should he apply more restrictions than the donor has wished for? The archivist and the repository should not accept the responsibility: 'Surely we should not martyr ourselves on the altar of privacy rights'(Greene 1993, 35). Greene favours an approach according to which the donors have control over access. They know the collection and therefore they are in the best position to decide dilemmas concerning access and privacy. By obliging archivists to actively protect privacy interests 'the manuals' (including ethical codes) ask the impossible and go against general archival practice. Referring to the North American situation, Greene is not impressed by the argument that his approach would increase the risk of litigation: there is little evidence of institutions that have actually faced legal action. Archivist should be aware of privacy interests but not become paralyzed: 'If we err, let us err on the side of access' (1993, 38).

The third approach, a case-by-case method, would mean that decisions could turn out to be different if taken by different archivists, and at different moments. 'Sensitivity' is a difficult notion, and an archivist would have a difficult time to defend his interpretation in the face of either blind-donors or researchers. Gaudette concludes that case-by-case decisions should not be tolerated. What is needed is that there are clear laws governing access to records with personal data. At the least, professional associations should adopt 'clearly delineated procedures that truly respect the right to privacy'(2003, 30-31).

7.4.4 Conclusion

This section has been concerned with the question how to balance access and privacy in the absence of clear legal guidance or institutional rules. Professional ethical guides have been found too vague to be of much assistance. Following Elena Danielson we see the following possible approaches to reconciling access and privacy:

1. *Shifting the responsibility to the donor or researcher*
   There are practical advantages connected to this approach. The archivist is not forced to police what researchers see or what donors offer. Yet archivists will always be responsible to some degree. This is largely the approach advocated by Mark Greene, discussed above, for private collections.
2. *Prior screening of collections by the archives staff*
   This approach might work well. The archivist could advise the donors on restrictions, and even refuse items. Danielson remarks that pre-screening would not work well in case of big volumes and a limited processing budget.
3. *Institutional review boards*
   This approach has also been advocated by MacNeil. It is labour-intensive and also time-consuming. Danielson considers it to be a very useful one, as it might help researchers more quickly and be fairer to the stakeholders. She observes that this approach is becoming increasingly common in Europe.
4. *Responding to complaints case by case*
   This is what Danielson calls the 'default approach'. A request to obscure a social security number could easily be agreed to; if complying would mean a form of censorship the requested change could not be made.

Which approach would suit an archive best? 'In the end', as Danielson says, 'each repository will need to construct its own combination of privacy safeguards in keeping with its mission, the nature of the collections, and its budget' (2010, 207-216).

7.5 New challenges and topical discussions

This section signals two trends, namely that digitization turns unstructured data into structured data, discussed in section 5.1, and challenges some of the traditional archival principles, touched upon in section 5.2. It also discusses a current dilemma, namely whether there should be an archival privilege (section 5.3).

7.5.1 Digitization turns unstructured data into structured data

'Historically the greatest privacy protection has been obscurity: the information is available, but hidden in heaps of dusty files, difficult and time consuming to access', says Danielson (2010, 202). Digitization changes all this. For archives in countries within the EU digitization of records can imply having to deal with a stricter legal regime. 'Scattered' personal data in traditional archival materials is considered not to be contained in a 'filing system' (GDPR recital 15). That means that they are not covered by the GDPR, and possibly neither by the national law of EU member states. Yet, as soon as records are digitized, data protection rules apply downright. If they are made available in this new form it is possible to search the records for names, addresses, and so forth with for instance OCR techniques (optical character recognition). Archives are under constant pressure to digitize their holdings, be it by their funders, the general public or researchers. Yet answering this demand makes privacy considerations more urgent and more complex, and not only in the European Union.

This changed legal status of archival activities and these new search possibilities might force archives to reconsider whether materials can be made accessible in digital format. More in particular, problems can arise with regard to sensitive data, which in the EU are governed by stricter rules. It will be prohibited there to process these personal data, unless there is an explicit legal exemption. Facts that normally are quite innocuous, for instance that a named person has once broken a leg, or is a member of a trade union, can qualify a record as sensitive.

With regard to digitizing private collections it might be questionable whether the original agreement with the donor actually covers this transformation and new uses. This depends on the wording of the agreement, but there is also an ethical dimension. An archivist might come to the conclusion that the exercise of balancing the interests of access and privacy has to be done once again.

'Availability' in the archive's reading room is of a different order than 'availability' to all via the Internet. Ferguson and others come to this conclusion. They investigated how library and information (LIS) professionals experience ethical dilemmas in the light of new technologies. ICT has qualitatively changed ethical dilemmas. 'It seems likely that, when considered in the abstract, new technologies do not appear to change ethical principles; however, when actually experienced in the workplace they do substantially change the factors the LIS professional has to weigh up' (Ferguson, Thornley, and Gibb 2016, 549).

7.5.2 Digitization, new technologies and archival principles

No doubt technology changes the balance of access and privacy. Privacy legislation, archival legislation, and archival ethics will have to deal with this developing playing field with its many interests. Archivists, to begin with, will try to uphold the archival principles which they see threatened because they seem to lose control of what can happen with data once these are made digitally available. Much cherished principles in the archival world like (historical) accuracy, documented provenance, original order, authenticity, and context that normally guide all archival processes, will lose much of their meaning once data enter the free flow of information. Anonymization, a much propagated way to enhance privacy, is, with regard to upholding the aforementioned principles, certainly not an attractive solution. There is also a practical and financial side to this: very often the sheer bulk of the material and its diversity form an unsurmountable challenge.

The expectations of users, under whom academic researchers, cannot be neglected. They expect more and more that materials are made available in a machine-readable format, to enable easy retrieval of relevant data and reuse. Archivist are willing to fulfil these expectations, but at the same time they must try to assess what these new methods could mean for privacy. The preservation of websites or other dynamic records, for instance, brings uncertainty of the extent persons may be identified. In an EU context this is decisive for the question whether the GDPR applies or not.

7.5.3 An 'archival privilege'? The Boston College subpoenas

The case of the Boston College subpoenas, also known under the names of The Boston Tapes and The Belfast Project has attracted much attention. The Belfast Project was an oral history project. People had been interviewed, under the explicit promise of confidentiality during their life. The case is generally associated with confidentiality being under threat, rather than privacy (although some consider confidentiality as a species of privacy). Yet there are good reasons to discuss The Boston Subpoenas also in the current context of privacy, data protection, and archives. There is little doubt that the testimonies in the case at hand qualify as personal data, and that the interviewees had reason to expect their privacy to be respected. The single fact of being interviewed already implied involvement of the interviewee in illegal activities and violence.

The Belfast Project consisted of interviews with people who had played a role as paramilitary fighter during the 'Troubles', a violent episode in Northern Ireland. They were promised that the tapes that were made during these interviews would remain sealed until after they died. The materials were stored at Boston College (US), which had sponsored the project. The police in Northern Ireland investigated murders during the Troubles, and had asked the American authorities to assist. As a result, Boston College was faced in 2011 with subpoenas, court orders to hand over materials. At first Boston College refused to hand over materials containing the testimony of persons that were still alive, but after a series of court proceedings Boston College decided not to continue the legal struggle and to turn over materials to the court (George 2013, 48-53).

Expounding this case, George argues that, in a United States legal context, archives should be able to protect their sealed holdings and thus to challenge subpoenas successfully. For that it is necessary to obtain an 'archival privilege', comparable to the privilege that doctors, attorneys and, to a certain extent, journalists already have. Their clients and sources are protected because of a relationship based on trust. A similar relationship exists between archives and donors. George calls upon archivists to do their best to obtain such an archival privilege (2013, 61). Future projects like the Belfast Project need to proceed cautiously in the absence of such a privilege, and researchers as well as repositories should inform participants that confidentiality cannot be guaranteed (King 2014, 37). There can be little doubt that the fact that confidentiality cannot now be guaranteed might have a chilling effect on future projects.

7.6 Conclusion

This chapter has been concerned with long-term archiving. Archives serve several goals, like providing information on government-linked activities, securing evidence that can be used to protect citizens' rights, and safeguarding collective memory and cultural heritage. The archival processes involve acquisition of information, and arranging, discarding, and providing access.

Archival holdings will often contain personal information, and decisions on providing access researchers and the general public imply that privacy aspects must be taken into consideration.

Privacy protection must be taken seriously in the archival field to prevent that individuals might experience harm because of the fact that certain materials will be stored for the long term, more in particular in the case that these materials will be accessed during their lifetime. The archival institution and the archivist have a responsibility in this respect, which will be the most clear at certain stages in the archival process. In a government setting privacy considerations must play a role at the time of formulating criteria for the selection process, and later in applying these criteria. Then it will be determined which records will be transferred to the repository and which will be discarded. At this same moment access restrictions will be decided upon. If archival collections are acquired, the materials have to be examined for privacy aspects, and access restrictions must be negotiated with the donor. Once materials are part of the collection archivists will have to deal with requests for access.

With regard to the role of privacy in society and the regulation of privacy there are differences between the US and Europe. Europe knows regulation for the protection of personal data that covers the private as well as the public sector. This legislation is a corollary of privacy seen as a basic human right. In the US privacy legislation is less comprehensive, and it is concerned for the larger part with public records. Access to public records is more valued than in Europe.

The tensions between access to government-held information and privacy are illustrated by a recommendation, issued by the Council of Europe: This document acknowledges privacy as a legitimate limitation of access to archives ('Council of Europe: Recommendation No. R [2000] 13 on a European Policy on Access to Archives' n.d.).

The European General Data Protection Regulation (GDPR) is a piece of legislation that is important for the protection of privacy and personal data in EU member states. The real consequences for European archives are difficult to measure, because member states have the freedom to enact complementary legislation with regard to exemptions and derogations to facilitate archives and historical research. A close scrutiny of the GDPR nevertheless has been essential for the subject of this chapter, as it has brought to light that there are numerous frictions between the principles of data protection and the objectives of archives and historical research. This becomes most clear at the level of data protection principles, but also with respect to data subjects' rights.

As a discussion of a sector code in the United Kingdom has shown, the European legislative model of general EU data protection legislation, complemented by national law, has left many questions in the archival sector unanswered. A sector or branch code could fill this gap ('Code of Practice for Archivists and Records Managers under Section 51[4] of the Data Protection Act 1998' 2007).

It might be the case that the law does not offer archivists with sufficient guidance to make reasoned decisions. In that case recourse can be taken to professional ethical codes. The three codes discussed in this chapter appeared to be rather vague. More in particular, the codes cannot effectively assist in solving access versus privacy dilemmas.

Outside Europe, and in a situation in which the law offers little guidance, there are a number of options to deal with privacy versus access dilemmas in an archival context. In general, arbitrary decisions by archivists are seen as unwelcome. Some authors find that archivists should not accept the responsibility to protect privacy, and instead leave this to either the donors or the users of archival materials. Some advocate institutional review boards, in order to reach fair decisions. Another solution that has been suggested, as a default mode, is just to deal with complaints. Assessing all materials for sensitive content could also be a line of action, but this is a time and money-consuming activity. An additional problem lies in the fact that the 'sensitiveness' of information is difficult to establish in an objective way.

The digitization of holdings, and the making available of records in digital format, augments the importance of privacy in the archival field. It means that balancing access and protecting privacy has to be regarded in quite a different light. Earlier decisions, taken in the analogue era, might need reconsideration. In the EU, personal data that used to be inaccessible through the application of specific criteria (e.g. alphabetically, or by an index) is not covered by personal data legislation. Digitization of holdings changes this: after digitization any personal data in a record will bring the file under data protection law. Information specialists, and archivists among them, experience that new technologies have qualitatively changed ethical dilemmas.

Archives acquire records, arrange and describe them, and make them accessible. There are archival principles that guide these processes, for instance guaranteeing authenticity of information, and presenting it in the right context. Recent technological and societal developments make it difficult to honour these archival principles, and protecting personal data is one of the complicating factors. Anonymization, which can contribute to data protection, might often prove incompatible with for instance maintaining the contextual integrity of records.

Information might be transferred to a repository under the promise of confidentiality. Archivists feel bound to honour all contracts with donors. However, the case of the 'Boston College subpoenas' has made clear that the promise of absolute confidentiality cannot always be upheld. Being faced with a court order a repository will see itself obliged to hand over the required documents. The introduction of an 'archival privilege', comparable to the privilege that doctors and attorneys have, could strengthen the position of archival institutions.

The case of the Boston subpoenas is yet another example of competing interests in the archival field. It is the archivist's duty to preserve documents that will serve as a witness of our history. A promise of confidentiality can be instrumental to acquiring documents, that otherwise might not be preserved. Promises of confidentiality, and more in general the demand to consider the protection of privacy and personal data, have to be balanced against the demands following from 'the right to know'. This chapter has set out to identify a number of relevant issues, and to discuss resources with which the archivist can manage to reconcile the principal interests.

Further reading

Bingo, Steven. (2011). 'Of Provenance and Privacy: Using Contextual Integrity to Define Third-Party Privacy'. *The American Archivist* 74(2), 506-21. https://doi.org/10.17723/aarc.74.2.55132839256116n4.

Cox, Richard J. (2008). 'Archival Ethics: The Truth of the Matter'. *Journal of the American Society for Information Science and Technology* 59(7), 1128-1133. https://doi.org/10.1002/asi.20852.

Hodson, Sara S. (2010). 'Ethical and Legal Aspects of Archival Services' in *Encyclopedia of Library and Information Sciences*, 3rd edition (online). City: Taylor & Francis.

Iacovino, Livia. (2006). *Recordkeeping, Ethics and Law: Regulatory Models, Participant Relationships and Rights and Responsibilities in the Online World*, vol. 4. The Archivist's Library. Dordrecht: Springer Netherlands.

Shepherd, Elizabeth. (2010). 'Archival Science'. *Encyclopedia of Library and Information Sciences*, 3rd edition (online). City: Taylor & Francis.

Szekely, Ivan. (2014). 'The Right to Be Forgotten and the New Archival Paradigm' in *The Ethics of Memory in a Digital Age*, 28-49. Palgrave Macmillan Memory Studies. London: Palgrave Macmillan. https://doi.org/10.1057/9781137428455_3.

References

Behrnd-Klodt, Menzi L., and Peter J. Wosh (eds.). (2005). *Privacy & Confidentiality Perspectives: Archivists and Archival Records*. Chicago: Society of American Archivists.

'Code of Practice for Archivists and Records Managers under Section 51(4) of the Data Protection Act 1998'. (2007). The National Archives, the Society of Archivists, the Records Management Society and the National Association for Information Management. www.nationalarchives.gov.uk/documents/information-management/dp-code-of-practice.pdf.

Council of Europe. (n.d.). 'Council of Europe : Recommendation No. R (2000) 13 on a European Policy on Access to Archives'.Accessed 31 January 2018. https://rm.coe.int/16804cea4f.

Danielson, Elena S. (2010). *The Ethical Archivist*. Chicago: Society of American Archivists.

Ferguson, Stuart, Clare Thornley, and Forbes Gibb. (2016). 'Beyond Codes of Ethics: How Library and Information Professionals Navigate Ethical Dilemmas in a Complex and Dynamic Information Environment'. *International Journal of Information Management* 36(4), 543-556. https://doi.org/10.1016/j.ijinfomgt.2016.02.012.

Gaudette, Marybeth. (2003). 'Playing Fair with the Right to Privacy'. *Archival Issues* 28(1), 21-34.

George, Christine Anne. (2013). 'Archives Beyond the Pale: Negotiating Legal and Ethical Entanglements after the Belfast Project'. *The American Archivist* 76(1), 47-67.

Greene, Mark A. (1993). 'Moderation in Everything, Access in Nothing?: Opinions about Access Restrictions on Private Papers'. *Archival Issues* 18(1), 3141.

Henttonen, Pekka. (2017). 'Privacy as an Archival Problem and a Solution'. *Archival Science* 17(3), 285-303. https://doi.org/10.1007/s10502-017-9277-0.

Iacovino, Livia, and Malcolm Todd. (2007). 'The Long-Term Preservation of Identifiable Personal Data: A Comparative Archival Perspective on Privacy Regulatory Models in the European Union, Australia, Canada and the United States'. *Archival Science* 7(1), 107-127. https://doi.org/10.1007/s10502-007-9055-5.

Ketelaar, Eric. (1995). 'The Right to Know, the Right to Forget?: Personal Information in Public Archives'. *Archives and Manuscripts* 23(1), 8.

King, James Allison. (2014). '"Say Nothing": Silenced Records and the Boston College Subpoenas'. *Archives and Records* volume(issue), 1-15. https://doi.org/10.1080/23257962.2013.859573.

MacNeil, Heather. (1992). *Without Consent: The Ethics of Disclosing Personal Information in Public Archives*. Metuchen: Society of American Archivists, Scarecrow Press.

Neuberger, Lord. (2015). '"What's in a Name?" – Privacy and Anonymous Speech on the Internet'. https://www.supremecourt.uk/docs/speech-140930.pdf.

Medical Privacy: Where Deontology and Consequentialism Meet

Robin Pierce

Privacy in the medical context is complex, not least because it requires operationalization of a concept that has variously been referred to as vague and indefinable (Thomson 1975), but also and primarily because it depends on the creation and maintenance of a sphere in which privacy and other norms function to achieve particular objectives relating to the effective delivery of healthcare. That is, medical approaches to privacy adopt a somewhat instrumental view to the role of privacy in the provision of healthcare. Even where a clear deontological imperative to respect persons is evident, for example, in the use of drapes and sheets to cover the body during sensitive examinations, the instrumental goal of treating patients in such a way that they will not feel humiliated and forgo necessary medical exams, operates in the background to motivate preservation of a private sphere. This is equally true of sensitive patient data such as a potentially stigmatizing diagnosis. The objective of the medical domain is to provide health-enhancing interventions at both the individual and population level (clinical care and public health, respectively). Privacy is essential to achieving this objective. As the examples of bodily and data privacy illustrate, the nature of privacy in the medical domain is multifaceted and consists not of a single type of privacy nor does it rest on a single legal basis. Rather, privacy in the medical setting is much like a protective encasement where the adage 'what happens here, stays here', has long been the prevailing norm.

I find it useful to think of medical privacy as a construct formed by intersecting strands of normative frameworks that create a conceptual sphere of privacy that is designed to protect a spectrum of privacy interests in the medical context. These privacy norms are derived from multiple sources. For example, the Article 8 of the EU Charter of Fundamental Rights recognizes a right to data protection, and Article 8 of ECHR, recognizes a right to respect for private life, which supports multiple dimensions of medical privacy. Article 9 of the General Data Protection Regulation (GDPR) provides for the protection of personal data and specifically classifies health data as sensitive data that merits heightened protection. Yet, the basis for privacy in the medical context extends beyond mere protection of health data. Privacy in the medical context seeks to protect and support a much broader array of interests that ultimately serve to enhance the provision of medical care.

The construct of medical privacy incorporates at least three of the common conceptions of privacy – informational, decisional, and local or spatial (Roessler 2005). Informational privacy is perhaps the most straightforward. Although codified in contemporary documents, protection of health information and the origins of informational privacy in the medical context can be found as far back as 300 BC in the Hippocratic Oath, which requires that physicians swear that whatever in (or not in) connection with their professional practice, that they 'see or hear, in the life of men, which ought not to be spoken of abroad, will not be divulged, as reckoning that all such should be kept secret' (British Medical Association 1993). This pronouncement about the importance of 'keeping patient secrets secret' has, over time, been incorporated into professional ethics, guidelines, and codes of conduct, as well as giving rise to the possibility of legal action in the case of violation.

A clear instance of what has come to be regarded as patient 'secrets' is health data. Health data occupies special status and must be handled in such a way that it does not expose the care seeker to further vulnerability on the basis of health status or characteristics. The protection of health data serves multiple purposes. Inseparable from the person (except through true anonymization), information about health can render the individual vulnerable in multiple ways. Health data can form the basis for stigma, social exclusion, embarrassment, and various forms of discrimination, for example. As a result, assurances of the privacy of health information are critical to creating the trust necessary to fostering an effective health system that encourages appropriate care-seeking. When I wrote *The Comparative Architecture of Genetic Privacy* in 2009, I explored how the relatively newly accessible genetic information exposed individuals and groups to new types of vulnerability and therefore needed specific protection. Consider the case of a soaring academic who learns that she has a set of genes highly correlated with Alzheimer's disease. Were this information to escape the clinical setting, an employer might be less willing to promote her even though the genetic risk may never result in disease onset. Without a strong privacy norm, a rising executive may refrain from undergoing screening for a treatable cancer for fear that it may affect his career. Yet at the same time, honouring privacy in some instances could be counter to the well-being of others (biological relatives for certain types of genetic risk or sexual partners in the case of STDs). While there are strict edicts about the need to protect privacy in the medical sphere, well-recognized exceptions exist and continue to emerge when non-disclosure is likely to result in harm to others. In addition to public health exceptions, the landmark case of *Tarasoff*, in which a psychiatrist was held to have a duty to warn on the basis of private clinical communications by a patient when it clashed with a need to disclose a probable imminent threat to a specific individual, illustrate that medical privacy, however essential, is not absolute.

Forming a long-standing edict of medical ethics, doctor-patient confidentiality stands as a core component of medical care. While confidentiality shares some overlap with privacy of health data, there are differences. In medical confidentiality, privacy finds yet another conceptualization in the explicit relationality of confidentiality. In other words, privacy, in principle, extends to the whole world and can be enforced by law or social norms. Medical confidentiality pertains to a protected communication between persons in a role-defined context. The doctor (healthcare team) is obligated to honour doctor-patient confidentiality in the non-disclosure of communications made within the context of this relationship. Another difference is in the type of communication that is protected. While privacy of health data pertains to a specific type of information, medical confidentiality pertains to the full spectrum of communication (verbal, physical, bodily, emotional), recorded or not, that occurs in the context of this relationship, subject to legally recognized exceptions. Fortified by professional codes of conduct, medical ethics, and implicitly in various legal instruments, medical confidentiality forms a critical component of effective provision of healthcare, not only as an operationalization of fundamental rights, but also for its instrumental value in assuring the kind of trust necessary for optimal functioning of the medical sphere.

Multiple types of interests are contained in the notion of decisional privacy in the medical context. First and foremost, consent by patients who possess capacity (the ability to understand the nature of an intervention and to make a reasoned decision regarding its use) is a requirement for the administration of treatment and care of patients. This stems in part from the foundational bioethical concept of respect for persons (Beauchamp and Childress 2001). Moreover, consent in the medical context also has roots in the legal concepts of tort (negligence/personal injury) and battery (bodily violation). As a result, patients enjoy a sphere of autonomy that guarantees that they will not be subject to treatment against their will. Such decisions exist within the 'private sphere' (see Pierce 2018). This seemingly simple concept frequently escalates in complexity when confronted with situations in which the decision would appear to lie with the patient, but the interests of others are implicated by the individual's decision. This occurs, for example, in the case of infectious disease and, in some jurisdictions, decisions regarding abortion. The operation of decisional privacy can also be seen in the use of consent-substitutes such as advance directives for care or research, a mechanism that is intended to perpetuate autonomy even after the person loses the capacity to consent (Pierce 2010).

Medical privacy must also attend to the sensitivity of the physical and psychological exposure that care seekers often experience in the medical setting. Even for healthy persons, undergoing bodily exams that form a part of routine clinical care, individuals necessarily expose themselves and become vulnerable in ways that rely heavily on a culture of trust. There is also a dignitary dimension in spatial or local privacy, which can draw on a range of sources from human rights to the earlier named foundational bioethics principle of respect for persons. Spatial privacy, in this way, can stand alone, apart from its instrumental value in facilitating the delivery of care in a way that supports and promotes appropriate care-seeking.

The scope and force of medical privacy is further supported by a secondary layer of protections based in sectoral and other forms of regulation. Anti-discrimination laws, disability rights, and employment law, for example, all recognize and operate to preserve medical privacy by restrictions of use of medical information outside of the medical setting.

These intersecting normative strands of medical privacy, derived from different sources, together form a set of norms designed to protect a bundle of interests that is essential to the maintenance of an effective healthcare system that encourages and protects appropriate care-seeking and treatment. Whether and how technological changes in the collection, storage, and processing of data affect the construct of medical privacy is a pressing question. Just as a bell cannot be unrung, erosion of the sphere of medical privacy is unlikely to be restored. The eager embrace of technological innovation such as big data, machine learning, AI, eHealth, data sharing, essentially forming a virtual explosion of connectedness is likely to present challenges to the construct of medical privacy. This essay seeks to emphasize the teleological basis for medical privacy and suggests that at least one aspect of evaluating and potentially remedying instances of erosion is assessing the impact on the ability of the current construct of medical privacy to achieve its objectives. If, indeed, the instrumental motivations of medical privacy become frustrated by the pervasiveness of connectedness, data sharing, and the co-mingling of social (non-sensitive) data with sensitive health data, solutions will need to be found, not only within the network of privacy protections, but also perhaps in the concept of medical privacy itself. In the presence of weighty justifications emphasizing improved delivery of healthcare, more effective personalized treatments, and advancing scientific research that may ultimately improve treatment options, privacy may seem to be a lightweight consideration. Interestingly, when instrumental considerations of privacy are placed on the scale, the balance tips sharply in favour of preserving whatever form of the private sphere is necessary to assure patients that seeking medical care will not make their lives worse and that going untreated is unlikely to be the better option.

References

Beauchamp, T. and J. Childress. (2001). *Principles of Biomedical Ethics*. New York: Oxford University Press, p. 309.

Pierce, R. (2018).'Honoring the Multiple Dimensions of Autonomy in all Phases of Treatment and Care'. *American Journal of Bioethics* (2018) Forthcoming.

Pierce, R. (2010)'A Changing Landscape for Advance Directives in Dementia Research'. *Social Science and Medicine* 70(4), 623-30.

Pierce, R. (2009)'Comparative Architecture of Genetic Privacy'. *Indiana International and Comparative Law Review*, 9(1), 89-127.

Roessler, B. (2005). *The Value of Privacy*. Cambridge: Polity Press.

Thomson, J. J. (1975). 'The Right to Privacy'. *Philosophy and Public Affairs*, 4(4), 295-31.

8. Privacy from a Medical Perspective

Wouter Koelewijn

8.1 Introduction

This chapter discusses the right to privacy and data protection in the context of healthcare, with a strong emphasis on the latter. In this sector, three spheres of privacy are necessarily interfered with. Firstly, private life and physical integrity are necessarily affected by medical actions. Secondly, local/spatial privacy concerns may rise during medical research, treatment, and stay in care institutions. Thirdly, informational privacy is at play. This concerns the processing of medical personal data, the control that patients can exercise over their medical data, and the obligation of doctors to keep medical personal data confidential. This section mainly focuses on informational privacy and data protection. It explains the concepts in the context of regulations.

First of all, section 2 describes the role and function of informational privacy protection within the healthcare sector. Subsequently, section 3 discusses the basic texts and authors. In section 4 the most important schools of thought and discussions about privacy in relation to the healthcare sector are briefly discussed. Subsequently, section 5 focuses on current and future changes and dilemmas and ends with a number of conclusions and recommended literature.

8.2 The role and function of informational privacy protection within the healthcare sector

The domain of healthcare is broad and hardly definable. It covers various sub-areas that vary from the treatment and care of patients (in hospitals and nursing homes) to the life science industry that is focused on the development of medical devices, medicines, and new technologies. Within these sub-areas, the focus of this chapter is on privacy protection within the field of nursing and care, also referred to as the care and cure sector.

8.2.1 Confidentiality vis-à-vis privacy in healthcare

Privacy protection within the healthcare sector is first of all/primarily associated with medical confidentiality. This is no surprise, as the concept of medical confidentiality addresses the sensitive nature of medical personal data, that obviously warrants protection. Nevertheless, medical-professional confidentiality or secrecy is only partially recognized in international treaties and is predominantly subject to national regulation. The most important international source of medical-professional confidentiality is the International Code of Medical Ethics of the World Medical Association. In this code, professional secrecy is defined as one of the doctor's obligations to the patient:

> A physician shall respect a patient's right to confidentiality. It is ethical to disclose confidential information when the patient consents to it or when there is a real and imminent threat or harm to the patient or to others and this threat can only be removed by a breach of confidentiality.[1]

Furthermore, the right of patients to confidentiality is recognized as such in the International Convention for the Protection of Economic, Social, and Cultural Rights, in which Article 12 sets out the right to health and the right to access to facilities of essential, necessary care. In General Comment No. 14 requires that healthcare be available and accessible and that healthcare is of good quality. In addition, the healthcare system must be in accordance with the medical ethical principles, whereby it is noted that healthcare facilities should be 'designed to respect confidentiality'.

In this way, medical-professional secrecy, serves the interest of the individual that affects and determines the relationship between the physician and the patient, and primarily guarantees the right to confidentiality. In addition, however, it is acknowledged that professional secrecy also serves a general, public interest, namely ensuring unhindered access to healthcare, so that the patient can invariably, and without hesitation or hindrance, seek assistance where the disclosure of confidential communication about his or her health is unavoidable.[2] When there are hesitations about the extent to which the patient's data are safe with a care provider or doctor, this may lead to the person concerned not seeking help or being (too) late in seeking help, with possible harmful consequences for the individual and public health in general.

The duty of secrecy contained in medical-professional, doctor-patient confidentiality first of all concerns the information entrusted to the doctor by the patient. This pertains to consciously communicated information. However,

1 WMA 2006.
2 Leenen 1987.

confidentiality stretches out to everything that the doctor comes to know about the patient. Think of data that emerge when taking anamneses, X-rays, laboratory result(s), data relating to diagnosis, therapy, consultation, prescription, notes in the medical administration. It can thus be assumed that all knowledge related to the patient that comes to the notice of a doctor, is directly covered by doctor-patient confidentiality as laid down in clauses, principles, and guidelines.

Doctor-patient confidentiality can only be breached with the consent of the patient and in very exceptional circumstances. Consideration should be given to weighty interests and emergency situations, in which the consequence of breaking the professional-secrecy code must be weighed against the individual interest of averting, for example, an emergency situation.

A physician is allowed under circumstances to breach confidentiality (for example where there is a strong indication of child abuse), but expressly does not have the obligation to do so. This independent responsibility of the physician is an outlier in the sphere of data protection rules, which are geared towards the exercise of control by the data subject. Explicit consent constitutes a clear basis for processing medical data both in laws on medical confidentiality and data protection. However, permission to breach professional secrecy does not trump the independent assessment by the physician – and the outcome of that assessment may be that the physician, in spite of the consent granted by the patient, decides does not proceed to breach his or her doctor-patient confidentiality.

8.2.2 Informational privacy principles

Within the healthcare sector, three informational privacy principles are of fundamental importance for the protection of the right to privacy. The three informational privacy principles formed the basis for the Data Protection Directive that was established in 1995 (Directive 95/46/EC). With this directive, the European Parliament envisaged on the one hand a far-reaching harmonization of the national rules and regulations of the member states for the processing and exchange of personal data. The European Parliament considered this important for the internal market and the safeguarding of the free movement of goods, people, services, and capital within the European Union.[3]

The first principle concerns the control that a patient can exercise over her medical personal data. For this purpose it is crucial, among other things, that the information and transparency obligations are met by healthcare

3 Hervey and McHale 2004.

institutions. A patient needs to know what data is collected and with whom the data are shared in the context of treatment.

The second principle is the subsidiarity principle. No more information may be collected, stored, and processed than is necessary for the treatment or care of the person concerned. The principle thus (and importantly) also governs medical examinations. The principle is of importance in situations in which weighty individual interests or circumstances give rise to breaching doctor-patient confidentiality. The principle of subsidiarity limits the damage in such circumstances.

The third principle is 'purpose limitation'. The collection, storage, and processing of medical personal data needs to be done with a clearly defined purpose and the personal data may not be used and processed for other purposes incompatible with the original purpose.

The principles also apply in the General Data Protection Regulation (hereafter: 'GDPR') which entered into force in 2016, to be applicable two years later, so as of May 2018. With the GDPR, the European legislator further harmonizes protection of personal data within the EU. The main difference with the privacy directive is that the GDPR has direct effect in the legal order of the European member states and does not need to be converted into national laws and regulations.

8.2.3 The prohibition on the processing of health data in EU law

The starting point for the standardization of the processing of medical personal data in the healthcare sector is laid down in Article 9 GDPR, which contains a general prohibition on the processing of genetic data, biometric data for the purpose of unique identification, and processing of health data in general. Within this chapter, I limit myself to regulations regarding personal data about health.

The first question that arises in this context is what exactly should be understood by personal health data. Recital 35 in the preamble to the GDPR broadly describes personal data concerning health. The definition includes all data relating to the health status i.e. condition of a data subject that provide information about the physical or mental state of health of the person concerned in the past, present, and future. The addition that this also includes data about the future is relevant with a view to the application of big data technologies that can make an assessment of the future health risks of a data subject on the basis of personal characteristics and genetic data in combination with a disease history. Such assessments and estimates are explicitly covered by the scope of data concerning health.

Furthermore, it is explicitly included that the information arising from the testing or research of a body part or body substance, including genetic data and biological samples and information on, for example, illness, disability, disease risk, medical history, clinical treatment, or the physiological or biometric state of the person concerned, falls within the scope of the GDPR. The source as such is thus not relevant for the qualification as health data. These data may come from a doctor or other health professional, a hospital, a medical device or, for example, an in-vitro diagnostics.

The preceding description of the scope of medical personal data under the GDPR shows that this scope is very broad indeed. It is broader than that of the preceding Data Protection Directive.

It may be noted that the European Court of Justice of the European Union (hereafter: ECJ) in its case law already applied a broad interpretation of the concept of health data. For example, the fact that someone has hurt his foot or is partially on sick leave was already considered by the ECJ as health data.[4]

The broad scope of the term 'data in relation to health' has a particularly profound impact on the application of data science based on technological innovations that are currently entering, and being designed for, the healthcare sector. How this poses a challenge will be discussed further in section 5.

8.2.4 Exceptions to the prohibition of processing

8.2.4.1 Explicit consent

The prohibition on the processing of data concerning health is not applicable if the person concerned has given his or her explicit consent. In this context, Article 29 Data Protection Working Party (meaning the Working Party ex Article 29 of the Privacy Directive 95/46/EC, [which is composed of the data protection authorities of EU member states, from here on Working Party]) has given a further interpretation to the requirements imposed on this consent. The Working Party has taken the position that the consent of the person concerned is only valid if it is a free, specific, and information-based expression of will and intent.

'Free' in this context means that consent must be an expression of will and intent by a person who possesses all his or her intellectual faculties and expresses this consent without any form of coercion, whereby one can think of coercion of social, financial, psychological, or other nature. This means that consent given under threat of non-treatment or less favourable treatment in a medical situation cannot be considered as free. This also applies to the

4 ECJ EU, 6 November 2003, case C-101/01 – Bodil Lindqvist.

consent of a data subject who has not been given the opportunity to make an informed choice or who is confronted with a fait accompli.[5]

The Working Party goes a step further by stating that when a care provider stores and processes personal data in an information system as a necessary and unavoidable consequence of the medical situation, this processing of health data cannot be considered as a legitimate processing basis of consent.

Furthermore, the WorkingParty states that, if the basis is found in consent, this consent must be specific and must therefore relate to a concrete situation in which the medical data will be processed. A general agreement of the data subject to allow, the collection of his or her medical data during the treatment relationship is insufficient for this. Additionally, consent must be based on information. A data subject must grasp the facts and implications of a course of action and, therefore, he or she must be fully and accurately informed of all relevant aspects of data processing in a clear and understandable manner.

These positions of the Working Party were already formulated during the DPD, and are recognizable in the stricter requirements that the GDPR stipulates. Consent, also outside of healthcare, cannot be inferred implicitly. There must be a clear and active act showing that the data subject agrees freely, specifically, and informed, and this act must be unambiguous. Next to these general rules, the additional requirement that the consent needs to be given 'explicitly' applies to consent as a processing basis for health-related data. According to the Working Party, this requirement must be seen in relation to the sensitivity and the nature of the data. In other words, in the context of health data, the data subject must be made aware that she renounces the special protection regime of the GDPR.[6]

8.2.4.2 *The protection of vital interests*

The second exception to the prohibition on the processing of data relating to health is that data processing is permitted if it is necessary to protect the vital interests of the data subject or of another natural person if the data subject is physically or legally incapable of giving his or her consent.

If this processing ground is applied in the context of healthcare, the processing of data must be necessary for, for example, a life-saving treatment in a situation where the person concerned is unable to express his or her will. Also in the health law context, it is generally accepted that, in order to protect the vital interests of a patient who is no longer able to express

5 Article 29 Data Protection Working Party 2007 and 2017.

6 Article 29 Data Protection Working Party 2007 and 2017.

his or her will due to an acute medical condition, the doctor may proceed to conduct an absolutely necessary life-saving treatment on a patient.[7]

The Working Party has pointed out that this processing basis for the processing of health data only applies in a limited number of cases and cannot be invoked as a basis for processing medical data for purposes other than the treatment of the health subject. The processing of medical data for general medical research that may only produce results in the future, is therefore not permitted on this basis.

In view of the Working Party, the legal grounds for the application of technological innovations (e-Health, big data, and artificial intelligence), for which large-scale medical personal data must be processed, cannot easily be found in the protection of a vital interest for the patient concerned.

8.2.4.3 Processing in the context of the treatment relationship

The third exceptive processing ground that is important for the healthcare sector is formed by Article 9, subsection h, GDPR, which provides, inter alia, that data relating to health may be processed if this is necessary for medical diagnoses, providing healthcare, or social services or treatments, or managing healthcare systems and healthcare services. In addition, the agreement with a health professional forms a basis for the processing of medical data.

This includes preventive care, diagnosis, treatment, therapy, and aftercare, as well as the management of healthcare where billing and assessment of insured claims are concerned.

Particular attention should be paid in this regard to the necessity requirement. It is not sufficient that the collected data are *useful* in view of the aforementioned purposes, but the data must be *necessary* for these purposes. This means that healthcare data providers within the healthcare sector must always check within this framework whether the nature of the collected health data is in reasonable proportion to the goal and whether less data could be sufficient for it.

The third paragraph of Article 9 GDPR provides the additional guarantee that data processing for the aforementioned purposes may only take place by or under the responsibility of a professional medical practitioner, who is bound by doctor-patient confidentiality pursuant to Union law or the law of the EU member states.

The foregoing also means that when the need arises that also non-medical professionals process personal medical information, they must also be

7 Leenen et al. 2017.

bound to secrecy. Responsibility for this obligation lies primarily with the one responsible for the data processing. In particular, this will impose requirements on the contractual sanctioning of breaches of confidentiality.

8.2.4.4 Overarching public interest

In paragraph 2 sub i of art. 9 GDPR an exception to the prohibition on the processing of health data is formulated that is linked to reasons of public interest or the common good. In this regard, public health and the protection of people against cross-border health hazards are mentioned. Furthermore, ensuring high standards of quality and safety of healthcare and of medicines and medical devices are explicitly mentioned as part and parcel of the common, public interest.

Based on the legislative wording, the processing of personal data to ensure high standards of quality and safety in the healthcare system expressly falls under the general-interest rubric. This seems to offer a starting point for technological innovations in healthcare, whereby large-scale processing of health data is (deemed) necessary. In particular, consideration should be given to the application of big data technology with medical personal data.

This provision expressly imposes requirements on the processing of data on this basis, involving appropriate and specific measures to protect the rights and freedoms of the data subject. In particular, professional secrecy i.e. doctor-patient confidentiality is again mentioned in this context as a privacy guarantee.

8.3 Classical texts and authors

In this section, five basic texts are discussed that specifically focus on the privacy issue in the healthcare sector.

Within the context of healthcare, a legal and a moral aspect is typically distinguished with respect to the right to privacy. On the one hand, it is internationally recognized that the right to privacy also includes the right to confidential use of the medical data, which is guaranteed in particular by medical-professional i.e. doctor-patient confidentiality. On the other hand, this confidentiality is under pressure because of the general public interest in making access to confidential medical data widely available for medical-scientific research. Various studies have been conducted internationally into this field of tension. I discuss here some of the most important studies.

In the collection *Privacy and Health Care*, six ethical and legal scientific essays edited by Humber and Almeder (2001) explain and discuss the aforementioned conflicting interests.[8] The authors describe the legal and ethical dilemmas primarily from the vantage point and experience of the American legal system. Because of the international nature of confidentiality in general and medical-professional confidentiality in particular, as well as the ethical dilemmas surrounding the breach of this confidentiality in the interest of the advancement of medical-scientific research, the study yields, in my view, important insights that are relevant for Western European jurisdictions as well.

In line with the study by Humber and Almeder, lies OECD's research on Health Data Governance. In this study, the member states of the OECD are offered guidance to find a balance between using and generating research data for medical scientific research and ensuring the privacy of those involved.[9]

For a useful exposition of the legal framework within the European Union of data protection and medical information, I refer to Hervey and McHale. In section 5 of their manual on health law within the European Union, they describe the way in which the processing of medical information is regulated under the privacy directive 95/46/EC. Because the regulation of medical personal data under the GPDR has not undergone fundamental changes to these components, this literature is also relevant for the current regulation under the GPDR.

Furthermore, I refer to the recent study by Munns and Basu on the aspects of sharing medical data within the healthcare sector. They place the right to privacy and medical professional secrecy or doctor-patient confidentiality in the context of technological developments within the information society and examine the sustainability of the principle that medical personal data are only shared in the (patient's) relationship with the physician.[10]

Finally, the *Medical Data Privacy Handbook*, edited by Gkoulalas-Divanis and Loukides, forms a standard work.[11] This study provides a comprehensive overview of new technologies in the field of data processing within the healthcare industry and the authors investigate, among other things, scenarios and new techniques for facilitating the anonymization of different types of medical personal data for a wide range of applications within the healthcare sector. These applications range from the classic data processing in the treatment relationship between the doctor and patient

8 Humber and Almeder 2001.

9 OECD 2015.

10 Munns and Basu 2016.

11 Gkoulalas-Divanis and Loukides 2015.

to the application of new big data technologies, e-Health applications and the use of medical data for medical-scientific research. The manual also discusses both legal and policy frameworks for the regulation of privacy in the United States and Europe.

8.4 Traditional debates and dominant schools

In literature, partly in the context of privacy in the healthcare sector, it is argued that the emphasis of privacy legislation in recent years has shifted from classical privacy protection to data protection. Mostert derives this from, among other things, the terminology used by the European legislator in the GDPR in which the term 'privacy' is hardly ever used any longer and has been replaced by the term 'data protection'.[12] These changes have occurred since the establishment of the right to data protection in Article 8 of the Charter of Fundamental Rights of the EU. This has shifted the accent within the European legal order, more than is the case under art. 8 ECHR, to positive obligations for the member states to subject the processing of personal data to appropriate regulations. In the academic legal discourse there exists a strong focus on requirements of data protection in relation to medical-scientific research. The use of medical personal data outside and beyond the doctor-patient treatment relationship will require the breach of confidentiality and secrecy.

In this respect, the Article 29 Working Party has taken the position that any breach of confidentiality is in principle only possible after explicit consent:

> This means that, in principle, scientific research projects can only include personal data on the basis of consent if they have a well-described purpose. Where purposes are unclear at the start of a scientific research program, controllers will have difficulty to pursue the program in compliance with the GDPR.[13]

Quinn[14] and other authors[15] point out that strict adherence to the consent requirement as intended by the Working Party can seriously impede

12 Mostert et al. 2016.
13 Article 29 Data Protection Working Party 2017.
14 Quinn 2016.
15 Tene et al. 2012. See also: Knoppers et al. 2012.

the effective use of personal data in data-intensive health research; they thus seek to solve this problem through anonymization, despite the fact that absolute anonymity can no longer be guaranteed according to some researchers.[16] In this context it can be pointed out that the GDPR does not require absolute anonymity but approaches this as a dynamic legal concept. The European legislator considers the following:

> To determine whether a natural person is identifiable, account should be taken of all the means reasonably likely to be used, such as singling out, either by the controller or by another person to identify the natural person directly or indirectly.[17]

In other words, data are only anonymous once direct or indirect identification is prevented, based on the resources that people can reasonably use. This should take into account not only current technology but foreseeable technological developments and (re)sources that may become available. The EU Court of Justice has offered further guidelines on what available information should be considered as a means that can reasonably be used for identification purposes. According to the ECJ, this is not the case if identification with the help of extra information is in practice 'impractical' or 'undoable', or at least if the danger of identification appears to be 'insignificant'. This is the case, for instance, if identification would take excessive effort in view of the required time, costs, and manpower.[18]

As Mostert et al. point out, however, when one meets the aforementioned standards of anonymization, this also results in a significant decrease in the usability of data for research purposes. This is caused firstly by the fact that many of the usable data have to be removed from a file in order to reasonably prevent redirection to the individual in question. Secondly, anonymization seems to prevent that data files can still be linked or updated. It is precisely these two disadvantages that, according to Mostert, hinder data-intensive research.[19]

Partly because of these reasons, Munns and Basu are looking for other solutions to be able to use medical data for scientific research. They argue that classical confidentiality in the relationship between physicians and patients needs to be reassessed, and that the current relationship is

16 Savage 2016.

17 GDPR, consideration 26.

18 ECJ, 19 October 2016, ECLI:EU:C:2016:779 – Breyer vs Bundesrepublik Deutschland.

19 Mostert et al. 2016.

unsustainable in the modern information society. According to the authors, there are two determining forces that should lead to changes in the views on confidentiality and information sharing. The first is the development of information technology in combination with medical technological innovations, with which the health of large populations of people can be improved on a large scale using medical data. Secondly, they point to the changes in the relationship between patients and medical professionals. The right to gain access to one's own medical data did not develop until the 1990s, and the medical professionals had to make a major change in their attitude towards, and relationship with, their patients as a result. The doctor no longer determines to a primary or decisive extent what is good and necessary for the patient, but the patient looks for information and medical knowledge him- or herself and forms a more fully fledged discussion partner for the medical professional. The sharing of all relevant medical information with the patient, and more control over this information by the patient, has become the norm. To give substance to this new role of the patient and the changing relationship with the medical professional, Munns and Basu argue that the concept of privacy protection will evolve more towards 'patient control' in healthcare information management. They argue that nowadays patients themselves must make new informed choices about sharing their health data, and they thus advocate a change in that regard in the relationship between doctors and patients.[20] This brings me to the current changes and future dilemmas when it comes to privacy in the healthcare sector.

8.5 New challenges and topical discussions

The evolving information society that prompted Munns and Basu to examine the changing role and relationship between the medical professional and the patient, is increasingly driven by technological developments. The historical emphasis on the classical right to privacy, as formulated by Warren and Brandeis, was predominantly a negative right, 'the right to be let alone'. Partly as a result of technological developments, the emphasis in law development over the coming decades will shift, including within the healthcare sector, to the right to data protection and more positive obligations for governments to ensure data protection through legislative measures. The key point here will be that the patient will be 'in control' of her medical personal data, far

20 Munns and Basu 2016.

more so than currently is the case. The following technological innovations are decisive in this trend.

8.5.1 E-Health

People increasingly collect and control data about their own health and well-being. This is also true for patients. In the field of healthcare, many mobile applications (hereafter: 'apps') are available that track or measure things like exercise, nutrition, and medication use. Often these apps are offered commercially. Their usability and application in medical practice by doctors is still limited, but increasingly, doctors and hospitals are involved in the development of such apps, hoping to increase their accuracy and thus their usefulness in the treatment process.

These developments are also referred to by the concept of 'e-Health'. In practice, a multitude of different technological applications are placed under and understood by that concept. Eysenbach (2001) formulated a more delineated definition of e-Health that more accurately indicates what this should mean:

> e-health is an emerging field in the intersection of medical informatics, public health and business, referring to health services and information delivered or enhanced through the internet and related technologies. In a broader sense, the term characterizes not only a technical development, but also a state-of-mind, a way of thinking, an attitude, and a commitment for networked, global thinking, to improve health care locally, regionally, and worldwide by using information and communication technology.[21]

According to this definition, e-health is ultimately about offering and improving health services and information through the Internet or related audiovisual and ICT technologies. The definition not only encompasses technological development in itself, but also particularly includes the attitude to improve the quality of healthcare at the local, regional, and global level by means of technological information and communication systems.

The increasing scale of development and application of e-Health technologies takes place in the care sector as well as the cure sector. It leads to a radical shift from the classical collection of medical data in the physician-patient relationship to automated data collection. These trends are also expected to have a major impact over the coming decades on the classical

21 Eysenbach 2001.

norms of confidentiality and privacy protection embedded in the current medical-professional i.e. doctor-patient confidentiality framework.

8.5.2 Big data in healthcare

The application of e-Health technologies in turn forms a driver of a second important technological development within the healthcare sector, namely the application of big data technologies. Big data can be seen as data collections that are so large and complex that traditional data processing tools, such as relational databases, are unable to process them within acceptable time and cost limits. Big data technologies increase possibilities to generate, combine, and analyse data, which in turn leads to new insights and new ways of reasoning.

The applications of big data technologies in healthcare are varied. It varies from its application in new drug research to measures aimed at cost reduction. Particularly in the United States, big data are employed to reduce costs in healthcare, by stimulating and furthering improvements in the effectiveness of medicines and treatments with the help of big data analyses. Research with very high healthcare costs is particularly relevant, because it turns out that about half of the costs of care are caused by only 5% of the patient population. As a result, substantial efficiency gains can be realized if this 5% group is timely identified i.e. anticipated so that healthcare practitioners can start the treatments earlier and the costs can be reduced at a later stage of the disease.

8.5.3 Artificial intelligence in healthcare

From the 1960s onwards, the healthcare sector has been working to programme medical knowledge into software using techniques in the field of artificial intelligence, like (advanced) machine learning. The aim is to create software that is capable of reasoning, analysing, and diagnosing. These applications are assumed to support or even replace physicians. One of the first applications in this field was the MYCIN software. This software focused on diagnosing and treating bacterial infections with antibiotics and was able to select the right therapy in 69% of cases (Buchanan and Shortliffe 1984). With this, the programme scored better than the average doctor. These 'artificial intelligence systems' were mainly rule-based in the early years of development, which means that the expert knowledge residing in the software consists of large amounts of decision rules. Analysing and disclosing expert knowledge in this way turned out to be a hindrance in practice for the large-scale development and application of these technologies in healthcare practice.

With the advent of big data analyses, artificial intelligence applications have received a new boost. Computational systems are fed with big data, and increasingly advanced techniques of machine learning generate correlation-based inferences that are seen to be of use for research and development in the sector. Worldwide, a multitude of companies are now active in this area within the healthcare sector.

8.5.4 Future privacy issues

The application of technologies in healthcare, in the field of e-Health and big data in combination with artificial intelligence, often lead to problems with the privacy of patients and the medical-professional confidentiality obligations of care providers. After all, information and software systems must have access to large amounts of patient data in order to be able to analyse these data and thus be able to extract useful knowledge from these. Moreover, due to the large scale of data processing and technology applications, it will become increasingly difficult to truly and effectively anonymize data.[22] This raises the question whether the current way in which the right to privacy is guaranteed is still adequate, and whether the current data-protection concepts will be sustainable in the long term. Moreover, multiple questions arise concerning the privacy risks and the risk of influencing people's behaviour by using personal medical information on a large scale by commercial parties.

8.6 Conclusion

In this chapter, the right to privacy in the context of healthcare was the central topic. I do not presume to give an exhaustive, complete overview of the way in which privacy law affects this sector, but have mapped out the most important themes and issues and indicated how the privacy-law frameworks affect this sector. It is certain that the importance of privacy in this sector is extremely high. This also applies to bridging the legal complications that entail the right to privacy as well as doctor-patient confidentiality in the development of electronic information systems for the storage of medical data – and in the technological developments of e-Health, big data, and

22 See Ohm et al. 2010.

artificial intelligence as applied in healthcare. Consequently, changes in the perceptions of patients and physicians vis-à-vis each other, and adaptations of the data-protection concepts, are almost certainly inevitable.

References

Article 29 Data Protection Working Party 2007.

Article 29 Data Protection Working Party 2007, 'Working Document on the processing of personal data relating to health in electronic health records (HER)', 00323/07/EN, WP 131, 15 February 2007.

Article 29 Data Protection Working Party 2017.

Article 29 Data Protection Working Party 2017, 'Guidelines on Consent under Regulation 2016/679', (17/EN WP259), 28 November 2017.

Buchanan, B.G. and E.H. Shortliffe. (1984). *Rule-Based Expert Systems. The Mychin Experiments of the Stanford Heuristic Programming Project.* City: Addicon Wesley.

Eysenbach, G. (2001). ' What is e-Health?', *Journal of Medical Internet research* 3(2), e20.

Gkoulalas-Divanis, A. and G. Loukides. (2015). *Medical Data Privacy Handbook.* London: Springer.

Hervey, T.K. and J.V. McHale (2004). *Health Law and the European Union.* Cambridge: Cambridge University Press.

Humber, J.M. and R.F. Almeder. (2001). *Privacy and Health Care.* New York: Springer-Science +Business Media.

Knoppers, B.M., M.H. Zawati, and E.S. Kirby. (2012). 'Sampling Populations of Humans across the World; ELSI issues'. *Annual Rev Genomics Hum Genet* 13, 395-413.

Leenen, H.J.J., J. Legemaate, J. Dute, E.J.C. de Jong, M.E. Gelpke, J.K.M. Gevers, and G.R.J. de Groot. (2017). *Handboek Gezondheidsrecht.* Den Haag: Boom Juridische Uitgevers.

Leenen, H.J.J. (1987). 'Patients' Rights in Europe', *Health Policy* volume(issue), 33-38.

Mostert, M., A.L. Bredenoord, M.C.I.H. Biesaart, and J.J.M. van Delden. (2016). 'Big Data in Medical Research and EU Data Protection Law: Challenges to the Consent or Anonymise approach', *European Journal of Human Genetics* volume(issue), 956-960.

C. Munns, C. and S. Basu. (2016). *Privacy and Healthcare Data, Choice of Control to Choice and Control.* London: Routledge.

OECD. (2015). *Health Data Governance: Privacy, Monitoring and Research,* OECD Health Policy Studies. Paris: OECD Publishing.

Ohm, P. (2010). 'Broken Promises of Privacy: Responding to the Surprising Failure of Anonymization'. *UCLA Law review* 57,1701-1777.

Quinn, P. (2016). 'The Anonymisation of Research Data – A Pyric Victory for Privacy that Should Not Be Pushed Too Hard by the EU Data Protection Framework?' *European Journal of Health law* 24,1-24.

Savage, N. (2016). ' The Myth of Anonimity'. *Nature* 537,70-72.

Tene, O. and J. Polonetsky. (2012). Privacy in the Age of Big Data: a Time for Big Decisions. *Standford Law Review Online* 64, 63-69.

World Medical Association (WMA). (2006). *International Code of Medical Ethics.* Pilanesberg: WMA.

Privacy Law – on the Books and on the Ground

Kenneth A. Bamberger & Deirdre K. Mulligan

Privacy law is at a crossroads. In light of the digital explosion, policymakers in Europe and North America are engaged in a wholesale process of revisiting the rules governing the treatment by the private sector of personal information.

For too long, such efforts have lacked critical information necessary for reform. Scholarship and advocacy around privacy regulation has focused almost entirely on law 'on the books'—legal texts enacted by legislatures or promulgated by agencies. By contrast, the debate has surprisingly ignored privacy 'on the ground'—the ways in which those who collect and control data in different countries have (or have not) operationalized privacy protection in the light of divergent formal laws, decisions made by local administrative agencies, and other jurisdiction-specific social, cultural, and legal forces.

For the two decades following a 1994 study that examined the practices of seven US companies,[1] no sustained inquiry was conducted into how corporations actually manage privacy in the shadow of formal legal mandates. No such work was ever done in Europe. And no one has ever engaged in a comparative inquiry of privacy practices across jurisdictions. Indeed, despite wide international variation in approach, even the last detailed comparative account of different countries' enforcement practices occurred over two decades ago.[2] Thus, policy reform efforts have often progressed largely without a real understanding of the ways in which previous regulatory attempts have actually promoted, or thwarted, privacy's protection.

A purely 'on the books' approach fails to recognize important attributes of the privacy landscape.

In the United States, despite a static statutory landscape characterized by a patchwork of privacy statutes, the absence of a dedicated data protection agency and a failure to provide across-the-board procedures empowering individuals to control the use and dissemination of their personal information, corporate

1 See H. Jeff Smith. (1994). *Managing Privacy: Information Technology and Corporate America*, 15-17.

2 This was a study of privacy in several North American and European countries. David H. Flaherty. (1989). *Protecting Privacy in Surveillance Societies: The Federal Republic of Germany, Sweden, France, Canada, and the United States.*

privacy management has undergone a profound transformation. Thousands of companies have created Chief Privacy Officer positions, a development often accompanied by prominent publicity campaigns. A professional association of privacy professionals boasts over 38,000 members and offers information-privacy training and certification. A robust privacy law practice has arisen to service the growing group of professionals and assist them in assessing and managing privacy. Leading firms conduct privacy audits across multiple sectors. And privacy seal and certification programs have developed.

Regarding Europe, moreover, much of the literature engaging in an 'on the books' approach focused on the existence of a 'European' data protection paradigm, characterized by omnibus privacy regulation and a dedicated data protection supervisor—masking, in turn, that there developed not just one European privacy regime, but many. The development of data protection in the various European Union member states reflected major variation across jurisdictions, in terms of administrative structure and behavior, social discourse, and corporate behavior[3] —all within the formal governance of a single legal framework that governed privacy practices for over two decades: the 1995 E.U. Privacy Directive.[4]

Privacy regulators and advocates increasingly recognize that misinformed public debates result, in part, from the paucity of research comparing various forms of privacy regulation—a deficit that our book, *Privacy on the Ground*,[5] attempted to address. To find out what actually does work, we surveyed people charged with protecting privacy inside companies, as well as government officials, scholars and privacy advocates in five countries—the U.S., England, France, Germany, and Spain.

Our interviews with privacy leads yielded startling findings about the limitations of formal laws' influence on corporate privacy practices, and the striking importance of legal institutions, as well as non-legal factors and actors. The research exposed a rift between the dominant story's emphasis on regulation on the books, and the reality of on-the-ground corporate behavior. Indeed, combining our work with a recent analysis of systematic government access to private sector data in 13 countries suggests that the chasm between privacy law on the books and on the ground is pervasive.

Driven by the findings in the U.S., we expanded our research to privacy leads in four other countries, including two—France and Spain—whose strictly

3 See infra Part IV.A–C; see also Abraham L. Newman. (2008). *Protectors of Privacy: Regulating Personal Data in the Global Economy*, 32-33, 94; Francesca Bignami. (2005). 'Transgovernmental Networks vs. Democracy: The Case of the European Information Privacy Network', 26 Mich. J. Int'l L. 807, 827-830 [hereinafter Bignami, *Transgovernmental Networks*].

4 Council Directive 95/46, 1995 O.J. (L 281) 31 (EC).

5 Kenneth A. Bamberger and Deirdre K. Mulligan, *Privacy on the Ground: Driving Corporate Behavior in the United States and Europe* (MIT Press: 2015)

defined and enforced privacy regulations seemed most consonant with the dominant story's prescriptions. Again, we found revealing results.

First, the countries such as France and Spain that have the strictest, most uniform, and most centralized regulatory processes—precisely what some American privacy advocates prescribe—had the *least* robust corporate privacy practices. According to those we interviewed, that's largely because companies operating under such regulations adopt a 'compliance mentality.' They interact mainly with regulators, not a wider community of stakeholders, and focus on doing just enough to satisfy government officials.

By contrast, the countries with the most robust corporate privacy practices, the United States and Germany, had broad similarities, despite very different cultural and legal frameworks. Ironically, these nations' more ambiguous rules encouraged stronger internal controls and facilitated broader conversations about the meaning of privacy among multiple stakeholders, including public advocates. Regulatory agencies established broad goals, set the table for these broader conversations (in Germany, the employee works councils play a crucial role), required greater transparency about privacy failures such as massive data breaches, and made the punishments fit the crimes—sometimes levying significant fines on outlier companies.

Largely as a result, the number and authority of privacy professionals inside companies has grown dramatically in recent years. In Germany, the role of a Data Privacy Officer (DPO) was mandated by the state, whereas in the U.S., more and more companies have elected to hire a Chief Privacy Officer (CPO). These professionals often have direct access to corporate leadership and work strategically to design and embed privacy concerns directly into business operations, rather than as a resented add-on. They typically engage more meaningfully with outside stakeholders, and join peers in national and global professional organizations that define and help enforce industry-wide standards. Paradoxically perhaps, the DPO/CPOs' internal independence, authority, and budgets grow with outside advocacy and media attention, government-mandated transparency requirements—especially in high profile breaches such as the J.P. Morgan, Sony, or Home Depot hacks—and enforcement actions.

Our study uncovered a set of broadly defined best practices largely shared by the U.S. and German companies, despite their quite different regulatory and political cultures. Distilling these findings, we summarized those best practices as:

- *Making the Board's Agenda*: a high level of attention, resources, access and prominence for the privacy function within the firm;
- *A Boundary-Spanning Privacy Professional*: a high-status privacy lead who mediates between external privacy demands and internal corporate privacy practices; and

- *The 'Managerialization' of Privacy*: the integration of privacy decision-making into technology design and business-line processes through the distribution of privacy expertise within business units and assignment of specialized privacy staff to data-intensive processes and systems.

Deriving the best practices in the U.S. and Germany in turn suggested ways in which regulatory policy could replicate, encourage, and expand those best practices on the ground. Defined as *properties* of the field rather than hard and fast rules, we distilled three that support the overall approach we've labeled 'bringing the outside in.'

1. *Ambiguity with accountability*: broad legal mandates and open regulatory approaches, activist regulators, and meaningful stakeholder scrutiny fostered dynamism in the face of changes and pushed more accountability onto firms;
2. *A boundary-spanning community*: U.S. and German,corporate privacy leads situated themselves in a broad and inclusive community of outside stakeholders, including other corporate privacy professionals as well as those from civil society and government, who both challenge the inside privacy officers and empower their role in the firm;
3. *Disciplinary transparency*: greater transparency around privacy failures, including data breach laws, enabled non-regulators, such as civil society groups and media, as well as the broader public, to become credible enforcers in the court of public opinion, leading corporations to invest greater resources and authority in internal privacy professionals and processes.

Happily, the General Data Protection Regulation (GDPR), approved in 2016 by the European Union Parliament and effective as of May, 2018, substantially incorporates many of the suggestions made in Privacy on the Ground. It envisions a new and enhanced role for Data Protection Officers, charged with advising the organization on their data protection obligations and monitoring compliance on the one hand, and facilitating interaction between the supervisory authority and the organization on the other. It further requires those who handle data to address privacy through design and defaults, reflecting the growing interest of regulators, privacy advocates, and privacy professionals in 'Privacy by Design.'

Yet the moment of the GDPR's implementation calls for a renewed commitment to an 'on the ground' approach. Despite the goal of continent-wide policy harmonization reflected in the Regulation, it has become increasingly clear that the GDPR will be implemented differently by individual member states, arising from specific derogations included in the legal instrument on the one hand, and divergent choices regarding regulatory behavior and enforcement practices on the other. It will be increasingly important to understand how these regulatory choices affect the ways those who collect, store and use data treat that information, what internal policies and practices they adopt, and how legal requirements like DPOs and a focus on privacy-by-design affect the actual ways that corporations, and their executives, managers and engineers, treat personal information in a digital age.

9. Privacy from a Media Studies Perspective

Jo Pierson & Ine Van Zeeland

9.1 Introduction

This chapter will discuss research within the domain of media and communication studies (MCS) in relation to privacy, of both individuals and groups, and society as a whole. MCS takes by definition an interdisciplinary perspective, as it discusses in an integrated way the relationship between media and/or communication processes on the one hand and society on the other hand. Mansell (2016, 719) posits as a starting point for MCS the 'critical inquiry into pressing social problems and the roles played by the media and communicative practice in a wide variety of contexts', which also fits the role and function of MCS for investigating issues like privacy.

Scholars in this field like to emphasize the link between media and other societal domains (politics, economics, culture, etc.) to demonstrate the relevance of the MCS viewpoint. Since MCS was introduced as a separate discipline, it has built up a corpus of theories, which is strongly influenced by developments and insights from other disciplines. The latter refers foremost to cross-fertilization with the social sciences (like social theory, sociology, political science, and economics), but also with the human sciences (like psychology, anthropology, philosophy, history, and linguistics) and even the 'exact' sciences (like engineering, cybernetics, and informatics) (Loisen and Joye 2018).

In addition to an ever-expanding number of research topics, MCS 'stands at the intersection of the social and human sciences, and is therefore open to 'immigrating' researchers from other more established disciplines. This certainly encourages interdisciplinarity, but it also leads to the blurring of the boundaries between specific disciplines' (Loisen and Joye 2018, 35). Nevertheless, this interdisciplinarity is perceived as a major strength, because it 'focuses scholarly attention on theory development concerning the material facets and the symbolic process of mediated communication that help to expose lived asymmetries of power at both the individual and the collective levels' (Mansell 2016, 720). In other words, studying the implications of communicating through media from different academic perspectives leads to more refined theories of the various power imbalances experienced by individuals, groups, and society as a whole. A number of examples of those implications will be highlighted in sections 2-4 below.

MCS became a separate area of research from the 1930s onwards, starting in the US and then taken up in the UK and in the rest of the world. The domain investigates human and social communication through media as well as from person to person. Generally, two main streams of research are identified: 'communication studies' and 'media studies'. The foundations can be found in social sciences (for communication studies, with an origin in the US in the 1930s) and in humanities (for media studies with an origin in the UK in the 1970s). Another influential input for MCS came from information science as developed by Shannon and Weaver (1949), dealing with the technological efficiency of communication channels for carrying information. The authors started from a simple transmission model, which defines communication as the intentional transfer of information from sender to receiver by way of physical channels which are influenced by noise and interference.

In the beginning MCS were foremost concerned with the sociology of mass communication in the US, pioneered by Paul Lazersfeld. The term 'mass' in 'mass communication' is essential, starting from the (monocausal) belief that masses could be easily manipulated via broadcasting media like radio and later television, based on the experiences of the Second World War and the propaganda of the Third Reich in Germany. Therefore one of the first aims was to investigate the linear effects that media messages had on a mass population. This perspective was later readjusted by the reception approach, by taking more into account the people themselves and the way they receive, process, and interpret the broadcasted messages. A next phase was to also take the sociocultural context more seriously, by exploring the production and reinforcement of culture through communication and interpreting media by people in everyday life. This was the start of the media studies stream in the UK, which was closely related with the so-called cultural studies tradition, with Stuart Hall and Raymond Williams as key scholars. From the 1960s onwards, also critical theories (based on Marxism) of the production side in the media industry and communication infrastructures became very influential. This perspective was pioneered by Max Horkheimer and Theodor Adorno of the Frankfurt School in the 1930s, and later followed up by scholars like Jürgen Habermas, with his work on the public sphere, and Dallas Smythe on the political economy of communication.

Given its diverse scientific background and relatively young age (established in the first half of the twentieth century), the name itself of the field is not fully fixed. Depending on the background and the (local institutional) setting in which the discipline is residing, the naming can vary from 'communication science', 'communication sciences' or 'communication studies' to 'media studies'. We will use the more generic label of 'media and communication studies' (MCS).

A major focus of MCS will continue to be on the 'mediation' of communication that takes place through technological media. We are living in times when technology is often employed for different forms of public and private surveillance, which affects personal interaction and the integrity of the communication context in a variety of ways. In section two we will explore these notions further in an introduction to the perspectives MCS takes on media and privacy. In the third section we will discuss foundational contributions to the discussion of privacy within MCS, while the fourth section will describe how the privacy debate within MCS has further developed. The fifth section will draw a line from these developments to foreseeable future developments, followed by a conclusion in section six.

9.2 Meaning and function of privacy

This chapter generally takes a 'Western' perspective on MCS, mainly discussing research traditions in Europe and the United States of America. Our focus will be on different forms of mediated communication, which means communication processes effected via a (technological) medium (Baran and Davis 2015, 6), while less attention will be given to interpersonal, face-to-face communication. This section of the chapter will clarify the specific MCS approach to privacy and data protection.

The formal introduction of the privacy notion in law by Samuel Warren and Louis Brandeis was spurred by the emergence of a medium that was 'new' at that time: portable photography. Warren and Brandeis's canonical essay, 'The Right to Privacy' (1890), was aimed at intrusive photojournalism in the 1880s, with a focus on the technology used: 'numerous mechanical devices threaten to make good the prediction that "what is whispered in the closet shall be proclaimed from the house-tops"' (Warren and Brandeis 1890, 195). The connection between media and privacy has since then always been pertinent, given that the use of interpersonal media and communication tools also signifies the separation between private versus public communication. Interpersonal media like letters and the telephone are seen as inherently private communication, and are also protected by law in that way. Mass media are positioned as public communication which are therefore heavily regulated, whereby the communicator can be held accountable for the public messages being broadcasted.

With the rise of digital media and technology like smartphones, social media, online platforms, drones, big data applications, and the Internet of Things, the boundaries between private and public are blurring. With the proliferation of these digital media and communication technologies, the relevance of approaching privacy issues from an MCS perspective has only increased.

Lievrouw and Livingstone (2002) define 'new' or digital media as information and communication technologies and their associated social contexts, incorporating three inextricable and mutually determining components:

1. Technology: artefacts or devices that enable and extend our abilities to communicate or convey information;
2. People: communication practices or activities we engage in to design and use these devices;
3. Society: social and economic arrangements or organizational forms that form around the artefacts and practices.

As this three-fold division of perspectives neatly clarifies the MCS approach to media and privacy, this division will be used as the main framework for discussion in this chapter. It should be noted, though, that the three perspectives are very much intertwined.

MCS explores the ways in which digital media configure communication and, at the same time, how they themselves are being shaped by society through artefacts, practices, and socioeconomic arrangements. As social life is more and more 'datafied' (Mayer-Schönberger and Cukier 2013) through the counting of clicks, 'likes', and 'friends' (thus changing the meaning of 'like' and 'friend'), so are technologies shaped by how we use them, e.g. the major online platforms continuously keep track of how subscribers use their online environments to pick up on trends and tendencies to incorporate in their systems. While tracking of user behaviour can in itself be considered a breach of privacy, these adaptations 'under the hood' of technologies also diminish transparency about their functions (curtailing users' control over what information about them is shared with whom), and agency for users, who are often unaware of how and when 'the user experience' is adapted to keep them engaged for longer or draw them in deeper. Users can become simultaneously empowered as well as disempowered with regard to privacy, as stated by Van Dijck (2013, 171): 'User empowerment is dependent on *knowledge of how mechanisms operate* and *from what premise*, as well as on the *skills to change them*' [emphasis added].

However, due to power asymmetries, even individuals who are aware of this often face a 'take-it-or-leave-it' situation, in which they stand to lose abilities they have shaped their lives around and which have changed social norms irreversibly. E.g. leaving a chat app out of informational privacy[1] concerns may mean complicating communications with friends and family, and being left out of certain types of interactions that one's social circle and society at large have become so accustomed to that they would be hard-pressed to change.

Each of the three interrelated perspectives – technology, people, and society – offers a particular added value for a more profound understanding and investigation of privacy in technologically mediated communications. Getting back to Warren and Brandeis, who sought legal provisions to protect privacy in view of a new technology they perceived as intrusive and laid the foundation of an increasingly prominent academic legal field, the MCS approach of privacy and data protection broadens the scope of study beyond a response to perceived intrusiveness. The mutual articulation between the perspectives from technology, people, and society improves our grasp on the context of privacy and data protection law, rather than regarding legal privacy provisions as an isolated phenomenon. The next section will discuss preeminent MCS research that illustrates the particular importance of context.

9.3 Classic texts and authors

In line with the previous section's division in perspectives, we start this discussion of influential privacy literature in the field of MCS from the perspective of how technological artefacts relate to people and society with regard to privacy. In a seminal article, Philip Agre (1994) discusses how information and communication technologies have an impact on privacy conceptions and how ideas about privacy impact technologies. For Agre, 'ideas about privacy are, among other things, cultural phenomena' (Agre 1994, 109) that are continuously reproduced and transformed. To demonstrate how ideas about privacy shape technical design-styles and how design practices change ideas about privacy, he contrasts two cultural models of privacy: the surveillance model and the capture model. These models are not explicitly present in anyone's mind or in society, but refer to coherent sets of metaphors that influence the design of technologies. Technical designs in turn feed more metaphors into the models. The models of privacy are

1 Informational privacy has been defined as 'the ability of the individual to control when, how, and to what extent his or her personal information is communicated to others' (Westin 1967) and as 'individuals' right to have control over the flow of information about them' (Nissenbaum 2009).

contingent, they change over time just like other cultural concepts. The surveillance model and the capture model are not mutually exclusive, but by contrasting them we can see more clearly how they change over time.

The surveillance model typically starts from visual metaphors (cf. 'Big Brother') and derives from historical experiences of secret police surveillance. This model is more pertinent than ever with the proliferation of mass state surveillance and surveillance capitalism, by means of CCTV, online tracking, and sensing devices in an Internet-of-Things world (Mosco 2017). We discuss this more in detail in the next section, from the society perspective.

The capture model is principally manifest in practices of information technologists. It is built upon metaphors that are linguistic rather than visual (e.g. speaking about a 'history' of a file). In the practices of applied computing, social and industrial work activities are systematically reorganized in order to allow computers to track them in real time. Consider for example how a product bought online is tracked technologically from the warehouse to the delivery address. Digital optimization systems have influenced this logistical process, enabling detailed control over the 'tasks' of the human workers involved in it and intertwining human activities with computerized mechanisms. Another example is how the user interface of a device allows only for specific keystrokes, menu selections, and settings manipulation, forcing people to adapt their actions to the technical design. The capture model fits in the Human-Computer Interaction (HCI) field of research, in particular in the more socially oriented subfield of Computer Supported Cooperative Work (CSCW) (Blomberg and Karasti 2013).

The value of the capture model is that it highlights the reorganization of activity to accommodate the tracking process. The concept of 'grammars of action' refers to the (socio-technical) way in which social activities are structured in and through information systems. The capture model then describes 'the situation that results when grammars of action are imposed upon human activities, and when the newly organised activities are represented by computers in real time' (Agre 1994, 109). In a simplified description, information systems impose grammars of action on humans through a five-stage cycle:

1. Analysis: empirical study of a real-life activity (e.g. 'post delivery'), discerning the fundamental units (post, postman, delivery van, walking to the door, etc.) and describing the interrelationship of those constituting entities.
2. Articulation: translating the real-life units into a grammar of the ways in which they can be woven together to form stretches of activity, creating a consistent whole.

3. Imposition: a normative force is articulated. People are induced to organize their actions in such a way that they can be easily understood by the system in terms of the proposed grammar. This can happen via social means (e.g. procedures suggested by management) or technical means (e.g. nudging via a user interface).
4. Instrumentation: provision of social and technical resources for continuous tracking of the ongoing activity. This stage and the next are where people's informational privacy loses out against the needs of the information system.
5. Elaboration: recording of the captured activity. These records can be stored, processed, audited, and used for other purposes. A current example of elaboration would be 'datafication', referring to the ability of social media and online platforms to convert online and offline interactions and activities into data (Mayer-Schönberger and Cukier 2013; Van Dijck, Poell, and de Waal, Forthcoming).

The capture model helps us to better understand the socio-technical challenges of the current media and communication environment. For one thing, omnipresent commercial social media and online platforms engage in ever broader and deeper capturing of people's social activities. This capturing, mapping people's social and work lives and imposing a 'new order', can come at significant cost to their informational and decisional privacy. This development has also led to a commodification of everyday activities: even your morning commute to work or the university is now for sale. Another result of this is an increased commensuration of very diverse human activities and interactions (Espeland and Stevens 1998), meaning that very different activities, interactions, or preferences are measured according to the same metric so they can be captured. Think of measuring 'online privacy concerns' so they can be factored (alongside other factors) in a metric of 'willingness to buy'. In addition, increased state surveillance has readily profited from the capture model for strengthening its grip on citizens.

We now move to the second perspective, that of the people. A key communications scholar who takes this perspective is Sandra Petronio. Her theory links the bottom-up interpretative perspective of media users with aspects of society and technology, fitting into the research that looks at social or contextual privacy.

Looking at privacy from an individualistic viewpoint, as promulgated by legal frames and technological implementations, has for long dominated public discourse. This is typically represented by Westin's classical definition of privacy: 'the claim of individuals [...] to determine for themselves when, how, and to what extent information about them is communicated to others' (Westin 1967, 7). However, many scholars in social science and MCS argue that privacy is first and foremost contextual (Altman 1977; De Wolf and Pierson 2014; Palen and Dourish 2003). Altman (1975, 24) defines privacy as 'the selective control of access to the self', complementing Westin's conceptualization by posing that privacy is a social process that not only functions on an individual level but also on a group level.[2] Altman states that each individual 'regulates privacy' via different behaviour mechanisms. This privacy regulation leads to a dynamic process of 'interpersonal boundary control' in which the physical and social environment in which the information disclosure takes place has a defining role.

The notion of 'boundary' as a metaphor for managing the communication of personal information (Margulis 2011, 12) is taken further by Petronio in her seminal work 'Boundaries of Privacy: Dialectics of Disclosure' (Petronio 2002). Petronio's Communication Privacy Management (CPM) theory highlights the necessity to juxtapose privacy with publicness, where both are in a dialectical tension defining each other's parameters (Petronio 2010, 178). She defines privacy 'as the feeling that one has the right to own private information, either personally or collectively' (Petronio 2002, 6). As people feel they own information about themselves, they also feel entitled to control 'their' information. They develop personal privacy boundaries, but also need to coordinate collective privacy boundaries when they have shared information about themselves with others. They may also need to implement corrective measures when those others have violated privacy rules. Privacy management implies strategies and tactics that are highly context-dependent.[3]

The contextuality of privacy management becomes even more incontestable in the social media environment. An influential study by Marwick & boyd (2014) of how teenagers navigate the amenities and pitfalls of social media, demonstrated that individual information control is not a workable approach in what they call a 'networked' environment. Instead, Marwick and boyd present a model of 'networked privacy'.

As we have seen in the discussion above about the perspective from technology, new technologies change human practices through their 'grammars of action', or how they organize human interaction with the technologies themselves and other (human or non-human) components

2 See also chapter 3 of this book.

3 CPM theory is more extensively explained in chapter 10 of this book.

in the socio-technical system. Social media platforms in particular change participants' practices of information-sharing through their 'affordances', configurations of the environment in ways that shape participants' engagement. Affordances that play a significant role in configuring social media environments are: the persistence of online expressions, the replicability of digital content, the scalability of shared information (social media content can travel far beyond its intended audience), and the searchability of content (boyd 2010, 36). As sharing is a key component of social media participation, participants have to share information about themselves, and this information may persist, be replicated, reach many unintended audiences, and may still be found years later.

Networked publics are defined as 'spaces constructed through networked technologies and imagined communities that emerge as a result of the intersection of people, technology, and practice' (Marwick and boyd 2014, 1052). Marwick and boyd's study of teenagers' online privacy management showed that they face many challenges in networked publics, often to do with privacy violations by others; snooping parents and teachers, angry exes, inappropriate sharing by friends or other family members, and so on. Many of these challenges are caused by the 'context collapse' that is typical of networked publics, which refers to the amalgamation of relations from distinct social contexts, all under the same header of 'friends' or 'followers'. Another typical aspect of networked publics consists of 'invisible audiences': the wider public (including government and companies), friends of friends, or future contacts, who may see personal information that was not originally intended for an audience of strangers.

The teenagers used a variety of strategies to regain control over their information, including tweaking (privacy) settings, content curation, and using linguistic and social cues to 'encode' content for specific audiences ('social steganography'). However, in a networked public it is impossible for participants to single-handedly control where information about them ends up and by whom it is seen. As the teenagers' challenges and strategies demonstrated, information-sharing cannot be regulated by technical means alone, but needs to be negotiated socially: 'networked privacy is the ongoing negotiation of contexts in a networked ecosystem in which contexts regularly blur and collapse' (Marwick and boyd 2014, 1063).

Marwick and boyd's model of networked privacy was partly a response to Helen Nissenbaum's ideas on contextual integrity (2010). Nissenbaum's contextual integrity approach is an example of the third perspective within MCS: the societal perspective. Where both Petronio and Marwick and boyd take a bottom-up approach, in which people interpretatively define and coordinate their private information boundaries, Nissenbaum's approach can be seen as a more top-down perspective focused on detecting privacy violations (De Wolf 2015, 35). She starts from the premise that society consists of different social contexts (Nissenbaum 2004, 119) in which privacy gets meaning and is enacted. Based on this premise she develops a justificatory framework for addressing privacy problems: ' people engage with one another not simply as human but in capacities structured by social spheres' (Nissenbaum 2010, 130). These contexts can be very diverse, like education, employment, healthcare, family or the marketplace. Reducing privacy to a particular class of information (e.g. 'sensitive data') or to one transmission principle (e.g. 'user control') overlooks important contextual factors of privacy.

The concept of contextual integrity consists of four essential claims (Nissenbaum 2016). The first claim is that privacy is achieved when the flow of information is appropriate. This contrasts with the idea that certain kinds of information flows (e.g. personal data collection, information leakage) would be by default privacy violations, or that a situation of no information flow or secrecy would signify privacy.

Second, an information flow is appropriate when it conforms with contextual informational norms; when it does, this constitutes a situation of contextual integrity. Hence, the theory of contextual integrity goes beyond looking at procedural outcomes, such as the legal requirement of informed consent or the Fair Information Practice Principles (FIPP).

Third, in order to understand contextual informational (privacy) norms, Nissenbaum identifies five independent parameters that need to be analysed: sender; subject; recipient; information type; and transmission principle. 'Information type' refers to the kind of information that is transmitted (e.g. financial, demographic, communications, medical status, etc.), while 'transmission principle' refers to the constraints under which the information flows (e.g. consent, buy, with a warrant, surreptitiously, required by law, etc.). If we wish to assess if an information flow is appropriate, our analysis must specify all parameters. For example, a patient (sender) talks to her doctor (recipient) about the medical condition (information type) of her husband (subject), under the assumption that the doctor keeps this information confidential to the outside world except towards possible other doctors for peer consultation (transmission principle). When all these parameters are upheld, the contextual informational norms are respected.

The fourth claim is that there are three levels in the evaluation of the ethical legitimacy of privacy norms. The first level concerns the evaluation of the interests and preferences of the affected parties, e.g. based on a stakeholder analysis. The second level of evaluation has to do with the ethical and political principles and values that are at the core of a society, e.g. fairness and justice. The third level is the evaluation of the context-specific functions, purposes, and values at stake. For example, in the development of wearable technologies for medical purposes, an evaluation should be made of whether these technologies promote medical ends (e.g. cure disease or alleviate suffering) and values (e.g. patient autonomy and equal access), and not merely commercial goals.

While Marwick and boyd (2014) acknowledge and expand on the importance of incorporating contextual factors in the analysis of media's influence on privacy, they warn that new technologies are increasingly collapsing contexts and blurring contextual lines, which may lead to confusion over privacy norms. Nissenbaum (2015) recognizes this confusion and calls these novel information flows in new technologies 'disruptive'. The disruption caused by new media should trigger an analysis and evaluation of the information types, actors (sender, recipient, subject) and transmission principles. This evaluation must focus on context-specific functions, purposes and values, in which 'context' should not merely refer to the platform, business or economic sector, but to the social domain, e.g. trusting families and friendships.

New technologies, specifically mediated communication through social media, remain an important orientation of MCS research, as will become clear in the next section.

9.4 Traditional debates and dominant schools

We will now discuss some key authors that have a background in or link to MCS, or that have mainly done research in this field, starting from one of the three perspectives: technology, people, and society. The classification of authors under one of the perspectives is based on the starting point or discipline of the authors, but is by no means intended as a demarcation: as the discussion of influential literature and authors in the previous section made clear the perspectives are very much intertwined. Finally, this is not meant as an exhaustive account of all scholars in the field, but more as a guideline to some 'capita selecta' offering a representative snapshot of the main ideas and schools of thought. For additional scholars and concepts, we wish to refer to the 'Further reading' part at the end of this chapter.

9.4.1 Technology

Looking at media as technological artefacts has been the focal point of many scholars in the MCS field. Artefacts like online social networks and social media need to be developed and programmed according to certain requirements. To capture the user and social needs and practices of mediated communication and privacy, a lot can be learnt from findings of research into human communication behaviour in the MCS field (Karahasanovic et al. 2008). In this way we avoid simplistic assumptions about privacy behaviour. Given that user effort is a limited resource, privacy and security settings must be intuitive to use, and thus attuned to people's communication patterns, in order to be effective (Sasse and Palmer 2014).

In 'Do Artifacts Have Politics?', Winner (1980, 127) stated: 'By far the greatest latitude of choice exists the very first time a particular instrument, system, or technique is introduced. Because choices tend to become strongly fixed in material equipment, economic investment, and social habit, the original flexibility vanishes for all practical purposes once the initial commitments are made. It comes as no surprise, then, that 'privacy by design' has attracted the attention of many policymakers today. 'Privacy by design' refers to embedding privacy into the design specifications of various technologies (De Wolf, Heyman, and Pierson 2012). 'Embedding', in a genuine sense, calls for an interchange of MCS research and technological insights in the early development stages of new technologies or services with privacy-sensitive characteristics, when more fundamental design choices are still possible (Gürses and del Alamo 2016).

In this regard, it is useful to identify different approaches in how to design privacy into technologies. Diaz and Gürses (2012) provide an overview of the landscape of privacy technologies, following the classification in privacy research paradigms proposed by Gürses (2010). They describe three interdisciplinary paradigms of engineering privacy:

- Within the 'privacy as confidentiality' paradigm, technologies aim to create an individual autonomous sphere free from intrusions from public authorities and private companies.
- The 'privacy as control' paradigm focuses on providing individuals with control and oversight over the collection, processing, and use of their data, with technologies aiming on the one hand to provide individuals with the means to exercise this control and on the other hand to provide organizations with purpose-based access control systems.
- The 'privacy as practice' paradigm has the closest links with social sciences as it starts from the assumption that privacy is not just an individual matter, but that it also has important social dimensions. Technologies within this paradigm aim to make data flows more transparent through feedback and raising awareness. Individuals and collectives are offered more insight into how information is collected, aggregated,

analysed, and used for decision-making. Enhancing transparency of systems also means enhancing the possibility to question, intervene, and renegotiate information practices. This paradigm will be discussed more extensively in the next section.

The MCS 'artefact perspective' is echoed in a number of related research fields, such as digital humanities. Framing the software code itself and the processes of software development as a distinct object of study, MCS has strong links with the field of 'software studies', an interdisciplinary research field studying software systems in combination with their social and cultural effects. Similar approaches can be found in 'platform studies', investigating the relationships between the hardware and software design of computing systems and the creative works produced on those systems, as well as in 'critical code studies', which are more closely attuned to studying the code itself. These approaches generate valuable input for research into how friendships and sociality, and related privacy and security issues in online social networks, are interacting with the underlying algorithms.

Associated with our discussion of Agre in the previous section, media scholars argue that the phenomenon of online social media shapes social behaviour, while social behaviour in turn shapes social media. A critical history of social media, Van Dijck's 'Culture of Connectivity' (2014) describes how "making the web social' in reality means 'making sociality technical" (Van Dijck 2014, 12). In the past decade and a half, people's experience of sociality has undergone rapid changes, as many social activities moved online and became mediated through Internet technologies. For example, chatting online differs from chatting 'in real life' in many ways, as a result of pre-coded restrictions: people communicating through a chat app have to use keyboards, are limited by app settings and functionalities, and have to forgo many prosodic and other nonverbal elements of communication – but they have gained emoticons, animated gifs, and direct Internet links, among other novelties.

While ordinary users may believe their social behaviour is simply supported by online technologies, those technologies steer users in certain directions (Sandvig 2015). For example, they suggest contacts ('People You May Know'), connections ('Customers who bought this also bought'), and actions ('See What's Next') based on probability computations and rankings. Social media technologies engineer and manipulate connections not just between one user and other users (connections labelled 'friends'), but also between a user and brands, events, celebrities, (political) ideologies, trends, and so on (connections labelled 'likes' and 'follows').

Social network sites like Facebook and Twitter have little interest in the real-life complexity of waxing and waning friendships; they have their own purposes for which they need to attract users' attention: 'data generation has become a primary objective rather than a by-product of online sociality' (Van Dijck 2014, 12). This leads back to the commensuration of diverse human activities and interactions discussed in section 3: social-media capture systems translate the various relationships with friends, family, classmates, co-workers, and so on – relationships the emotional content of which appears to humans as something intimate and ungraspable in its nuance – into the single metric of 'friend' or 'follower', with a certain value (unknown to the user) that is run through probability computations to predict the extent to which users can be influenced by these 'friends' or 'followers' for commercial purposes. Heyman and Pierson (2015) suggest that Facebook limits users' communicative capabilities to only those aspects that drive profit for Facebook, Inc.

Privacy concerns come into play when we look at the commodification of data collected by social media platforms: many users are only dimly aware of how online sociality generates behavioural and profiling data that is monetized by commercial social media platforms. Creating an account and logging onto a social media website almost by default requires surrendering personal data for mining and reselling, but few users read the terms of service or the privacy policy. Facebook changes both the service itself and its terms of service on an ongoing basis, mostly without explicit notification of the changes to users, as a result of which users are hardly capable of knowing what they are consenting to, and have 'limited agency to reject and leave or to accept and stay' (Heyman and Pierson 2015, 8). Their informational privacy is thus unwittingly (rather than intentionally) exchanged for connectedness.

Nevertheless, users are not merely targets of manipulation; they employ tactics of their own to negotiate the social media environment in order to incorporate it into their everyday lives and make it meet their needs. This tactical behaviour of users is, in turn, fed back into the social media algorithms and incorporated into the technologies. 'It is a common fallacy [...] to think of platforms as merely *facilitating* networking activities; instead, the construction of platforms and social practices is mutually constitutive' (Van Dijck 2014, 6). As Marwick and boyd (2014) show, social media users also develop tactics to negotiate their informational privacy within the social network environment. Additionally, 'users can enjoy connective media and still be critical of their functioning, for instance by taking a vocal stance on privacy issues or data control' (Van Dijck 2014, 18). However, although both the social media platforms and their users can manipulate the technologies to align with their preferences, there is an important difference in power that needs to be taken into account.

The aim of Van Dijck and other media scholars is not to identify how the technical co-construction of social media platforms comes about, nor how social media violate privacy laws, but to trace changes in norms for what counts as private or public, and to analyse the clash between user tactics and platform strategies in order to distinguish implications for society. Van Dijck (2014, 23) specifically notes the growing influence of algorithmic decision-making and automated connectivity on society as 'platformed sociality' matures.

9.4.2 People

Given the intensified integration of digital media and communication technologies into everyday life, it becomes essential to investigate to what extent and how people are still aware of communication mediated by technology and hence can critically reflect on possible consequences like privacy intrusion. For this we need to complement media literacy with data and privacy literacy.

According to Deuze (2012) the key challenge of MCS in the 'media life' of the 21st century is, or will be, the disappearance of media, where people increasingly are living 'in' media instead of living 'with' media, by the merging of online and offline world.[4] This has also been indicated as the evolution towards an 'onlife' world, which is the increased convergence of offline and online in a hyperconnected world (Floridi 2015).[5] Because our society increasingly becomes an infosphere, mixing physical and virtual experiences, we are acquiring an onlife personality. The latter is different from who we innately are in the 'real world' alone (Botsman 2017) which requires us to rethink traditional privacy ideas. This perspective fits the

4 The perspective of 'media life' has been criticized by Couldry and Turow (2014, 1720-1721) as undercutting any further normative critique.

5 The dissolving of borders that formerly separated online and offline spaces fits in a much larger societal development of so-called 'digital seepage'. We observe how characteristics, events, and experiences from one context seep into another context, which makes it increasingly difficult to distinguish and reflect upon different contexts. This crumbling of boundaries and the merging of (once contrasted) worlds happens on many levels: not only between online and offline experiences, but also between private and public spaces, between editorial content and advertising, between home and work, between free and paying business models, between regulation and self-regulation, and in many other contexts (Pierson 2013; 2014).

notion of 'mediation' referring to the idea that 'mediated connection and interconnection' are part of the infrastructure of most people's lives in the Internet age (Mansell 2012; Silverstone 2006). This is also in line with similar schools of thought that have been developed in social sciences, by scholars like Bauman (2000) on 'liquid modernity', Wellman and Haythornthwaite (2002) on 'networked individualism', Orgad (2007) on 'online and offline', and Couldry (2011) on 'media practices'.

One of the first ways to conceptualize the merger between media technology and everyday life can be found with Roger Silverstone, Eric Hirsch, and David Morley (1990), proposing the notion of 'domestication'. This refers to the gradual process by which digital media are consumed and 'tamed' within the sociocultural context of the home context and beyond (Lie and Sorensen 1996; Silverstone and Haddon 1996). People adapt their media consumption behaviour according to the requirements of the (new) technology, while they simultaneously aim to adapt the technology to their wishes as users. The end result is that domesticated digital media technologies disappear and dissolve into the everyday life of people, as they are not perceived as technologies anymore, but as natural extensions of personal interactions and communication. This gives these platforms a self-evident character, making it difficult to reflect on them, and hence to critique how they operate and the role they have. For example, if the steering aspect of social media technologies 'disappears' (from observation) as they become fully domesticated, it becomes increasingly hard to observe and criticize their possible misuse of personal data for disputed social engineering (boyd 2014), unethical political micro-targeting (The Economist 2018) or illegal behavioural advertising via cookies (Gibbs 2018).

In order to thoroughly understand the consequences for users and their privacy in this changing socio-technological environment of digital media, various scholars take a critical perspective (Mansell 2004; Röhle 2005), as opposed to an administrative, instrumental perspective. In this 'social imaginary' of the critical perspective: 'attention is drawn to the potential of innovations in technologies to be associated with people's empowerment and their disempowerment, depending on the extent to which they are able to master or control the innovation process' (Mansell 2012, 37). In order to investigate the development of digital media from vantage points that make issues of power explicit in the analysis of mediated experience, an interdisciplinary as well as a critical research agenda is required (Mansell 2004, 102). This also relates to safeguarding and embedding public values like privacy and data protection, but also transparency, autonomy, and non-discrimination.

In order to operationalize this critical research agenda from the perspective of people, users need to become sufficiently aware of and reflect on their 'onlife' context while domesticating digital media. This requires a particular form of media literacy among citizens, that includes a data analytics perspective (Couldry 2014). Media literacy is generally defined as 'the ability to access, analyze, evaluate and create messages in a variety of forms' (Livingstone 2004, 5). Due to the proliferation and domestication of digital media, we need to enhance media literacy with data literacy. The latter involves increasing awareness, building attitudes, enhancing capabilities, and adjusting behaviour among users of social media technologies and online platforms regarding the impact of (personal) data collection, processing and (re)use on fundamental public values like privacy (Pierson, Forthcoming). The concept of data literacy goes beyond media literacy by for example integrating understandings of the material conditions and technological affordances of the proprietary control of personal data (Morozov 2013; Naughton 2017).

In times when much of the Internet is fuelled by advertising, various academic and scientific disciplines have researched Internet users' response to online advertising, often with a focus on studying consumers' privacy concerns from a marketing or management perspective. Online behavioural advertising is becoming more salient as a concern for Internet users, who often experience targeted, personalized ads as 'creepy' (Ur et al. 2012). When it comes to media literacy research, a study by Smit, Van Noort, and Voorveld (2014) shows that many Internet users do not understand how online behavioural advertising works and feel worried that their personal privacy is at stake. However, most Internet users do not read privacy statements, they rather use ad blockers and anti-spyware applications or clear their browser history.

9.4.3 Society

When it comes to current MCS perspectives from society on privacy and data protection, MCS helps to understand and counter how human connectedness in social media is increasingly converted into 'automated connectivity' for commercial aims, especially by platforms publicly listed on the stock exchange (Van Dijck 2013).

One aspect that was not yet fully addressed in the discussion of people's response to new technologies in the previous section, is surveillance. Leaving aside targeted surveillance, which so far has not been a main theme in MCS research, mass surveillance through new technologies is carried out by both private and public actors. As social media platforms, their partners, and other third parties capture and use personal data from social media, challenges to users' privacy are not limited to the relationships between people who know each other. Again, 'surveillance studies' is an emerging research discipline in itself, but there is still a specific MCS approach that adds to academic knowledge accumulation in this field. This was already highlighted earlier by positioning the surveillance model and the capture model in MCS (Agre 1994).

Surveillance is defined by David Lyon (2007, 14) as 'the focused, systematic and routine attention to personal details for purposes of influence, management, protection or direction'. Routine means that it occurs as a 'normal' part of everyday life in all societies depending on bureaucratic administration and some kind of information technology. In a surveillance society, precise details of our personal lives are collected, stored, retrieved, and processed every day within huge computer databases belonging to large companies and government departments (Lyon 1994, 3). It was Gary T. Marx (1985) who first introduced this concept, referring to computer technology. Roger Clarke emphasized the central position of (personal) data for surveillance by introducing the notion of 'dataveillance' as the 'systematic monitoring of people's actions or communications through the application of information technology' (Clarke 1988, 500). According to Oscar Gandy, this type of monitoring leads to a panoptic sort: 'The panoptic sort is a difference machine that sorts individuals into categories and classes on the basis of routine measurements. It is a discriminatory technology that allocates options and opportunities on the basis of those measures and the administrative models that they inform' (Gandy 1993, 15). By relating the panoptic sort to computer technology, Internet, and data, Lyon develops the idea of social sorting: 'The surveillance system obtains personal and group data in order to classify people and populations according to varying criteria, to determine who should be targeted for special treatment, suspicion, eligibility, inclusion, access, and so on' (Lyon 2003, 20).

Developments in social media and digital technologies have only exacerbated the risks of surveillance, social sorting, and related privacy violations. The latter is perspicuously demonstrated by Bauman and Lyon (2013) by introducing the concept of 'liquid surveillance' as a way of situating surveillance developments in the fluid and unsettling modernity of today, where social forms are melting and power and politics are splitting apart. They connect this with the mutual relation between digital media and fluid relationships (Bauman 2013). In a similar way, Mosco discusses the major transformations of the primary technological systems that make up the so-called 'Next Internet', resulting from the convergence of cloud computing, big data analytics, and the Internet of Things (Mosco 2017, 4).

For Gandy in particular corporations and the state conduct surveillance: 'The panoptic sort is a technology that has been designed and is being continually revised to serve the interests of decision makers within the government and the corporate bureaucracies' (Gandy 1993, 15 and 95). Hence, according to Fuchs (2012, 141), privacy is permanently undermined by corporate and state surveillance into human lives. Fuch's main focus is on how capitalism protects privacy for the rich and companies, but at the same time legitimates violations of consumers' and citizens' privacy. He therefore proposes a so-called socialist concept of privacy, as opposed to the liberal individual notion of privacy. The aim of this critical perspective on privacy is to strengthen the protection of consumers and citizens from corporate and governmental forms of surveillance. Fuchs (2012) has applied the study of corporate or economic surveillance extensively to social media, by for example analysing the political economy of Facebook. He starts from Toffler's notion of 'prosumer', referring to a new type of consumption where formerly distinct roles of producers and consumers are increasingly merging together (Toffler 1980, 267). According to Fuchs, this notion describes important changes in media structures and practices which need to be adopted in critical studies. For this he indicates how 'prosumer commodity' is used for capital accumulation and how – by those means – users of social media are exploited by selling them as commodities to advertisers. 'Internet prosumer commodity' consists of user-generated content, transaction data about browsing and communication, personal data, and virtual advertising space and time. This is different from the traditional audience commodity for traditional mass media, as now users are not only watching but also producing content as well as being tracked. In that way Fuchs' perspective brings together the surveillance model and the capture model, discussed earlier.

9.5 New challenges and topical discussions

This section draws the lines set out in the previous section further to topical questions in MCS research. New forms of media and communication in the form of digital intermediaries and online platforms are becoming deeply entrenched in social and public activities. In that way they are becoming essential for the realization of fundamental public values associated with these activities. In addition to privacy and data protection, this concerns values like freedom of expression, diversity, transparency, and non-discrimination. 'Online platforms' are defined as socio-technical architectures that enable and steer interaction and communication between users through the collection, processing, and circulation of user data (Van Dijck, Poell, and de Waal, Forthcoming). These intermediaries can be general-purpose platforms for social (media) communication (e.g. Facebook, Twitter, etc.) or specific platforms in various sectors like mobility, health, education, and housing (e.g. Uber, PatientsLikeMe, edX, Airbnb, etc.).

The previous section described how Van Dijck and other media scholars studied how mediation of communication by social media platforms affected social norms regarding sociality (Van Dijck 2013). As platform ecosystems advanced beyond sociality into other domains (transportation, education, hospitality, etc.) MCS scholars moved their focus to the mechanisms of 'platformization' and the effects of platformization on societal norms and values (Van Dijck, Poell, and de Waal, Forthcoming). MCS research scrutinizes how social media platforms are not merely neutral conduits or facilitators of social interaction, but rather communication intermediaries which reflect certain perceptions of social interaction and pursue ulterior profit-driven motives. The same goes for other types of online platforms: they 'govern' human interaction through explicit provisions such as the terms of service and (privacy) policies, as well as through implicit norms embedded in user interfaces and inscrutable algorithmic sorting (Gillespie 2017). Despite platform owners' rhetoric about openness, impartiality, and transparency (Gillespie 2010), platformization may be causing societal structures to become more opaque, since the socio-technical processes that orchestrate human interaction on platforms evade democratic control (Sandvig 2015). Moreover, as platforms increasingly feel the need to 'police' user-generated content – both to comply with legislation and to promote a friendly atmosphere for users – they become invisible arbiters of public values (Gillespie 2017).

The particular role online platforms play in society is closely related to long-term changes in mediated communication (Schroeder 2018). Traditionally, mass media, like radio and newspapers, control communication over their channels in line with certain professional standards and values, related to objectivity, pluralism, commercial motives etc. This type of 'curating' is reflected in the selection and positioning of news items, scheduling, advertising placement, must-carry obligations, and other activities. Both public broadcasting media and commercial media can be held accountable through media regulation and government supervision for how they curate and edit their content. The traditional media landscape also includes interpersonal media, such as post and telephone, which typically only 'facilitate' communication between sender and receiver. Interpersonal media have no right to interfere with communication, except on specific legal grounds (e.g. criminal investigation).

Based on digitization and convergence of media, we have now entered the age of 'mass self-communication' (Castells 2009). On the one hand, this is 'mass communication' because users can potentially reach the global Internet audience through online platforms. On the other hand, this is 'self-communication' because the content can more easily be self-generated, the potential receiver(s)'s definition is self-directed and the content retrieval is self-selected. Hence the borders between mass and interpersonal communication have blurred, which has led to the gradual folding of the roles of 'curator' and 'facilitator'.

Focal points of MCS research related to privacy and data protection currently and for the near future are therefore the 'platformization' of society, the (lack of) transparency of automated decision-making, and the (privacy) effects of mass self-communication on the development of technology, people and society.

9.6 Conclusion

Given the gradual transition of the main focus of research from social media to online platforms, the MCS perspective generates a uniquely interdisciplinary insight into how these digital media and society mutually articulate each other. This is particularly relevant as these media and technologies are penetrating in all fibres of society, from social communication to domains like health, education, mobility, urban life, and smart cities. Consequently, the need to investigate and address fundamental public values like privacy and data protection from a media and communications perspective will only increase. Media are thereby interpreted in a broad sense, namely as technological tools that mediate the interaction between people. For example – related to mobility and location privacy – autonomous vehicles also become part of this discussion, given their darker side as potential 'panopticons on wheels', surveilling the environment and the movements people make. The latter will even intensify when human-driven cars are gradually banned on safety grounds, with passengers losing the freedom to go anywhere they choose and thereby opening the door to segregation and discrimination (The Economist 2018).

These new and upcoming issues can best be addressed by taking the threefold perspective that structured our chapter. First we need to build in technological affordances that avoid unregulated surveillance and safeguard privacy and data protection, by applying 'privacy by design' in the way that it has a meaningful impact. In addition to mitigating the exposure towards privacy invasions, privacy enhancing tools (PETs) should further be developed and supported in order to help citizens. Ideally we aim for so-called 'empowerment by design', which refers to building infrastructures and systems (like the Internet of Things, participatory sensing platforms, open data systems, government databases, etc.) in such a way that citizens and activists have agency to safeguard and strengthen public values (like privacy) in society (Pierson and Milan 2017).

For people to better cope with these new challenges of digital media and their threats for privacy, additional efforts on the level of data literacy are required. As mentioned earlier, this means that media literacy needs to be complemented with awareness, attitudes, and capabilities with respect to technologies, data, and privacy. This is particularly urgent as we see how people are resigned to giving up their data, which is even more prominent among more knowledgeable parts of the public (Turow, Hennessy, and Draper 2015). The growing power imbalance between those who generate data and those who convert these data into value also creates a need to open up new forms of digital literacies, such as privacy literacies, algorithmic literacies, and code literacies. Kennedy, Poell and van Dijck (2015, 5) state that it is only through these types of literacy that citizens can act with agency in the face of data power. These efforts can further be strengthened by initiatives of collective action and by involving civil society organizations on digital rights.

However, the attention for data literacy does not mean that the burden of data privacy should be for the largest part on the shoulders of individuals. It is crucial that also public authorities and industry take their part of the responsibility. The European General Data Protection Regulation (GDPR) at least sets forth a promising international legal framework for better protecting personal data and enforcing privacy. Yet, it remains to be seen what this legislation's real impact will be on society and economy. The industry has an important role to play. Besides of course complying with privacy regulations, companies also need to create conditions so that individual users can easily interact in a privacy-friendly way, by for example creating clear and accessible privacy settings. Creating awareness and informing and educating users about data privacy also helps in empowering users.

Finally, industry should also anticipate by way of 'prospective design responsibility' (Thompson 2014). The latter notion refers to shifting the perspective from responsibility for outcomes to responsibility for privacy-preserving design of the infrastructures and services.

Allocating responsibilities based on a dynamic interaction between the different stakeholders as described enables a situation of 'cooperative responsibility' (Helberger, Pierson, and Poell 2018). This multi-stakeholder approach would offer a realistic, balanced, and fair prospect of safeguarding privacy and data protection in the context of media and communications.

Further reading

Internet and digital media

Castells, Manuel. (2009). *Communication Power.* Oxford: Oxford University Press, 608.

Couldry, Nick. (2012). *Media, Society, World: Social Theory and Digital Media Practice.* Cambridge: Polity Press, 242.

Lievrouw, Leah A. & Sonia Livingstone. (eds.). (2002). *The Handbook of New Media: Social Shaping and Consequences of ICTs.* London: Sage, xxiv, 564.

Mansell, Robin. (2012). *Imagining the Internet: Communication, Innovation, and Governance.* Oxford: Oxford University Press, 289.

Media and technologies

Berker, Thomas, Maren Hartmann, Yves Punie, and Katie Ward. (2005). *Domestication of Media and Technology.* Berkshire: Open University Press, 255.

Feenberg, Andrew. (1999). *Questioning Technology.* London: Routledge, 243.

Gillespie, Tarleton, Pablo J. Boczkowski, and Kirsten A. Foot. (eds.). (2014). *Media Technologies: Essays on Communication, Materiality, and Society.* Cambridge: MIT Press, 325.

Morozov, Evgeny. (2013). *To Save Everything, Click Here: the Folly of Technological Solutionism* (First edition). New York: Public Affairs, 415.

Oudshoorn, Nelly and Trevor J. Pinch, (2003). *How Users Matter: the Co-construction of Users and Technologies.* Cambridge, MA/ London: MIT Press, vii, 340.

Media, privacy, and everyday life

boyd, danah. (2014). *It's Complicated: the Social Lives of Networked Teens.* New Haven: Yale University Press.

Cohen, Julie. (2012). *Configuring the Networked Self: Law, Code, and the Play of Everyday Practice.* New Haven: Yale University Press, 352.

McStay, Andrew. (2017). *Privacy and the Media.* London: Sage.

Nissenbaum, Helen. (2010). *Privacy in Context: Technology, Policy, and the Integrity of Social Life.* Stanford: Stanford Law Books, 288.

Petronio, Sandra. (2002). *Boundaries of Privacy: Dialectics of Disclosure.* Albany: State University of New York Press, 225.

Social media

Fuchs, Christian. (2014). *Social Media: a Critical Introduction.* City: Publisher, 293.

Pierson, Jo. (2012). 'Online Privacy in Social Media: a Conceptual Exploration of Empowerment and Vulnerability'. *Communications & Strategies* (Digiworld Economic Journal) 4(88), 99-120.

Pierson, Jo. (forthcoming). 'Media and Communication Studies, Privacy and Public Values: Future Challenges' in Fuster, Gloria González, van Brakel, Rosamunde and De Hert, Paul (eds.), *Research Handbook on Privacy and Data Protection Law: Values, Norms and Global Politics*. Cheltenham: Edward Elgar Publishing.

Van Dijck, José. (2013). *The Culture of Connectivity: a Critical History of Social Media*. Oxford: Oxford University Press, 228.

Online platforms

Couldry, Nick. (2014). 'Inaugural: A Necessary Disenchantment: Myth, Agency and Injustice in a Digital World'. *The Sociological Review* 62, 880-897.

Gillespie, Tarleton. (2010). 'The Politics of Platforms'. *New Media & Society* 12(3), 347-364.

Van Dijck, José, Thomas Poell, and Martijn de Waal. (Forthcoming). *The Platform Society: Public Values in a Connective World.* Oxford: Oxford University Press.

Big data, algorithms, and smart media

Beer, David. (2009). 'Power through the Algorithm? Participatory Web Cultures and the Technological Unconscious'. *New Media & Society* 11(6), 985-1002.

boyd, danah and Kate Crawford. (2012). 'Critical Questions for Big Data'. *Information, Communication & Society* 15(5), 662-679.

Barocas, Solon and Andrew D. Selbst. (2016). 'Big Data's Disparate Impact'. *California Law Review* 104, 671-732.

Bucher, Taina. (2012). 'Want to Be on the Top? Algorithmic Power and the Threat of Invisibility on Facebook'. *New Media & Society* 14(7), 1164-1180.

Couldry, Nick & Alison Powell. (2014). 'Big Data from the bottom up'. *Big Data & Society* 1(2), 1-5.

Crawford, Kate. (2016). 'Can an Algorithm be Agonistic? Ten Scenes from Life in Calculated Publics'. *Science, Technology & Human Values* 41(1), 77-92.

Mosco, Vincent. (2017). *Becoming Digital: Towards a Post-Internet Society* (1st ed.). Bingley: Emerald Publishing Limited, 227.

Pariser, Eli. (2011). *The Filter Bubble: What the Internet is Hiding from You.* New York: Penguin Press, 294.

Pasquale, Frank. (2015). *The Black Box Society: The Secret Algorithms That Control Money and Information*. Cambridge, MA: Harvard University Press, 320.

Sandvig, Christian, Kevin Hamilton, Karrie Karahalios, and Cedric Langbort. (2014). 'An Algorithm Audit' in Seeta Peña Gangadharan (ed.), *Data and Discrimination: Collected Essays*, 6-10. Washington: New America Foundation.

References

Agre, Philip E. (1994). 'Surveillance and Capture: Two Models of Privacy'. *The Information Society* 10, 101-127.

Altman, Irwin. (1975). *The Environment and Social Behavior: Privacy, Personal Space, Territory, Crowding.* Monterey: Brooks/Cole Pub. Co., 256.

Altman, Irwin. (1977). 'Privacy Regulation: Culturally Universal or Culturally Specific?' *Journal of Social Issues* 33(3), 66-84.

Baran, Stanley J. and Dennis K. Davis. (2015). *Mass Communication Theory: Foundations, Ferment, and Future* (Seventh edition). Stamford: Cengage Learning, 408.

Bauman, Zygmunt. (2000). *Liquid Modernity*. Cambridge: Publisher, 228.

Bauman, Zygmunt and David Lyon. (2013). *Liquid Surveillance*. Cambridge: Polity Press, 182.

Blomberg, Jeanette and Helena Karasti. (2013). 'Reflections on 25 Years of Ethnography in CSCW'. *Computer Supported Cooperative Work* 22(4-6), 373-423.

boyd, danah. (2010). 'Social Network Sites as Networked Publics: Affordances, Dynamics, and Implications' in Zizi Papacharissi (ed.), *Networked Self: Identity, Community, and Culture on Social Network Sites*. City: Publisher, 39-58.

boyd, danah. (2014). 'What Does the Facebook Experiment Teach Us?' Message posted to zephoria.org (1 July 2014, www.zephoria.org/thoughts/archives/2014/07/01/facebook-experiment.html), Last accessed on 27 August 2015.

Botsman, Rachel. (2017). 'Big Data Meets Big Brother as China Moves to Rate Its Citizens'. *Wired* (21 October 2017). https://www.wired.co.uk/article/chinese-government-social-credit-score-privacy-invasion

Castells, Manuel. (2009). *Communication Power*. Oxford: Oxford University Press, 608.

Clarke, Roger. (1988). 'Information Technology and Dataveillance'. *Communications of the ACM* 31(5), 498-512.

Couldry, Nick. (2011). 'The Necessary Future of the Audience… and How to Research It' in V. Nightingale (ed.), *The Handbook of Media Audiences*. Malden: Wiley-Blackwell, 213-229.

Couldry, Nick. (2014). 'Inaugural: A Necessary Disenchantment: Myth, Agency and Injustice in a Digital World', *The Sociological Review* 62, 880-897.

Couldry, Nick and Joseph Turow. (2014). 'Advertising, Big data, and the Clearance of the Public Realm: Marketers' New Approaches to the Content Subsidy', *International Journal of Communication* 8, 1710-1726.

De Wolf, Ralf. (2015). *Privacy in a Networked Life: Collective Privacy Practices, Audience Management and Visualizations in Social Network Sites*. PhD thesis Vrije Universiteit Brussel, 263.

De Wolf, Ralf, Rob Heyman, and Jo Pierson. (2012). 'Privacy by Design through Social Requirements Analysis of Social Network Sites from a User Perspective'. Paper presented at the 5th International Conference CPDP (Computers, Privacy and Data Protection), 25-27 January 2012, Brussel (Belgium), 19.

De Wolf, Ralf and Jo Pierson. (2014). 'Who's My Audience Again? Understanding Audience Management Strategies for Designing Privacy Management Technologies'. *Telematics & Informatics* 31(4), 607-616.

Deuze, Mark. (2012). *Media Life*. Cambridge: Polity, 305.

Diaz, Claudia and Seda Gürses. (2012). 'Understanding the Landscape of Privacy Technologies'. Presentation at Information Security Summit, 20 November- 6December, Hong Kong.

Espeland, Wendy Nelson and Mitchell L.Stevens. (1998). 'Commensuration as a Social Process'. *Annual Review of Sociology* 24, 313-343.

Floridi, Luciano (ed.). (2015). *The Onlife Manifesto: Being Human in a Hyperconnected Era*. Cham: Springer Open, 264.

Fuchs, Christian. (2012). 'The Political Economy of Privacy on Facebook'. *Television & New Media* 13(2), 139-159.

Gandy, Oscar H. (1993). *The Panoptic Sort: a Political Economy of Personal Information*. Boulder: Westview, 283.

Gibbs, Samuel. (2018). 'Facebook ordered to stop collecting user data by Belgian court', The Guardian, 16 February.

Gillespie, Tarleton. (2010). 'The Politics of "Platforms"'. *New Media and Society* 12(3), 347-364.

Gillespie, Tarleton. (2017). 'Governance of and by Platforms' in J. Burgess, T. Poell, and A. Marwick (eds.). *The SAGE Handbook of Social Media* Los Angeles: Sage.

Gürses, Seda. (2010). *Multilateral Privacy Requirements Analysis in Online Social Networks.* Leuven Publisher.

Gürses, Seda and Jose M. del Alamo. (2016). 'Privacy Engineering: Shaping an Emerging Field of Research and Practice' in *IEEE Symposium on Security and Privacy.* City: Publisher, 2-8.

Helberger, Natali, Jo Pierson, and Thomas Poell. (2018). 'Governing Online Platforms: from Contested to Cooperative Responsibility' *The Information Society* 34(1), 1-14.

Heyman, Rob and Jo Pierson. (2015). 'Social Media, Delinguistification and Colonization of Lifeworld: Changing Faces of Facebook'. *Social Media + Society,* 1(2), 11.

Karahasanović, Amela, Petter Bae Brandtzæg, Jan Heim, Marika Lüders, Lotte Vermeir, Jo Pierson, Bram Lievens, Jeroen Vanattenhoven, and Greet Jans. (2009) 'Co-Creation and User-Generated Content – Elderly People's User Requirements'. *Computers in Human Behavior,* 25 (3), 655-678.

Kennedy, Helen, Thomas Poell, and José van Dijck. (2015). 'Data and Agency'. *Big Data & Society* July-December, 1-7.

Lie, Merete and Knut H. Sorensen. (1996). 'Making Technology Our Own? Domesticating Technology into Everyday Life' in M. Lie and K.H. Sorensen (eds.), *Making Technology Our Own? Domesticating Technology into Everyday Life.* Oslo: Scandinavian University Press, 1-30.

Lievrouw, Leah A. and Sonia Livingstone. (2002). 'Introduction: the Social Shaping and Consequences of ICTs' in L.A. Lievrouw and S. Livingstone (eds.), *The Handbook of New Media.* London: Sage, 1-15.

Livingstone, Sonia. (2004). 'Media Literacy and the Challenge of New Information and Communication Technologies'. *The Communication Review* 7(1), 3-14.

Loisen, Jan and Stijn Joye. (2018). *On Media and Communication: An Introduction to Communication Sciences.* Leuven: Acco, 457.

Lyon, David. (1994). *The Electronic Eye: the Rise of Surveillance Society.* Oxford: Polity Press.

Lyon, David. (2003). 'Surveillance as Social Sorting: Computer Codes and Mobile Bodies' in D. Lyon (ed.), *Surveillance as Social Sorting.* New York: Routledge, 13-20.

Lyon, David. (2007). *Surveillance Studies: an Overview.* Cambridge: Polity Press.

Mansell, Robin. (2004). 'Political Economy, Power and New Media', *New Media & Society* 6(1), 96-105.

Mansell, Robin. (2012). *Imagining the Internet: Communication, Innovation, and Governance.* Oxford: Oxford University Press, 289.

Mansell, Robin. (2016). 'Recognizing 'Ourselves' in Media and Communications Research'. *The International Communication Gazette* 78(7), 716-721.

Margulis, Stephen T. (2011). 'Three Theories of Privacy: An Overview' in S. Trepte and L. Reinecke (eds.), *Privacy Online: Perspectives on Privacy and Self-disclosure in the Social Web* (Chapter 2). Heidelberg: Springer, 9-17.

Marwick, Alice and danah boyd. (2014). 'Networked Privacy: How Teenagers Negotiate Context in Social Media' *New Media & Society* 16(7), 1051-1067.

Marx, Garty T. (1985). 'The Surveillance Society: the Threat of 1984-style Techniques', *The Futurist* (June), 21-26.

Mayer-Schönberger, Viktor and Kenneth Cukier. (2013). *Big Data: a Revolution that Will Transform How We Live, Work and Think.* London: John Murray.

Morozov, Evgeny. (2013). *To Save Everything, Click Here: the Folly of Technological Solutionism* (First edition). New York: PublicAffairs, 415.

Mosco, Vincent. (2017). *Becoming Digital: Towards a Post-Internet Society* (1st ed.). Bingley: Emerald Publishing Limited, 227.

Naughton, John. (2017). 'Why We Need a 21st-century Martin Luther to Challenge the Church of Tech'. The Guardian, 29 October.

Nissenbaum, Helen. (2004). 'Privacy as Contextual Integrity' *Washington Law Review* 79(1), 119-158.

Nissenbaum, Helen. (2015). 'Respecting Context to Protect Privacy: Why Meaning Matters'. *Science and Engineering Ethics*, 1-22.

Nissenbaum, Helen. (2016). 'At the Intersection of Ethics and Technology: Contextual Integrity and Other Values'. Presentation at Interdisciplinary Summerschool on Privacy, 12 July, Berg en Dal, the Netherlands.

Nissenbaum, Helen. (2010). *Privacy in Context: Technology, Policy, and the Integrity of Social life.* Stanford: Stanford Law Books, 288.

Orgad, Shani. (2007). 'The Interrelations between Online and Offline: Questions, Issues, and Implications' in R. Mansell, C. Avgerou, D. Quah, and R. Silverstone (eds.), *The Oxford Handbook of Information and Communication Technologies*. Oxford: Oxford University Press, 514-536.

Palen, Leysia and Paul Dourish. (2003). 'Unpacking 'privacy' for a networked world'. Presentation at SIGCHI conference on human factors in computing systems, organized by ACM, 5-10 April, Fort Lauderdale.

Petronio, Sandra. (2002). *Boundaries of Privacy: Dialectics of Disclosure*. Albany: State University of New York Press, 225.

Petronio, Sandra. (2010). 'Communication Privacy Management Theory: What do We Know about Family Privacy Regulation?' *Journal of Family Theory & Review* 2(3), 175-196.

Pierson, Jo. (2013). 'Conclusie: sociale media en empowerment' in R. Heyman, A. Daems, D. Baelden, and J. Pierson (eds.), *Hier vloekt men niet, Facebook ziet alles – Sociale netwerken ontrafeld.* Leuven: Davidsfonds Uitgeverij, 131-138.

Pierson, Jo. (2014). 'Interdisciplinary Perspective on Social Media, Privacy and Empowerment: the Role of Media and Communication Studies in Technological Privacy Research' in K. O'Hara, S.L. David, D. De Roure, and M.H.C. Nguyen (eds.), *Digital Enlightenment Forum Yearbook 2014 – Social Networks and Social Machines, Surveillance and Empowerment.* Amsterdam: IOS Press, 265-274.

Pierson, Jo. (forthcoming). 'Media and Communication Studies, Privacy and Public Values: Future Challenges' in Fuster, Gloria González, Rosamunde van Brakel, and Paul De Hert (eds.), *Research Handbook on Privacy and Data Protection Law: Values, Norms and Global Politics*. Cheltenham: Edward Elgar Publishing.

Pierson, Jo and Stefania Milan. (2017). 'Empowerment by Design: Configuring the Agency of Citizens and Activists in Digital Infrastructure'. Presentation at IAMCR Conference 'Transforming Culture, Politics & Communication: New Media, New Territories, New Discourses 16-20 July', 17 July, Cartagena, Colombia.

Röhle, Theo. (2005). 'Power, Reason, Closure: Critical Perspectives on New Media Theory', *New Media & Society* 7(3), 403-422.

Sandvig, Christian. (2015). 'The Social Industry'. *Social Media + Society* 1, 1-4.

Sasse, Martina Angela and Charles C. Palmer. (2014). 'Protecting You', *IEEE Security & Privacy* 12(1), 11-13.

Schroeder, Ralph. (2018). 'Towards a Theory of Digital Media' *Information, Communication & Society* 21(3), 323-339.

Silverstone, Roger. (2006). *Media and Morality: on the Rise of the Mediapolis*. Cambridge: Polity press, 215.

Silverstone, Roger and Leslie Haddon. (1996). 'Design and Domestication of Information and Communication Technologies: Technical Change and Everyday Life' in R. Mansell and R. Silverstone (eds.), *Communication by Design: the Politics of Information and Communication Technologies*. Oxford: Oxford University Press, 44-74.

Silverstone, Roger, Eric Hirsch, and David Morley. (1990). 'Information and Communication Technologies and the Moral Economy of the Household' in A.-J. Berg (ed.), *Technology and Everyday Life: Trajectories and Transformations*. Trondheim: University of Trondheim, 13-46.

Smit, Edith, Guda van Noort, and Hilde A.M Voorveld. (2014). 'Understanding Online Behavioural Advertising: User Knowledge, Privacy Concerns and Online Coping Behaviour in Europe'. *Computers in Human Behavior* 32, 15-22.

The Economist. (2018). 'Who Is Behind the Wheel? Self-driving Cars Offer Huge Benefits – But Have a Dark Side', 3-9 March, 426, 14.

The Economist. (2018). 'Digital Privacy – The Facebook Scandal Could Change Politics as well as the Internet', 22 March, 426, 39-40.

Thompson, Dennis F. (2014). 'Responsibility for Failures of Government: The Problem of Many Hands' *The American Review of Public Administration* 44(3), 259-273.

Toffler, Alvin. (1980). *The Third Wave* (1st ed.). New York: Morrow, 544.

Turow, Joseph, Michael Hennessy, and Nora Draper. (2015). *The Tradeoff Fallacy: How Marketeers are Misreprenting American Consumers and Opening Them up to Exploitation*. Philadelphia: Annenberg School for Communication, 24.

Ur, Blase, Pedro G. Leon, Lorrie F. Cranor, Richard Shay, and Yang Wang. (2012). 'Smart, Useful, Scary, Creepy: Perceptions of Online Behavioral Advertising'. *Proceedings of the Eighth Symposium on Usable Privacy and Security*, ser. SOUPS'12. New York: ACM, 4:1-4:15.

Van Dijck, José. (2013). *The Culture of Connectivity: a Critical History of Social Media*. Oxford: Oxford University Press, 228.

Van Dijck, José, Thomas Poell, and Martijn de Waal. (Forthcoming). *The Platform Society: Public Values in a Connective World*. Oxford: Oxford University Press.

Warren, Samuel & Louis D. Brandeis. (1890). 'The Right to Privacy'. *Harvard Law Review* 4(5), 69-83.

Wellman, Barry and Caroline A. Haythornthwaite. (2002). *The Internet in Everyday Life*. Oxford: Blackwell, xxxi, 588.

Westin, Alan F. (1967). *Privacy and Freedom* (1st ed.). New York: Atheneum, 487.

Winner, Langdon. (1980). 'Do Artifacts Have Politics?' *Daedalus* 109, 121-136.

Diversity and Accountability in Data-Rich Markets

Viktor Mayer-Schönberger

For millennia the market has been an amazing mechanism for humans to coordinate with each other, without them having to share the same goals. How well the market coordinates, depends not just on the liquidity of the market, but crucially also on the ability to find a suitable match. This requires easy discoverability, comparability, and an effective way to translate such information into a transaction decision. Unfortunately, sharing information with everyone (or at least many) on the market has long been too cumbersome and costly.

Price acted as a suitable 'workaround': according to classical economics, all our preferences get condensed into price, greatly reducing the amount of information that needs to be communicated. Price also enables straightforward comparison, easing decision-making. In short, price (and hence money) have long greased the market.

Price, however, is not without weaknesses as an informer on markets. By condensing all available preferences into price, a lot of detail gets lost, leading to erroneous decisions, and thus an inefficient market. Fundamental human biases hinder our ability to translate information gleaned through price into decisions. The result is a market that is far from matching buyers and sellers well. But it's also a market, in which individuals' detailed preferences often stayed private, because they are only shared in the aggregate.

In contrast to these money-based markets, recently a new breed of markets has arisen. These are markets teeming with information about preferences and product (or service) qualities, as well as buyer (or seller) behaviour. These data-rich markets, when combined with the appropriate technical tools (data ontologies, matching algorithms, and especially data-driven machine learning systems for decision assistance), offer matches that are vastly superior compared with conventional markets.

Often today, we read about how consumers lose control over their data online, and thus their preferences become transparent to sellers. But often overlooked has been the opposite dynamic: how online markets have made sellers and their products and services more transparent to consumers, leading to better matches through consumer empowerment. Better matches mean more value for the money consumers spend, creating significant additional consumer welfare, and an improved, more efficient, and more sustainable economy.

The flip side of these advances is that more detailed consumer data has to be shared for the superior matching to work. Transparency translates into transactions that are better for both sides, but transparency also makes consumers more predictable; they feel more vulnerable as a result.

The fundamental policy issue is how to help consumers succeed being empowered in such data-rich markets.

Many have recently argued for a stronger empowerment of individuals, either through better enforcement of (and education about) individual rights, or even through the enactment of a property right in personal data. Such suggestions appeal to our idealistic representation of the empowered, self-determined individual. But they are far removed from actual reality, in which individuals routinely trade away their rights. With regards to a property right, information economists have cautioned that this may lead less to an empowerment than a commercialization of personal data, or – equally troubling – not have much practical impact at all.

Moreover, the empowerment of the individual regarding her personal data will do little to address the fundamental structural challenge we face in data-rich markets: the decision collusion.

Today's data-rich markets are mostly online (because of very low transaction costs of information flows online) and run by private companies. Amazon operates a data-rich market, and so do Google, Apple, Facebook, Alibaba, etc., but also niche players such as Airbnb or Spotify. Consumers prefer such marketplaces, because of the superior matching experience compared to most conventional markets. But for the superior matching to happen, they have to share information not only with other market participants, but also with the market provider, i.e. the company running the market. The market provider is the central conduit and knows everything about everyone on the market that can be gleaned online. This is a tremendous (and potentially troubling) concentration of power. But it gets worse. The market provider also operates the data-driven decision assistants – whether they are called Siri or Alexa or are less anthropomorphic recommendation engines. In other words: the super powerful central information intermediary also operates the leading decision assistant.

Some may see this not as immediately troubling. It's in the interest of the market provider to offer excellent matches to its customers. After all, if the decision assistant is biased to the detriment of the consumers, consumers will no longer use the market place. This is true, but whether or not a match is optimal is not obvious to consumers. Market providers could thus avoid detection by either only selectively altering the decision suggestions, or by doing so in small ways that most consumers fail to notice.

Worse, even if a market provider does not intend to shape decision-making on its market, the fact that it operates the leading or even the only decision assistant on the platform will create a 'single point of failure', a single weakness that could affect everyone's decisions and cripple the market. It would undo decentralized decision-making, the very quality that makes the market such a resilient system of coordination to begin with.

Put in positive terms, the task in data-rich markets is to create a diverse ecosystem of decision assistants that are *independent* of the market provider and that are fully accountable to the consumers as their principals. Hardly any of the leading online markets today permit for that. But without such diversity and accountability, data-rich markets suffer from a structural defect that makes them and their users singularly vulnerable.

To avoid such a fundamental weakness and to guarantee diversity and accountability requires most likely a combination of regulatory and technical elements.

On the technical side, sharing preference information to decision assistants must be possible easily and at low cost, but also in a trustful manner. Distributed digital ledger technology, such as blockchain, could provide some of the necessary technical elements, but it's no panacea. Other technologies, such as homomorphic encryption or differential privacy, may be part of the mix as well.

On the regulatory side, decision assistants would need to be subject to tough accountability standards, including perhaps ex ante certification and strict liability. This is necessary because these decision assistants on data-rich markets would 'see' lots of personal data, and work with other similar digital assistants to find appropriate matches, but do so in a way that's largely non-transparent to the consumer (as the principal). Here, transparency is not what consumers want or need – they need a helpful (digital) intermediary that they can trust because it works *for them*. To an extent, it's like consumers needing food safety experts to work *for them* without the need to be informed of all the details of their work.

This puts the onus on the user of personal data – the digital decision assistants first and foremost, but not necessarily just them. This will entail additional efforts (and thus cost) by the providers of (digital) decision assistance. But this cost is justified, because in a data-rich market with diverse decision assistants, these providers of assistance (rather than the market platform providers) create the value-add, and thus will reap a part of it.

Much remains to be done to put the elements of diversity and accountability in place, but as markets transform, now is the time to act.

Over the coming years, we'll shift from conventional markets based on price as the lead informer to data-rich markets. Because they offer significant advantages over conventional markets, the shift will only accelerate.

But as this happens, and we enjoy some additional consumer welfare – our digital dividend – we must be alert and aware of the need to shape such data-rich markets so that individuals feel both protected and treated well. Diversity and accountability are the two key elements through which we may be able to achieve this.

10. Privacy from a Communication Science Perspective

Sandra Petronio

10.1 Introduction

Privacy has emerged as a prominent topic of inquiry and has made individuals evermore mindful of its presence in everyday life. Privacy is a value in many cultures, contexts, academic disciplines, and within legal domains. Yet, privacy is often difficult to unpack. In each sphere, there are often a multiplicity of ways to think about privacy as is represented in the chapters found in this volume. This chapter discusses privacy from a communication science perspective focused on a theoretical understanding of privacy through the lens of Communication Privacy Management theory (CPM) (Petronio 1991, 2002, 2016; Petronio and Durham 2008). Communication privacy management research ascribes to using social science methodologies in juxtaposition to the historic disciplinary base of rhetoric and public speaking that reflects early developments of the communication discipline in the United States (National Communication Association 2018). Communication science typically uses both quantitative and qualitative methods rather than rhetorical or humanistic inquiry. However, the scope of communication studies in the United States sits side by side with the historic approaches focused on the power of communicating through public speaking in everyday life.

Communication between and among individuals spans many conditions and situations. Understanding the nature of human communication requires such issues as knowledge about societal issues, psychological conditions, message generation, the nature of conflict, the way groups function, organizational issues, persuasion, and communication that surrounds political issues.

The importance of having a broad-based understanding of how human communication functions allows for a deeper grasp of the complexities inherent in communication interactions. This approach provides a more complete way of understanding the nature of human communication. National and regional communication studies organizations in the United States, such as the National Communication Associations, encompass many areas of expertise and theoretical approaches. As an outcome, the discipline of communication, in which communication science resides, has many different emphases and theories. For example, foci such as planning theory (Berger 1997) that is devoted to understanding issues of goal attainment through communicative action and the theory of imagined interaction that focuses on such issues as rumination (Honeycutt 2003). By contrast, Communication Privacy Management theory emphasizes the communicative importance of understanding the *management* of private information. The development of Communication Privacy Management theory exemplifies the utility of incorporating a broad spectrum of knowledge to understand communicative issues such as privacy management.

10.2 Meaning and function of privacy management

Unpacking the paradoxical sense of privacy management seems an increasingly difficult task. People worry about their private information and are often not sure how to deal with decisions to tell or protect their information. With the Internet introducing endless examples of privacy breakdowns, it is often unclear whether privacy is still possible. Decisions to disclose or conceal private information can be tricky. For example, when an individual makes a choice to disclose a secret and that friend posts the information without asking the individual's permission, there is a relational price to pay. Communication Privacy Management theory and research offer a more informed understanding of the place privacy management has in today's world. The predictive nature of CPM allows individuals to recognize the behaviours of individuals when granting or denying access to their information. CPM identifies ways to learn how privacy management functions and ways individuals can effectively take charge of their private information.

Rather than focusing on typologies or assuming the definition of private information is the same for all people, CPM theory places an emphasis on information that individuals 'themselves' define as private. In addition, CPM theory focuses on how individuals make decisions to reveal or conceal, disclose or protect, and grant or deny access to their information when others are involved. Correspondingly, CPM theory takes into account decisions that recipients make regarding how they will or should care for the owner's information once told. While there are many other ways to consider privacy, CPM theory offers a targeted approach factoring in relationships with others through communicative actions. The way individuals manage and regulate these social and communicative encounters is the nucleus of CPM theory.

Communication privacy management theory, therefore, provides a roadmap that furthers the understanding of judgments made by both information owners and recipients of the owners' information in regulating disclosure and protection of private information with others (Petronio 2002). CPM theory is evidence-based, meaning that the theoretical concepts have been tested for viability and validity. In the years since the initial publication of CPM in 2002, researchers have used this theory in multiple contexts. For example, these contexts include applications in healthcare (e.g. Broekema and Weber 2017), in social networking (e.g. Child, Petronio, Agyeman-Budu, and Westermann 2011), exploring organizational domains (e.g. Gordon 2011) in family studies (e.g. Petronio 2010), within the context of personal and interpersonal relationships (e.g. Ngcongo 2016), conducting LGBT studies (e.g. McKenna-Buchanan 2015), in exploring educational issues (e.g. Sideliger, Nyeste, Madlock, Pollak, and Wilkinson 2015), social work issues (e.g. Cohen, Leichtentritt, and Volpin 2012), group interactions (e.g. Petronio, Jones, and Morr 2003), and in finance (e.g. Allen 2008).

In addition, researchers are using CPM theory and research in a number of countries, for example, Malaysia (Badrul, Williams, and Lundqvist 2016), Hong Kong (Hawk 2017), South Africa (Ngcongo 2016) Latvia (Peterson and Khalimzoda 2016), Kenya (Miller and Rubin 2007), Scandinavia (Heikkinen, Wickstrom, and Leino-Kilpi 2007), United States (Scharp and Steuber 2014), and Beligum (De Wolf, Willaert, and Pierson 2014).

While there is more to achieve, these data show interest in, if not promise of, continued expansion and applications of ideas about privacy management from a CPM perspective. In doing so, the volume and nature of CPM-based research has the potential to develop systematic ways to isolate communalities and differences across contexts and countries.

The next segment illustrates the fundamentals of Communication Privacy Management theory. Two main areas characterize the tenets of CPM theory. First, the underlying foundation of Communication Privacy Management theory is presented. Next, the operational system of CPM theory is explained.

10.2.1 Understanding the underlying foundation of communication privacy management theory

Communication Privacy Management theory is based on three major assumptions framing the CPM theoretical system of managing private information, they include: (1) dialectics, (2) centrality of others, (3) meaning of private information (Petronio 2002).

1. *Dialectical Tensions.* For CPM theory, the concept of dialectics is fundamental to the theory and overall privacy management system (Altman 1975; Altman, Vinsel, and Brown 1981; Petronio 2002). In other words, a

dialectical tension underpins choice making about revealing and concealing private information. These choices are typically between wanting to connect interpersonally through disclosing to others and at the same time being mindful about retaining a sense of autonomy to protecting the individual's private information. For example, Jane is recently diagnosed with breast cancer. She is new in town and she feels uncomfortable about talking to people she does not know well about this diagnosis. Because she needs some support, she takes a chance and confides in her neighbour. Nevertheless, she feels a sense of caution because she is not sure whether her neighbour will keep the information to herself. This example illustrates a push and pull of needing or wanting to disclose and at the same time worrying about whether the private information could be compromised. This type of tension prevails in most privacy situations to greater or lesser degrees. In this regard, the tensions become mindful when individuals have to make decisions about disclosing or protecting an individual's private information. Both protection and access of information enter the calculus reflective of managing private information.

2. *Centrality of Others.* Others play a significant role in understanding the mission of communication privacy management theory. Only when others are involved is there a need to *manage* private information. CPM theory considers communicative interactions among and between individuals and groups where private information is concerned. The way individuals make choices about how and to whom they communicate private information represents one half of the structure by which privacy management occurs. How recipients of private information handle the revealed private information is the second half of the equation. There are varying reasons individuals construct ways to control choice making about revealing and concealing, disclosure and protection, and granting or denying access to owners' private information. CPM theory and research provide tools to investigate what people do when faced with these issues (e.g. Kennedy-Lightsey, and Frisby 2016; Petronio 1991, 2002, 2010, 2013).
3. *Meaning of Private Information.* Over the years, many different ways of defining privacy have emerged, each portraying privacy and private information somewhat differently. After a considerable number of observations, CPM theory and research advocate that there is likely not one consistently held definition of private information. Individuals tend to define something as private information when there is a reason to do so, with the possibility that the nature of that information will change once the need for defining the information as private dissipates. These shifts in defining information as private can be held for a short time or a long time depending on the need

> to have the information defined as private. Variations in what constitutes private information likely occur across the lifespan with changing privacy needs. Often, what is considered private information to one person differs for others. The character of private information for individuals also shifts when circumstances change. For example, the nature of private information that adolescents hold may dramatically change as the adolescent moves into adulthood (Petronio 2002).

Because there is a changeability in how individuals define information as private, CPM theory argues there is a likelihood that privacy indicators can be detected through actions taken in communicative encounters (both verbal and non-verbal). For example, when a friend discloses information and states, 'don't tell anyone about this', the act of making this statement signals the information is likely to be private and expectations can be discerned regarding how the owner wants the recipient to treat the information (Petronio and Bantz 1991). Many types of behavioural enactments give rise to identifying a person's information as private. Obviously, more consideration is needed to work out these issues.

While there is a variability in how individuals define information as private, the underlying factor of vulnerability is, likely, inherent in the nature of the information deemed as private. In the cyber world, privacy risks play an important role because they highlight a level of presumed vulnerability (Ezhei and Ladani 2017). Although the notion of vulnerability is not a definition per se, there is a sensitivity to experiencing degrees of vulnerability. When that occurs, this state can trigger the need for exercising levels of ownership and control over information considered private by the owner. Degrees of vulnerability can range from high to low. When the vulnerability is high, there can be more intense management processes working to protect information perceived as private to the individual. This state suggests that privacy boundaries will likely be more impenetrable. When the sense of vulnerability is low, individuals are likely less concerned about sharing private information. The privacy boundaries surrounding the information are likely more permeable (Golden 2014; Millham and Atkin 2016).

10.2.2 Operations of communication privacy management

CPM theory proposes an operational structure that captures the scope of how this privacy management system functions (Petronio 2002, 2010, 2012, 2013). These highlight the component parts of the CPM private information management system that work in conjunction with each other forming a way to grasp how people make decisions about their private information and the underlying issues that drive the management process. Five aspects are discussed in more detail below, namely: (1) issues of private information ownership and privacy boundaries, (2) privacy rules and privacy control, (3) coordination operations when others are involved, (4) collective privacy boundaries reflecting multiple privacy relationships, (5) *privacy turbulence in privacy relationships.*

10.2.2.1 Private information ownership and privacy boundaries

Individuals believe they own their private information and it belongs to them individually. Individuals thus create a metaphoric 'boundary' to represent where individuals house this information. Individuals consider themselves rightful owners of their information. When information owners grant access to others, thereby sharing the information, the recipients become 'authorized co-owners' (e.g. Petronio and Durham 2008). Intended access by the owner transforms the privacy boundary from personal to collective thereby creating a 'privacy relationships' between the information owner and authorized co-owner or co-owners.

When individuals disclose or reveal their private information to selected others, the 'information owner' presumes the recipient understands the 'fiduciary' responsibilities for the information. Thus, the act of sharing prompts expectations regarding assumptions about how recipients should care for the owner's private information. The expectations that 'information owners' have about recipient responsibilities stem from a clear sense that they own and should have the right to control how the authorized co-owners handle their information. These expectations remain even after individuals reveal their private information. Thus, individuals believe they own rights to their private information and they feel justified in believing they should be the ones controlling their privacy, regardless of the fact others are privy to the information.

The notion of privacy boundaries also play a part in identifying the level of information access owners grant 'authorized co-owners' through identifying permeability levels of the privacy boundary walls. Thus, privacy boundary walls can be thick and at times impermeable when the information is restricted, such as secrets, and thin when the information is fluid or permeable where owners tend towards allowing more openness. Thick and thin walls reflect the anchor points on a continuum regarding access to private information. In addition, privacy boundaries are often layered. For example, families tend to have boundaries that regulate a family's collectively held information to others outside the family. These mark what members can or cannot discuss with people outside of the family (Petronio 2002). Families also have internal privacy boundaries that focus on how the family members are expected to treat the privacy regulation within the family (Petronio 2002).

10.2.2.2 Privacy rules and privacy control

Individuals control their private information by using 'privacy rules' (Petronio 2002). Privacy rules represent the engine of this privacy management system where choices about access, protection, and how authorized co-ownership is managed with others. Privacy rules are often constructed and reconstructed depending on the needs of the owner and the extent to which authorized co-owners adhere to the owner's expectations (see privacy turbulence). Once others are involved, successful and continued control post-access is accomplished through *coordinating and negotiating privacy rules* with authorized co-owners regarding third party access (Petronio 2002).

Individuals use two types of criteria to determine privacy rule selection, '*core criteria*' and '*catalyst criteria*' (Hammonds 2015; Petronio 2013).

1. *Core criteria* tend to remain stable and often work in the background when determining privacy rule usage. For example, '*privacy orientations*' as developed by CPM theory illustrates that groups, such as families, socialize members to use certain types of privacy rules consistent with the expectations of the family as a whole (Morr Serewicz, and Canary 2008; Petronio 2002). Core criteria include the notion of *cultural expectations* reflecting values of privacy anchored in cultural tendencies. Thus, all aspects of culture, including societal, ethnic, and regional, influence an individual's expectations about the nature of privacy and impacts choices of management (Yep 2000). In addition, a person may have *gendered tendencies* toward the kind of information that is held private. Gendered tendencies evolve out of the gender identity one adopts and the socialization one experiences (Manning 2015). In addition, *privacy orientations* to privacy management emerge when individuals are socialized to regulate their private information in a particular way, thereby becoming routinized in their decision-making about private information (Petronio 2013).
2. *Catalyst criteria* tend to be triggered when there is a needed change in the established privacy rules a person uses. Three examples can be provided:
   a. When '*motivational goals*' shift and change, privacy rules may need altering to accommodate a desired outcome. The goal of knowing more about someone a person finds attractive can for example

trigger a change in a person's privacy rules. The person may disclose more about him or herself than is typical, thereby modifying the privacy rules to accommodate the *motivational goals.* In unfamiliar circumstances, the individuals tend to weigh '*risks against benefits*' of granting access or concealing their private information. These cases trigger the need for new or different privacy rules.

b. '*Situational conditions*' also act as a catalyst for privacy rule changes. As the context or situation calls for different rule structures, it propels the need for modifications or alterations of the current set of rules typically used for privacy management. For example, divorced couples necessarily need to change the privacy rules they have established in their marriage (Miller 2009).
c. '*Emotional needs*' often change the privacy rules a person might typically use. For example, when a partner is uncharacteristically critical of a loved one, catching that person off-guard, the result may lead the loved one to unwillingly disclose hidden feelings that otherwise would not be discussed (e.g. Hesse and Raunscher 2013; McLaren and Steuber 2013).

As these circumstances show, there are catalysts that trigger the need for change in the privacy rules people use to make decisions about revealing or concealing private information where others are concerned.

10.2.2.3 Coordination operations when others are involved.
CPM research illustrates that when the information owner wishes to grant access to their private information, three types of operation work to coordinate privacy rules so that a smooth co-management of the owner's private information occurs (Petronio 2002). These operations include decisions about privacy boundary *linkages* reflecting the owners' selection of individuals as co-owners of their private information. Coordination regarding privacy boundary *permeability* determines how much the owner tells the recipient and how much the 'authorized co-owner' is able to tell others (see Liu and Fan 2015). Thus, the information owner sets parameters for the authorized co-owners regarding such issues as to who they may tell, if they can tell, and how much they are permitted to tell others. The third operation includes privacy boundary *judgments about co-ownership control.* This operation reflects the level of propriety rights the authorized co-owner is granted concerning independent rights to make judgments about the control over how the owner's private information is handled.

10.2.2.4 Collective privacy boundaries reflecting multiple privacy relationships.

CPM theory argues that authorized co-ownership status can include multiple people leading to jointly held and operated collective privacy boundaries (Petronio 2002). Within these co-constructed boundaries, all members are considered co-owners; however, there are different forms of coordination processes that determine privacy rules for access and protection. CPM argues at least three types of privacy boundary coordination that take place with collectively held privacy boundaries. First, these are situations where one or more of the members appropriate control over the collective information held in the privacy boundary. These are defined as *power privacy relationships*. For example, in sexual child abuse situations, the perpetrator manipulates control over the child to keep the incidents secret. CPM also argues that there are collective privacy boundaries that reflect *equitable privacy relationships*. In these cases, all parties in the collective share responsibility for the ownership and control over the private information through negotiating the privacy rules that are used in the group. Finally, there are collective privacy boundaries where there is a unified agreement among the co-owners where everyone understands that the information within the privacy boundary is co-owned by all members of the group. This type of coordination is defined as representing a *participative privacy relationship.* For example, joining Alcohol Anonymous, a self-help group for alcohol dependency, illustrates this type of privacy boundary coordination where the information the members share are well-kept secrets outside the group. Recently, this type of unified boundary coordination has been insightfully applied to issues found in social networking (De Wolf and Pierson 2014).

10.2.2.5 Privacy turbulence in privacy relationships

CPM theory argues that efforts to coordinate privacy rules can be problematic. Because individuals do not live in a perfect world, breakdowns in privacy management will likely occur. CPM theory identifies the notion of 'privacy turbulence' to reflect the assumption of change in privacy management that has the potential to ultimately sustain the privacy management system. There are two categories of privacy turbulence, 'privacy miscalculations' and 'privacy transgressions'. 'Privacy miscalculations' reflect unintentional mishaps that occur in privacy management. For example, information owner's privacy rule choices are left unsaid at times. In so doing, the co-owner is in the dark about what privacy rules the owner wants him or her to use. Consequently, the authorized co-owner may second-guess which rules seem acceptable triggering a potential for miscalculating the privacy rules the owner expects (Hewes and Graham 1989). When the choice of privacy rules is problematic, there is a potential for awkward interactions and possible challenges to the privacy relationship in the future. Regardless of how the authorized co-owner deals with these ambiguities, the lack of 'privacy rule coordination' can lead to mistakes and misunderstandings for both the recipient and the information owner.

While 'privacy miscalculations' have challenges, CPM theory also points out that there are incidents of 'privacy transgressions' that erupt (Petronio 2002). These violations are more serious in nature. Instances of 'privacy transgressions' further increase the complexity of managing private information. For example, the notion of betrayal reflects a circumstance where the actions taken are deliberate and are frequently complex. President Clinton's affair with Monica Lewinsky illustrates situations where privacy transgressions took place. A newspaper account indicated that one of Lewinsky friends, Linda Tripp, betrayed her confidence and revealed damaging information (Petronio 2002). Betrayals such as this are often a surprise to the aggrieved individual. 'Trust credit points' are lost and are difficult to regain. Consequently, the privacy relationships are significantly compromised and thorny to overcome.

CPM theory points out that the *ramifications* for the information owners in these turbulent incidences is often perplexing and problematic. However, the authorized co-owners can also experience turbulence when they encounter situations that become difficult.

There is a wide berth of reactions to receiving someone's private information. Among them are incidents where confidants are reluctant to accept the burden of knowing a person's private information. The notion of a 'reluctant confidant', as identified in CPM theory, speaks to the need for better understanding the role of authorized co-owners (Petronio 2002; Petronio and Reierson 2009). Receiving unwanted private information, whether the information owner is a relative, friend, or stranger can negatively affect a privacy relationship. Being asked to keep confidences when an individual knows others might benefit from having the information or encountering situations where knowing creates a dilemma because the confidant finds out information that could negatively affect a relative can be difficult to manage (McBride and Bergen 2008; Petronio 1991, 2013).

As these issues illustrate, privacy turbulence disrupts the privacy management system. However, these disruptions call into question the viability of the privacy rules used in these circumstances. Discovering that the current privacy rules do not address the needs of the owner or compromise co-ownership brings about the impetus to make changes in the management system. The recognition of the need for recalibrating privacy rules allows the privacy management system to sustain itself and provides a viable way to sustain control and ownership of private information (Child and Petronio 2015; Child and Petronio 2011; Child, Petronio, Agyeman-Budu, and Westermann 2011).

As this discussion points out, the framework of CPM has many facets as indicated in this segment of the chapter. The foundation of the theory gives depth and breadth of understanding to the assumptions and the platform upon which this theory stands. The operations of CPM theory identify core apparatus with which individuals gain insight into this type of communicative actions. Overall, these fundamentals of CPM guide us toward a more comprehensive understanding of how privacy management works in everyday life.

10.3 Classic texts and authors

The evolution of CPM theory development started with testing the viability of concepts and working to validate the ideas about managing private information. While CPM theory is primarily grounded in the discipline of communication, several lines of inquiry in communication science and other social science disciplines made a significant contribution to understanding issues of privacy management. Namely, authors in the discipline of psychology have influenced the development of many ideas in CPM theory. The history starts with Jourard (1958) introducing the concept of self-disclosure and a few years later publishing a book entitled *The Transparent Self* (1964) where he expanded his notion of self-disclosure. Interestingly, his introduction of this concept brought to the forefront the utility of communicating about one's self and significantly informed the nature of decision-making regarding revealing and concealing aspects of one's self to others. As mentioned in the preface of Chelune's (1979) edited book, 'self-disclosure has come a long way in its relatively short history' (p. ix). Although Jourard passed away early in his life, the legacy of his quest to understand self-disclosure has a rich history and continues to grow with many branches expanding the scope of understanding.

A bridge was erected between self-disclosure issues and privacy through the scholarship of Derlega and his colleagues. In particular, Derlega and Chaikin (1977) highlighted the synergy between disclosure and privacy by introducing the concept of dyadic boundaries. Their insights contributed to identifying ways that self-disclosure and privacy are integrated. Derlega,

Metts, Petronio, and Margulies (1993) added to the dialogue in their book on self-disclosure merging several important aspects about the role privacy plays in human interaction.

The seminal work on secrecy by Bok (1982) opened up additional considerations regarding the tension between revealing and concealing information considered secret. Her insights into reasons for secrecy and the way individuals treat secrets helped provide a set of comparisons. The insights Bok offers in her book broadened the scope of understanding the nature of privacy in relation to secrecy.

Privacy issues in communication received early attention from Burgoon (1982). Among other insights, she presents four states of privacy that capture different dimensions of privacy. She suggests that these states include the notion physical privacy such as public territory and home territory. Privacy states also include the notion of social privacy where Burgoon notes 'in asmuch as privacy presupposes the existence of others, a fundamental facet of privacy is the ability to withdraw from social intercourse' (Burgoon, 1982, p. 216). She also includes 'psychological privacy' that 'concerns one's ability to control affective and cognitive inputs and outputs' (Burgoon, 1982, p. 224). Burgoon also includes the state of informational privacy that 'is closely allied to psychological privacy but its legalistic and technological implications coupled it significance beyond the individual to the society as a whole is treated separately' (Burgoon, 1982, p. 228).

In addition to Burgoon's work on privacy, her research on interpersonal and family communication contributed to a better understanding of the way these relationships help contribute to the development of privacy management (e.g. Baxter 1988; Duck 1994; Rawlins 1989).

Although each of these inquires offer a useful way to understand communicative aspects of privacy management, Altman's foundational work built a platform that has inspired the emergence of new ideas and ways to understand the notion of privacy. Early in his search for understanding privacy, Altman was intrigued to discover that there was 'almost no empirical research' that had been done on privacy (Altman 1975, 6). He further stated, 'that social and behavioral scientists have generally not seen the issue of privacy as central or worthy of their empirically directed energies' (Altman, 1975, 6).

Altman's (1975) work on the environment and social behaviour charts a path to investigating privacy in a broader set of considerations. For example, Altman's inquiries regarding privacy issues incorporate processes that accommodated cultural issues, groupness, and the significance of dialectics. His ideas further explored the interface of disclosure and privacy among other significant inquiries (Altman 1975; Altman 1987; Altman 1992; Altman 1993a; 1993b; Altman 1977; Altman, Vinsel, and Brown 1981). His seminal books, such as *The Environment and Social Behavior* (Altman 1975) and *Social Penetration: Development of Interpersonal Relationships* (Altman and Taylor 1973) have been the focus of attention for several generations of students and researchers. As these articles and books illustrate, Professor Altman's vision opened the door to a more comprehensive way of considering the notion of privacy.

Clearly, Professor Altman's work has influenced the development of Communication Privacy Management theory as he points out in the foreword to the book introducing this theory (Petronio 2002, xiii-xix). CPM theory benefited from the insightfulness of Professor Altman. In addition, part of the CPM journey included a need to gain a more comprehensive understanding about the relationship between self-disclosure and private information. During the late 1960s and early 1970s, there was a proliferation of very good self-disclosure research (e.g. Jourard 1971). Curiosity was a main reason for working through the way that self-disclosure and private information could be seen as integral. After much thought and work testing this hypothetical relationship, it seemed that the act of disclosure could be understood as a process of revealing and that the information revealed could be identified as an individual's private information. Treating the relationships between disclosure and private information in this manner proved to be important to the framing of CPM theory.

10.4 Traditional debates and dominant schools

The most traditional debate regarding privacy issues occurred between Altman (1975) and Westin (1967). These foundational theories of privacy have both commonalities and differences that lead to challenges. Margulis (2003, 2011) discusses a comparative analysis of the underlying differences and similarities between Westin's focus on privacy and that of Altman's position. Margulis' (2003) assessment, in general, argues that Westin's focus is on how people protect themselves by limiting access to others. Westin states that

> privacy is the claim of individuals, groups, or institutions to determine for themselves when, how, and to what extent information about them is communicated to others. Viewed in terms of the relation of the individual to social participation, privacy is the voluntary and temporary withdrawal of a person from the general society through physical or psychological means, either in a state of solitude or small group intimacy of, when among large groups, in a condition of anonymity or reserve. (Westin 1967, 7).

Margulis (2003) points out that the states of privacy, according to Westin's perspective, focus on the 'hows' of privacy that include solitude, intimacy, anonymity, reserve, and the 'whys' of privacy that include personal autonomy, emotional release, self-evaluation, and limited and protected communication.

In the assessment of Altman's perspective on privacy, Margulis (2003) points out that Altman emphasizes the importance of individual and group levels of analysis and ways privacy is regulated. He also brings a dialectical approach to privacy regulation and a commitment to social and environmental psychology where social interaction is the underlying focus of the theory (Margulis 2003). Margulis (2003) argues that Altman's theory has five properties (p. 418). First, that privacy involves a temporal/dynamic of process of interpersonal boundary control. Second, that there is a differentiation of desired and actual levels of privacy. Third, where there can be an optimal desire for privacy or too much privacy. Fourth, privacy is bidirectional involving inputs and outputs. Fifth, Altman advocates that there can be individual and group levels of analysis. Margulis (2003) notes that 'Altman has challenged us to consider a number of important aspects of privacy' (p. 419). For example, Altman contributed the needed apparatus to illustrate how privacy is fundamentally a social process. At the time of Altman's theoretical breakthroughs, he challenged psychologists to recognize that where privacy was concerned, there needed to be an interplay among individuals, the social world in which they live, and account for cultural as well as contexts in which people navigated issues of privacy.

While there are specific differences in the way Altman and Westin envision the notion of privacy as Margulis (2011) points out, there are overlaps in some fundamental ways. However, the importance of these legendary leaders who have carved important paths of understanding cannot be overstated. Certainly, the advances made in the development of Communication Privacy Management theory has significantly benefited from the insightfulness of their work.

10.5 New challenges and topical discussions

In today's world, there are many challenges concerning privacy to consider. Clearly, social media is producing a number of issues that confront the continued efforts of privacy researchers and theorists. These challenges centre most certainly on capabilities to sustain the perceptions of privacy and directly call for more theoretically driven ways to capture behaviours in everyday life. For example, the relationship between privacy and security will need a new and more effective type of framework.

From a CPM theory vantage point, privacy and security, though sharing fundamental issues, tend to have essential differences. When people talk about their private information, they act on their assumption that they are in charge. People do not stop assuming control even after a disclosure is made. Research shows that people work to sustain their control over what happens to their private information by recalibrating their privacy rules to right the system (e.g. Child, Petronio, Agyeman-Budu, and Westermann 2011).

However, in the context of security, these situations seem to have a different calculus. Security arises as an issue when individuals must provide private information to access services. For example, when someone needs hospitalization in the United States, HIPAA privacy forms must be signed in order to receive healthcare. There appears to be a transfer of control over specific types of private information with the assumption that the information will be considered 'confidential'. To ensure that, people are asked to sign a form that is considered a promissory note to protect the owner's private information.

However, patients know that not signing this form can potentially mean they will not receive healthcare. A HIPAA form is not the only circumstance people are asked to sign a document to insure the parameters of responsibility are identified. Likewise, when people want a loan, private information must be provided to achieve this goal. When someone opens a bank account, banking procedures require disclosure of private information. The nature of security in these situations implicitly assumes that individuals will trade off access to their private information for a particular outcome. As these examples illustrate, at times, relinquishing private information is utilitarian to achieve a specific goal (Pastalan 1974).

In these cases, responsibility for promising protection of the individual's private information falls to the entity caring for the person's information. There are dialectic tensions between the assumed level of privacy protection and the expectation of security. For instance, when there is a data breach at a bank, the patrons hold the bank responsible; they have essentially entered into a 'contract'. Yet, the patrons cannot necessarily dictate how the bank should handle the loss of their private information. Thus, in this example, the notion of 'security' has to do with an entity being responsible for an individual's privacy, but the management of the private information needed for using a bank is limited to what the bank perceives is reasonable. A person can change banks, but still has limited control over the information disclosed. Though this is true, the trade-off is limited for the private information owner;

however, there is likely a judgment of the risk-benefit ratio by the information owner that calculates how much control over privacy management they want to relinquish in order to gain access to services they need or want.

With the continued investigations regarding the use of big data, CPM theory points out that until the collection of private information is personal for individuals, they tend to feel less engaged or bothered about new technologies they do not understand or to which they do not have direct access. The technology seems to be viewed as complicated and often inaccessible, thus out of their control. Only when individuals are directly impacted by a breach do they become more mindful of the potential ramifications.

However, consistent with CPM theory and research, when the use of data from data banks directly affects a person's life, individuals do not necessarily expect they will be negatively affected. For example, the New York Times reported on a case involving the use of targeted advertising by the Target Corporation (Duhigg 2012). The incident concerned a privacy breach discovered by a father. This father was upset that his teenage daughter was receiving a number of coupons that related to pregnancy. He called the local store and asked the manager why his daughter was receiving these coupons. He said, 'she is still in high school, and you are sending her coupons for baby clothes and cribs? Are you trying to encourage her to get pregnant?' The manager apologized and then called a few days later apologizing again. On the phone the father said, 'turns out that there's been some activities in my house I haven't been completely aware of. She's due in August, I owe you an apology' (Duhigg 2012, 2-19). This case illustrates that, in general, individuals give little notice to

Discerning potential similarities and differences in privacy management regarding personal relationships as opposed to corporate or public services offers an intriguing research opportunity. Examining how individuals conceive of the trade-off with their information in order to obtain services, products, or resources and comparing the findings with the research on choices about revealing or concealing private information with others would add to both research areas. Similarly, identifying how individuals treat corporate or public entities when security breaches occur is a useful line of inquiry.

10.6 Conclusion

The nature of privacy has long been a part of the human condition (Veyne 1987). Yet, our attention to this important aspect of life, where individuals need both privacy and the ability to be social with others is in constant need of new discoveries. A mission of communication privacy management theory is to bring about new insights into this phenomenon. The mission is to push these ideas further and help others to advance their interests in privacy inquiries. The Communication Privacy Management Center (www.cpmcenter.iupui.edu) at Indiana University-Purdue University, Indianapolis, has been recently started to provide resources, such as citations of research using this theory. We have harvested over 1000 citations thus far. There are over fifteen countries where researchers have been applying CPM theory, thus enabling cross-cultural research opportunities. There are also many different contexts and methodologies used in CPM research allowing for cross comparisons. Our team is working on teaching tools and devising ways to translate research into meaningful practice to help others.

Learning about privacy is a mission, yet, watching human behaviours unfold is remarkably entertaining and enriching. This volume offers a multitude of voices, opinions, and challenges. I appreciate the opportunity to take part in this mission.

Further reading

Beck, C., S. Chapman, N. Simmons, K. Tenzek, and S. Ruhl. (2015). *Celebrity Health Narratives and Public Health*. City: McFarland & Company Publishers (ebook).

Greene, K., V.J. Derlega, G.A. Yep, and S. Petronio. (2003). *Privacy and Disclosure of HIV in Interpersonal Relationships: A Sourcebook for Researchers and Practitioners*. Mahwah: Lawrence Erlbaum Associates.

Child, J.T., and S. Petronio. (2015). 'Privacy Management Matters in Digital Family Communication' in C.J. Bruess (ed.), *Family Communication in the Age of Digital and Social Media*. New York: Peter Lang, 32-54.

References

Allen, M.W. (2008). 'Consumer Finance and Parent-Child Communication' in J.J. Xiao (ed.), *Handbook of Consumer Finance Research*. New York: Springer New York, 351-361.

Altman, I. (1975). *Environment and Social Behavior: Privacy, Personal Space, Territory, and Crowding*. Belmont: Wadsworth Publishing.

Altman, I. (1977). 'Privacy Regulation: Culturally Universal or Culturally Specific?' *Journal of Social Issues* 33(3), 66-84. doi:10.1111/j.1540-4560.1977.tb01883.x.

Altman, I. (1987). 'Centripetal and Centrifugal Trends in Psychology'. *American Psychologist* 42(12), 1058-1069. doi:10.1037/0003-066X.42.12.1058.

Altman, I. (1993a). 'Dialectics, Physical Environments, and Personal Relationships'. *Communications Monographs* 60(1), 26-34. doi:10.1080/03637759309376291

Altman, I. (1993b). 'Challenges and Opportunities of a Transactional World View: Case Study of Contemporary Mormon Polygynous Families'. *American Journal of Community Psychology* 21(2), 135-163. doi:10.1007/BF00941618.

Altman, I., & Taylor, D. A. (1973). *Social Penetration: The Development of Interpersonal Relationships*. New York: Holt, Rinehart, & Winston.

Altman, I., A. Vinsel, and B.B. Brown (1981). 'Dialectic Conceptions in Social Psychology: an Application to Social Penetration and Privacy Regulation'. *Advances in Experimental Social Psychology* 14, 107-160. doi:10.1016/S0065-2601(08)60371-8.

Badrul, N.A., S.A. Williams, and K.O. Lundqvist. (2016). 'Online Disclosure of Employment Information: Exploring Malaysian Government Employees' Views in Different Contexts'. *Computers and Society* 45(3), 38-44. doi:10.1145/2874239.2874245.

Berger, C.R. (1997). *LEA's Communication Series. Planning Strategic Interaction: Attaining Goals through Communicative Action*. Mahwah: Lawrence Erlbaum Associates Publishers.

Baxter, L.A. (1988). 'A Dialectical Perspective on Communication Strategies in Relationship Development'. In S. Duck, D.F. Hay, S.E. Hobfoll, W. Ickes, and B.M. Montgomery (eds.), *Handbook of Personal Relationships: Theory, Research and Interventions*. Oxford: John Wiley, 257-273.

Baxter, L.A., and B.M. Montgomery. (1996). *Relating: Dialogues and Dialectics*. New York: Guilford Press.

Bok, S. (1989). *Secrets: On the Ethics of Concealment and Revelation*. New York: Pantheon Books.

Broekema, K., and K.M. Weber. (2017). 'Disclosures of Cystic Fibrosis-related Information to Romantic Partners'. *Qualitative Health Research* 27(10), 1575-1585. doi:10.1177/1049732317697675.

Burgoon, J.K. (1982). 'Privacy and Communication'. *Annals of the International Communication Association* 6(1), 206-249. doi:10.1080/23808985.1982.11678499.

Chelune, G.J. (1979). *Self-disclosure: Origins, Patterns, and Implications of Openness in Interpersonal Relationships*. San Francisco: Jossey-Bass.

Child, J.T., and S. Petronio. (2011). 'Unpacking the Paradoxes of Privacy in CMC Relationships: The Challenges of Blogging and Relational Communication on the Internet' in K.B. Wright and L.M. Webb (eds.), *Computer-mediated Communication in Personal Relationships*. New York: Peter Lang, 257-273.

Child, J.T., and S. Petronio. (2015). 'Privacy Management Matters in Digital Family Communication' in C.J. Bruess (ed.), *Family Communication in the Age of Digital and Social Media*. New York: Peter Lang, 32-54.

Child, J.T., S. Petronio, E.A. Agyeman-Budu, and D.A. Westermann. (2011). 'Blog Scrubbing: Exploring Triggers that Change Privacy Rules'. *Computers in Human Behavior* 27(5), 2017-2027. doi:10.1016/j.chb.2011.05.009.

Cohen, O., R.D. Leichtentritt, and N. Volpin, N. (2014). 'Divorced Mothers' Self-perception of their Divorce-related Communication with their Children'. *Child & Family Social Work* 19(1), 34-43. doi:10.1111/j.1365-2206.2012.00878.x.

Collins, N.L., & L.C. Miller. (1994). 'Self-disclosure and Liking: a Meta-analytic Review'. *Psychological Bulletin* 116(3), 457-475. doi:10.1037/0033-2909.116.3.457.

De Wolf, R., and J. Pierson. (2014). 'Who's My Audience Again? Understanding Audience Management Strategies for Designing Privacy Management Technologies'. *Telematics and Informatics* 31(4), 607-616. doi:10.1016/j.tele.2013.11.004.

De Wolf, R., K. Willaert, and J. Pierson. (2014). 'Managing Privacy Boundaries Together: Exploring Individual and Group Privacy Management Strategies in Facebook'. *Computers in Human Behavior* 35, 444-454. doi:10.1016/j.chb.2014.03.010.

Derlega, V.J., and A.L. Chaikin. (1977). 'Privacy and Self-disclosure in Social Relationships'. *Journal of Social Issues* 33(3), 102-115. doi:10.1111/j.1540-4560.1977.tb01885.x.

Derlega, V., S. Metts, S. Petronio, and S. Margulis. (1993). *Self-disclosure*. Newbury Park: Sage Publications.

Duck, S. (1994). 'Stratagems, Spoils, and a Serpent's Tooth: on the Delights and Dilemmas of Personal Relationships' in W.R. Cupach and B. Spitzberg (eds.), *The Dark Side of Interpersonal Communication*. Hillsdale: Lawrence Erlbaum Associates, 32-54.

Duhigg, C. (2012). 'How Companies Learn Your Secrets'. *The New York Times*, February 16. Retrieved from www.nytimes.com/2012/02/19/magazine/shopping-habits.html.

Ezhei, M., and B.T. Ladani. (2017). 'Information Sharing vs. Privacy: a Game Theoretic Analysis'. *Expert Systems with Applications* 88, 327-337. doi:10.1016/j.eswa.2017.06.042.

Gordon, M.E. (2011). 'The Dialectics of the Exit Interview: a Fresh Look at Conversations about Organizational Disengagement'. *Management Communication Quarterly* 25(1), 59-86. doi:10.1177/0893318910376914.

Golden, A.G. (2014). 'Permeability of Public and Private Spaces in Reproductive Healthcare Seeking: Barriers to Uptake of Services among Low-income African American Women in a Smaller Urban Setting'. *Social Science & Medicine* 108, 137-146. doi:10.1016/j.socscimed.2014.02.034.

Hammonds, J.R. (2015). 'A Model of Privacy Control: Examining the Criteria that Predict Emerging Adults' Likelihood to Reveal Private Information to their Parents'. *Western Journal of Communication* 79(5), 591-613. doi:10.1080/10570314.2015.1083117.

Hawk, S.T. (2017). 'Chinese Adolescents' Reports of Covert Parental Monitoring: Comparisons with Overt Monitoring and Links with Information Management'. *Journal of Adolescence* 55, 24-35. doi:10.1016/j.adolescence.2016.12.006.

Heikkinen, A.M., G.J. Wickstrom, and H. Leino-Kilpi. (2007). 'Privacy in Occupational Health Practice: Promoting and Impeding Factors'. *Scandinavian Journal of Public Health* 35(2), 116-124. doi:10.1080/14034940600975740.

Hesse, C., and E.A. Rauscher. (2013). 'Privacy Tendencies and Revealing/concealing: the Moderating Role of Emotional Competence'. *Communication Quarterly* 61(1), 91-112. doi:10.1080/01463373.2012.720344.

Hewes, D.E., and M.L. Graham. (1989). 'Second-guessing Theory: Review and Extension'. *Annals of the International Communication Association* 12(1), 213-248. doi:10.1080/23808985.1989.11678720.

Honeycutt, J.M. (2008). 'Imagined Interaction Theory' in L.A. Baxter and D.O. Braithwaite (eds.) *Engaging Theories in Interpersonal Communication: Multiple perspectives*. City: Publisher, pages.

Imber-Black, E. (1993). *Secrets in Families and Family Therapy: an Overview*. New York: W.W. Norton & Co.

Jourard, S.M. (1958). 'A Study of Self-disclosure'. *Scientific American* 198(5), 77-86.

Jourard, S.M. (1964). *The Transparent Self: Self-disclosure and Well-being*. New York: Van Nostrand.

Jourard, S.M. (1971). *Self-disclosure: an Experimental Analysis of the Transparent Self*. New York: Wiley.

Kennedy-Lightsey, C.D., and B.N. Frisby. (2016). 'Parental Privacy Invasion, Family Communication Patterns, and Perceived Ownership of Private Information'. *Communication Reports* 29(2), 75-86. doi:10.1080/08934215.2015.1048477.

Liu, Y., and J. Fan. (2015). 'Culturally Specific Privacy Practices on Social Network Sites: Privacy Boundary Permeability Management in Photo Sharing by American and Chinese College-age Users'. *International Journal of Communication* 9, 2141-2060.

Manning, J. (2015). 'Communicating Sexual Identities: a Typology of Coming Out'. *Sexuality & Culture* 19(1), 122-138. doi:10.1007/s12119-014-9251-4.

Margulis, S.T. (2003). 'On the Status and Contribution of Westin's and Altman's Theories of Privacy'. *Journal of Social Issues* 59(2), 411-429. doi:10.1111/1540-4560.00071.

Margulis, S.T. (2011). 'Three Theories of Privacy: an Overview'. In S. Trepte and L. Reinecke (eds.), *Privacy Online: Perspectives on Privacy and Self-disclosure in the Social Web*. Berlin: Springer Heidelberg, 9-18.

McBride, C.M., and K.M. Bergen. (2008). 'Becoming a Reluctant Confidant: Communication Privacy Management in Close Friendships'. *Texas Speech Communication Journal* 33(1), 50-61.

McKenna-Buchanan, T., S. Munz, S., and J. Rudnick. (2015). 'To Be or Not to Be Out in the Classroom: Exploring Communication Privacy Management Strategies of Lesbian, Gay, and Queer College Teachers'. *Communication Education* 64(3), 280-300. doi:10.1080/03634523.2015.1014385.

McLaren, R.M., and K.R. Steuber. (2013). 'Emotions, Communicative Responses, and Relational Consequences of Boundary Turbulence'. *Journal of Social and Personal Relationships* 30(5), 606-626. doi:10.1177/0265407512463997.

Miller, A.E. (2009). 'Revealing and Concealing Postmarital Dating Information: Divorced Coparents' Privacy Rule Development and Boundary Coordination Processes'. *Journal of Family Communication* 9(3), 135-149. doi:10.1080/15267430902773287.

Miller, A.N., and D.L. Rubin. (2007). 'Factors Leading to Self-disclosure of a Positive HIV Diagnosis in Nairobi, Kenya: People Living with HIV/AIDS in the Sub-Sahara'. *Qualitative Health Research* 17(5), 586-598. doi:10.1177/1049732307301498.

Millham, M.H., and D. Atkin. (2017). 'Managing the Virtual Boundaries: Online Social Networks, Disclosure, and Privacy Behaviors'. *New Media & Society* 20(1), 50-67. doi:10.1177/1461444816654465.

Morr Serewicz, M.C., and D.J. Canary. (2008). 'Assessments of Disclosure from the In-laws: Links among Disclosure Topics, Family Privacy Orientations, and Relational Quality'. *Journal of Social and Personal Relationships* 25(2), 333-357. doi:10.1177/026540750708796.

Ngcongo, M. (2016). 'Mobile Communication Privacy Management in Romantic Relationships: a Dialectical Approach'. *Communicatio: South African Journal for Communication Theory & Research*,42(1), 56-74. doi:10.1080/02500167.2016.1140666.

National Communication Association. (2018). https://www.natcom.org/about-nca.

Pastalan, L.A. (1974). 'Privacy Preferences among Relocated Institutionalized Elderly'. *Man-environment Interactions: Evaluations and Applications* 2, 73-100.

Pennebaker, J.W. (1990). *Opening Up: the Healing Power of Confiding in Others*. New York: William Morrow.

Petersons, A., and I. Khalimzoda, I. (2016). 'Communication Privacy Management of Students in Latvia'. *Problems and Perspectives in Management* 14(2), 222-227. http://essuir.sumdu.edu.ua/handle/123456789/49812.

Petronio, S. (1991). 'Communication Boundary Management: a Theoretical Model of Managing Disclosure of Private Information between Marital Couples'. *Communication Theory* 1(4), 311-335. doi:10.1111/j.1468-2885.1991.tb00023.x.

Petronio, S. (2002). *Boundaries of Privacy: Dialectics of Disclosure*. Albany: SUNY Press.

Petronio, S. (2010). 'Communication Privacy Management Theory: What Do We Know about Family Privacy Regulation?' *Journal of Family Theory and Review* 2(3), 175-196. doi:10.1111/j.1756-2589.2010.00052.x.

Petronio, S. (2013). 'Brief Status Report on Communication Privacy Management Theory'. *Journal of Family Communication* 13(1), 6-14. doi:10.1080/15267431.2013.743426.

Petronio, S. (2016). 'Communication Privacy Management Theory' in K.B. Jensen, R.T. Craig, J.D. Pooley, and E.W. Rothenbuhler (ed.), *The International Encyclopedia of Communication Theory and Philosophy*. Hoboken: Wiley-Blackwell, 278-286. doi:10.1002/9781118766804.wbiect138.

Petronio, S., and C. Bantz. (1991). 'Controlling the Ramifications of Disclosure: 'Don't tell anybody but...''. *Journal of Language and Social Psychology* 10(4), 263-269. doi:10.1177/0261927X91104003.

Petronio, S., and W. Durham. (2008). 'Understanding and Applying Communication Privacy Management Theory' in L.A. Baxter and D.O. Braithwaite (eds.), *Engaging Theories in Interpersonal Communication: Multiple perspectives.* Thousand Oaks: Sage Publications, 309-322.

Petronio, S., and J. Reierson. (2009). 'Regulating the Privacy of Confidentiality: Grasping the Complexities through Communication Privacy Management Theory'. In T. Afifi, and W. Afifi (eds.), *Uncertainty, Information Management, and Disclosure Decisions: Theories and Applications.* New York: Routledge, 309-322.

Petronio, S., S.M. Jones, and M.C. Morr. (2003). 'Family Privacy Dilemmas: Managing Communication Boundaries within Family Groups' in L. Frey (ed.), *Group Communication in Context: Studies of Bona Fide Groups* (2nd ed.) (pp. 23-56). Mahwah: Lawrence Erlbaum Associates, 23-56.

Rawlins, W.K. (1989). 'A Dialectical Analysis of the Tensions, Functions, and Strategic Challenges of Communication in Young Adult Friendships'. *Annals of the International Communication Association* 12(1), 157-189. doi:10.1080/23808985.1989.11678717.

Scharp, K.M., and K.R. Steuber. (2014). 'Perceived Information Ownership and Control: Negotiating Communication Preferences in Potential Adoption Reunions'. *Personal Relationships* 21(3), 515-529. doi:10.1111/pere.12046.

Sidelinger, R.J., M.C. Nyeste, P.E. Madlock, J. Pollak, and J. Wilkinson. (2015). 'Instructor Privacy Management in the Classroom: Exploring Instructors' Ineffective Communication and Student Communication Satisfaction'. *Communication Studies* 66(5), 569-589. doi:10.1080/10510974.2015.1034875.

Yep, G.A. (2000). 'Explaining Illness to Asian and Pacific Islander Americans: Culture, Communication, and Boundary Regulation' in B.B. Whaley (ed.), *Explaining Illness: Research, Theory, and Strategies.* Mahwah: Lawrence Erlbaum Associates, 271-284.

Westin, A. (1967). *Privacy and Freedom.* New York: Atheneum.

Still Uneasy: a Life with Privacy

Anita LaFrance Allen

1. Feminism and privacy

Thirty years ago, I published a book about privacy that focused on the problems of American women: imposed domesticity, reproductive autonomy, harassment, sexual violence, and sexual liberty.[1] *Uneasy Access: Privacy for Women in a Free Society* was not only the first book-length treatment of privacy by a philosopher to focus on women, it was the first book-length treatment by an academic philosopher to focus on *any* aspect of privacy.

My book was a response both to the academic debates about the meaning and value of privacy found in analytic-style philosophy journals; and to feminist critiques of privacy emanating from many disciplines, well represented in legal scholarship by Catharine MacKinnon. Writing about abortion privacy doctrines in US constitutional law, MacKinnon had dismissed privacy as 'an injury got up as a gift' – a patriarchal value representing and facilitating the continued subordination of women in inferior social, political, and economic roles.[2]

While conceding the women have historically lived their lives as ancillaries and inferiors, I argued in *Uneasy Access* that they have had 'too much of the wrong kind of privacy'. Women have suffered, for example, isolation, confinement, and imposed domestic roles. What they merit morally and politically are 'the right kinds of privacy', namely, meaningful opportunities for voluntary seclusion, intimacies, and legal rights of decision about personal life and health. As a counterpoint to 1980s feminism, which over-disparaged privacy, I turned to 19th-century utopian writer Charlotte Perkins Gilman,[3] who understood that true privacy for women would further equality and entailed radical transformation. Subsisting without alternatives in separate family houses to cook, clean, and take of others is not meaningful privacy.

To deepen the legal dimensions of the story, a few years later I published 'How Privacy Got its Gender', a law review article co-authored with one of my students.[4] The article was based on a realization that the development of the US tort and constitutional law of privacy was in important respects driven by

1 Anita L. Allen. (1988). *Uneasy Access: Privacy for Women in a Free Society. City:* Rowman and Littlefield.

2 Catharine MacKinnon. (1987). *Feminism Unmodified.* City: Publisher.

3 Charlotte Perkins Gilman. (1989). *Women and Economics.* City: Publisher.

4 Anita L. Allen and Erin Mack. (1991). How Privacy Got its Gender, *N. Ill. U. L. Rev.* 10(441).

concerns about the proper regard for and place of women. Problematically, in the 19th and early 20th centuries, the impetus behind the expansion of privacy law was often to protect ideals of women's modesty and domesticity. Fortunately, by the final third of the 20th century, the motivating ideals had changed to women's decisional autonomy and independence. I showed through detailed textual analysis that the famous Warren and Brandeis article,[5] 'The Right to Privacy', was a florid testament to male privilege, and depended on outmoded gender norms for its rhetorical authority.

In the course of this work closely examining the early US legal cases, I became so interested in the discourse of privacy law and how it evolved that I put together increasing comprehensive legal textbooks on the subject.[6] Moreover, concerned that it might appear to other feminist scholars that I believed accountable social relationships were of lesser value than secrecy and privacy, I wrote a book illustrating why, from an ethical point of view, privacy is not always the paramount consideration.[7] Reflecting my interest in how privacy regimes impact or fail to impact specific groups, my work often identifies the winners or losers of privacy rules and practices – be they women, corporations, the LGBTQ+ community, inmates, African Americans, or children. Whether privacy is a good thing for the people who have it is a question with normative and empirical dimensions.[8]

2. Unpopular privacy and ethical duty

At the invitation of the *Stanford Law Review* I critically revisited the central themes of *Uneasy Access* in light of the growth of the Internet and web.[9] In the preface to *Uneasy Access*, I had observed that '[t]he felt need of recent generations to demarcate the limits of intervention into the privacy and private lives of women has done more than even the information technology boom to inspire analysis of privacy and the moral right to it'. My observation no longer holds. Rather, starting in the 1990s many of the most visible and novel efforts to understand privacy been driven by developments in information technology, digital

5 Samuel D. Warren and Louis D. Brandeis. (1890). *The Right to Privacy, Harv. L. Rev.* 4(193).

6 The most recent edition is, Anita L. Allen and Marc Rotenberg. (2016). *Privacy Law and Society.* City: West Thomson Reuters.

7 Anita L. Allen.(2003). Why Privacy Isn't Everything: Feminist Reflections on Personal Accountability. City: Rowman & Littlefield.

8 Anita L. Allen. (2015). 'Compliance-Limited Health Privacy Laws', in *Social Dimensions of Privacy: Interdisciplinary Perspectives*, Beate Roessler and Dorota Mokrosinka (eds.). Cambridge: Cambridge University Press; Anita L. Allen. (2010). 'Privacy Torts: Unreliable Remedies for LGBT Plaintiffs', *Cal. L. Rev.* 98(1711).

9 Anita L. Allen. (2000). 'Gender and Privacy in Cyberspace', *Stan. L. Rev.* 52(1175).

communications, data protection, cybersecurity, big data, and the media. Still, although concerns about gender are no longer the driving forces, new debates about sexual harassment, stalking, violence against women, shaming, and unequal employment have been spawned by the digital economy. The Internet opened new opportunities for women to communicate, collaborate, learn, express themselves, and make money; but the status quo of gender inequality persists.

In the past decade or so my attention has shifted away from privacy and gender to broader fundamentals. A major theme has been 'coercing privacy', a phrase I coined to help raise concerns about the proper limits of regulation and self-regulation. I have come to view privacy as a 'foundational good', a necessary resource for a liberty-lover's successful life. A nation committed to personal freedom must be prepared to mandate inalienable, liberty-promoting privacies for its people, whether they eagerly embrace them or not. The human rights and natural rights discourses around privacy are consonant with my views.[10]

In *Unpopular Privacy: What Must We Hide*,[11] a book about seclusion, concealment, confidentiality, and data-protection, I focus on privacies disvalued by their intended beneficiaries and targets. The book outlined the best reasons for imposing them, and the worst. It looked at laws designed to keep website operators from collecting personal information from young children, anti-nudity laws that force strippers to wear pasties and thongs, and the myriad of employee and professional confidentiality rules – including insider trading laws – that require strict silence about matters whose disclosure could earn us small fortunes. I tried to show that such laws -- and ethical rules of concealment currently strained by trends in media and technology – recognize the extraordinary importance of dignity, reputation, and trust, and help to preserve social, economic, and political options throughout a lifetime.

Following this work, I went further to argue that individuals have a strong moral obligation to protect their own privacy,[12] that privacy is not a take it or leave it good, but one people ought to embrace as an act of self-respect, and for the sake of virtues of reserve and prudence. I was among the first to recognize that there are not only rights of privacy but duties, obligations, and virtues of privacy as well. The foundational status of the privacy good supports both state paternalism and duties not to throw privacies away on, for example, Instagram,

10 Anita L. Allen. (2012). 'Natural Law, Slavery, and the Right to Privacy Tort', *Fordham L. Rev.* 81(1187).

11 Anita L. Allen. (2011). *Unpopular Privacy: What Must We Hide*? Oxford: Oxford University Press.

12 Anita L. Allen. (2013). *'An Ethical Duty to Protect One's Own Information Privacy?', Ala.. L. Rev.* 64(845).

Alexa, or Twitter. Individuals have a moral obligation to respect not only other people's privacy but also their own.

Individuals are generously feeding Big Data. If the experience of privacy is important to human dignity and well-being, it is something individuals with a choice should not choose to carelessly discard or give away. Our ethical responsibility to do so could entail circumspect use of social media and credit cards, along with diligent use of passwords, encryption, and security software to limit access to devices.

We can close the blinds and cover the web cam lens. Yet as the Internet of things and AI overwhelm, it is harder and harder for typical individuals to exercise responsible control over digital information about themselves.[13] Ascribing an obligation of protecting our privacy seems to require something exceedingly difficult or impossible: the eschewal of activities that contribute to the production of massive data sets and analysis. The moral obligation to protect one's own privacy remains a meaningful concept so long as one recognizes that the obligation requires participating in the political process and supporting consumer activism and advocacy, as well as making adjustments to one's own individual behaviour and family education. Collectively, individuals can push for reforms and be critical of government.

13 Anita L. Allen. (2016). 'Protecting One's Own Privacy in a Big Data Economy', *Harv.. L. Rev. F.* 130(71).

11. Privacy from an Anthropological Perspective

Sjaak van der Geest

11.1 Introduction

Before writing about the anthropological perspective on privacy I will need to briefly explain what constitutes an anthropological perspective and how it comes into being. The first 'article' of my anthropological *Credo* is *context*. Anthropologists study people, practices, words, thoughts, objects, traditions, institutions, and so on *in their context*, while many other disciplines do the opposite. No spoken or written word has a fixed meaning but derives its meaning from the sentence or the wider context in which it occurs. This insight forms the basis of the anthropological research approach. We cannot be sure of the meaning of people's words or actions if they are delivered to us out of context. If we have not seen the expression on the speaker's face and the situation in which s/he was when speaking, we cannot be sure of the intended meaning of the spoken words. Was the person at ease when s/he spoke or did s/he rather feel uncomfortable or annoyed? Was s/he perhaps ironic, was s/he lying or did s/he try to flatter or just to get rid of the visitor who asked him or her impertinent private questions? Was s/he impatient or did s/he rather enjoy the conversation with this visitor? Of course, we can never be absolutely certain what someone has in mind during a conversation, but being with him or her in the same context is the best we can do to capture someone's intentions.

The necessity of knowing the context implies that the anthropologist is *present at the spot where the research takes place*. Not as an interrogator and distant observer, but preferably as an engaged participant in the conversation and a respectful and empathic observer. Doing this kind of research requires that the researcher does not present himself as someone who knows but rather as a learner. Why should anyone tell me about their life if they believe that I already know everything? The conversation (I prefer that less formal term to 'interview') would turn into an exam or interrogation, the worst that can happen in fieldwork. Presenting oneself as a learner, a not-knowing visitor is not so much a clever trick by the researcher; it is the reality. The local 'interlocutors' (an unfortunate and clumsy term) are indeed the knowers while the visiting researcher is the not-knower. Moreover, for a researcher, not-knowing and the curiosity accompanying not-knowing are the natural and most valid motives to undertake the research.

Another crucial aspect of the anthropological perspective is that the researcher is not so much interested in drawing outsider conclusions about a certain group of people but rather wants to capture what is on these people's minds, the *emic perspective*. Anthropology not only wants to know *what* people are doing and thinking but also *why*. What is important to them? What matters? And yet, the ambitions of anthropologists are modest. They want to *understand* others, *not to explain* and predict their acting and thinking. We will see, therefore, that anthropologists rarely attempt to explain people's behaviour with respect to privacy by linking it to particular features or historical antecedents that are typical for a specific society.

Interestingly, the challenge to understanding has shifted over the years. In the early days of the discipline anthropologists thought that they had to travel far away to 'exotic' cultures to find people who posed a challenge to their understanding; they took everything happening at home for granted, as 'normal'. Gradually, however, they realized that understanding others is also a challenge in their own society. The study of privacy in the Netherlands is a splendid example: how to make sense of people who leave their curtains wide open at night but loudly protest against camera's in the street?

Finally, trying to understand others always implies introspection on the part of the researcher. We can only ask sensible questions and understand the answers, if we know from our own experience what we are talking about. This subjectivity of anthropological research is often regarded as suspect and disapproved of by other ('exact') disciplines. The overall opinion says that scientific research must be objective. But we see subjectivity as an indispensable asset rather than an obstacle to good research. The implicit comparison between 'my' and 'their' experience is a prerequisite for understanding 'them' and a sine qua non for a fruitful conversation. Does this make anthropology a subjective discipline? Yes, in the sense that we use ourselves (as subjects) to make sense of what people do, say, and think. If science requires excluding this 'subjectivity' and basing the study of human behaviour on 'objective' observations (as we do with mice in a laboratory) or on short responses to a questionnaire by respondents we have never seen, anthropology is not 'science'.

If we do not recognize anything from ourselves in them, our data will remain stale and meaningless. It would be like reading a novel about people and events which do not touch us in any way; if there is nothing we can share with the characters of the story, we will take little interest in them and fail to understand them. Instead of suppressing personal views and feelings, therefore, the anthropological researcher should examine them carefully and try to use them in conversation, observation, and participation. Through personal exposure to an interlocutor, a deeper level of mutual understanding and appreciation will be reached. When the anthropologist Desjarlais (1991, 394) asked an old man in Nepal what happens when one's heart is filled with grief, the man smiled and gave the best possible answer: 'You ask yourself'. The need for this type of introspection becomes clear when we try to describe 'privacy', as we will see in the next section.

11.2 Meaning and function of privacy

After the above introduction about 'context', 'emic perspective', and 'introspection', it will not come as a surprise that anthropologists are reluctant to use fixed definitions of privacy in order to analyse privacy in other cultures or subcultures. They rather try to observe and discuss emotions and practices that appear akin to what they call 'privacy' in their own society. To put it differently, they use introspection to arrive at their own privacy experiences and use these to engage in a dialogue with the people they study. It implies that they do make use of a temporary working definition of privacy as a tool to explore how (and why) others think and act in situations where personal and social concerns are at play. It does make sense, therefore, to investigate descriptions and definitions of privacy in publications on Western society.

Interestingly, anthropologists have to look outside their own discipline for solid and useful discussions about privacy. Alan Westin (1970), an American law professor with a broad view on culture and society, discerns four types or aspects of privacy and four functions. The four aspects are *solitude* (being alone), *intimacy* (being alone with only one or a few close others), *anonymity* (being with others but unknown to them and unobserved, 'lost in a crowd'), and *reserve* (being with others but having erected a 'psychological barrier against unwanted intrusion') (Westin 1970, 32). The four functions or effects of privacy, mentioned by Westin, are *personal autonomy* (which includes self-identity and the ability to control communication and interaction with others); *emotional release* (the option of withdrawing and being free from observation by others); *self-evaluation* (the possibility of reflecting on one's position vis-à-vis others); and *protected communication* (sharing confidential things with selected others).

Irwin Altman (1975), a social psychologist, largely follows Westin but places more emphasis on 'the dialectic quality of privacy, the optimization nature of privacy, and privacy as a boundary regulation process' (1975, 21).

Privacy, in other words, is not only about excluding but also about including others. Altman (1975, 22) quotes George Simmel:

> We become what we are not only by establishing boundaries about ourselves but also by a periodic opening of these boundaries to nourishment, to learning, and to intimacy (Simmel 1971, 81).

A recent typology of privacy by Bert-Jaap Koops and colleagues (2017), loosely based on Westin's (1970) types and functions, shows the multilayered, multifunctional, and multifocal character of the privacy concept (see the figure below). They distinguish eight basic types of privacy occurring in four zones (personal, intimate, semi-private, and public) with a ninth type (informational privacy) that partly overlaps all eight basic types. The typology must serve as an analytic tool to understand what privacy is and does but it also shows the extreme complexity and variability of privacy.

Moreover, the concept of privacy defies a precise definition because it refers to experiences that are too close to look at objectively. The Stanford Encyclopedia of Philosophy (DeCew 2013) lists a number of attempts to capture the meaning of privacy: 'control over information about oneself', 'required for human dignity', 'crucial for intimacy', 'necessary for the development of varied and meaningful interpersonal relationships', 'the value that accords us the ability to control the access others have to us', 'a set of norms necessary not only to control access but also to enhance personal expression and choice', or 'some combination of these'. I will add my anthropological attempt to grasp what privacy is, or rather does.

Privacy is the condition of life in which a person feels comfortable, safe, and secure. The metaphor of a house presents itself: a place where one can live, protected against unwanted elements from outside such as cold and heat, wind and rain, against spies, authorities, thieves, and other unwelcome visitors. A house offers the possibility to allow some people and elements in while keeping others out. Usually it accommodates love and intimacy and is a base from which we engage with others in meaningful relationships. It provides freedom and creates room for self-control, self-reflection, and self-expression, according to Smith (2004, 11250). Monitoring one's privacy can be compared to keeping one's house open to some and closed to others. This metaphor provides 'feel-knowledge' of privacy that may be more effective in defining it than a conventional definition. Privacy is the realization of security in life, a condition that forms the ground for living the type of life one wants to live, a comfortable balance between intimacy and publicity. Examples of how this security and comfortable balance is achieved in various

Fig. 11.1

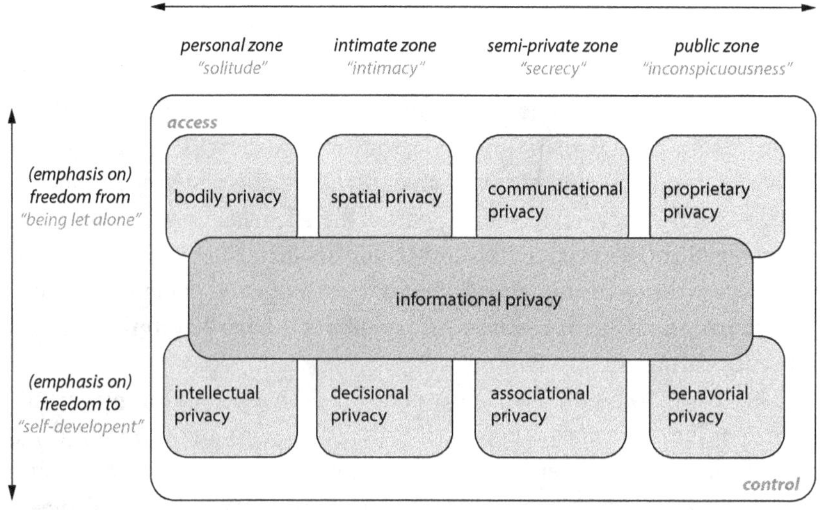

Source: Koops et al. 2017, 484.

contexts (poor or affluent, low or high class, gender-equal or -unequal, authoritarian or liberal) will be seen in the remainder of this chapter.

11.3 Classic texts and authors

Since privacy and feeling secure and comfortable appear universal human values and since modes of privacy and security seem to vary enormously between and within cultures and societies, one would expect anthropologists to be deeply interested in privacy. Surprisingly, however, there is no classic anthropological ethnography that takes local ideas and practices of privacy as its central theme of research. Anthropological observations concerning privacy are mainly found somewhat hidden in wider ethnographic studies, often 'between the lines'. There are however three anthropological texts that discuss aspects of privacy on a more general and theoretical level.

In 1959, Edward T. Hall published his bestseller *The Silent Language*, followed in 1969 by *The Hidden Dimension*. Hall was an anthropologist who had done fieldwork in 'reservations' of Amerindians in the 1930s. During his life, first in the American army and later in the US Foreign Service Institute, his attention was drawn to problems in international communication in politics and trade relations. He taught intercultural communication skills and wrote several books on this topic. The two above-mentioned books are the most well-known.

The Silent Language starts with a complaint about the cultural ignorance and ethnocentrism of his citizen Americans in their contacts with people from other cultures. The book is an attempt to teach his readers what culture is and does to human relations interlacing his message with examples from his earlier fieldwork and his extensive travelling. Privacy is not explicitly discussed in the book but it is clearly part of the 'silent language'. The publication can be seen as a preparation for his 1969 book. In *The Hidden Dimension* he proposes the term 'proxemics', the study of space and personal territory that people in different cultures and contexts use while interacting and communicating with each other.

Hall distinguishes different sorts of distance that people maintain: intimate (0–15 cm), personal (4–125 cm), social (125–365 cm), and public (365–750 cm or more). The measurements in centimetres sound too exact and absolute to the average anthropologist, but his point is well taken: people differ in what they consider a comfortable distance in the company of different types of people. These differences are not only culturally inscribed but depend on countless other aspects of their identity and the specific situation. He first discusses proxemics in animal behaviour and then moves to human beings. Hall points out that the perception of distance is not solely based on metrical space; vision, hearing, and smell also determine what distance is comfortable and when the proximity of other people is felt as invasive, threatening, or simply unpleasant. Two chapters of the book are about cross-cultural proxemic experiences of Americans versus German, British, French, Japanese, and Arabic people. They are the most ethnographic part of the book including some intriguing – but rather generalizing – observations. Two examples:

> Germans who come to America feel that our doors are flimsy and light. The meanings of the open door and the closed door are quite different in the two countries. In offices, Americans keep doors open; Germans keep doors closed. In Germany, the closed door does not mean that the man behind it wants to be alone or undisturbed, or that he is doing something he doesn't want someone else to see. It's simply that Germans think that open doors are sloppy and disorderly. To close the door preserves the integrity of the room and provides a protective boundary between people. Otherwise, they get too involved with each other. (pp. 135-136)

> Arabs look each other in the eye when talking with an intensity that makes most Americans highly uncomfortable. (p. 161)

A third classic is Barrington Moore's (1984) monograph *Privacy: Studies in Social and Cultural History*. The first chapter 'Anthropological perspectives' draws on observations by anthropologists who worked among various Amerindian communities and the Mbuti Pygmees in Central Africa. The next three chapters, which are the main body of the study, try to tease out information on ideas and practices concerning privacy in three societies in the distant past: classic Athens, Hebrew society as recorded in the Old Testament, and ancient China. The result – unavoidably – is a rather speculative description of juridical, political and philosophical data and some rare conjectural glimpses of everyday life, mostly among the urban elite. In a concluding chapter Moore asks 'what these investigations have revealed about the factors that promote or inhibit the growth of rights against intrusion, an expression that includes both personal aspects of privacy and private rights against holders of authority' (p. 267). The limitations of his sources do not allow Moore to draw solid conclusions about the different shades of privacy. He undoubtedly makes valuable comments regarding the advances of privacy and its enemies in his time (the 1970s and 1980s), for example about the opportunities and threats regarding privacy in a modernizing bureaucratic and industrialized society but he could have made these as well without the extensive data he draws from his various sources.

Anthropologists will find the handbook by social psychologist Irwin Altman (1975) the most useful introduction to the cross-cultural study of privacy. Altman builds upon Hall's concepts of proximity and personal territory. He discusses functions, meanings, mechanisms, and dynamics of privacy and focuses on personal space and crowding. Throughout the book he relates his observations to multidisciplinary social theory and research. Cultural variations in privacy mechanisms and coping with lack of (physical) privacy receive a fair amount of attention and constitute the main focus in a separate publication (Altman 1977).

In the 1977 article Altman rightly remarks that the use of ethnographic materials to 'test' his concept of privacy is problematic.

> Many cultural descriptions are not sufficiently explicit and were not developed with our particular model of privacy in mind. Thus, there may be instances in which a culture is described as having 'no privacy', examples are provided, and the situation is left at that. If we use such material are we to conclude that our hypothesis is invalid, and/or that it is not adequately testable because the ethnography may have been incomplete in its description of the total range of privacy regulation mechanisms? (p. 71)

The problem of interpretation is closely related to this. It is often impossible to deduce from the anthropological description if certain practices that seem privacy-motivated are in fact evidence of concern about what Altman calls 'privacy'. Clearly, more detailed and context-rich ('thick') ethnography is needed to draw conclusions about the meaning of privacy from a cross-cultural perspective. In the next section I will present and discuss some examples of anthropological research related to privacy, mostly, however, in an indirect and implicit way.

11.4 Traditional debates and dominant schools

In this section, eleven different topics of discussion will be presented that I regard relevant to the anthropological perspective on privacy. These are 'Guilt and shame', 'Elias' civilizing process', 'Witchcraft', 'HIV/AIDS', 'Physical and social privacy', 'Gossip', 'Secrecy', 'Lying', 'Privacy of the anthropologist', 'Undesired intimacy', and 'Thoughts'. In most of these debates, however, the concept of privacy is only indirectly referred to, for reasons stated in the previous section.

11.4.1 Guilt and shame

The oldest debate in anthropology related to privacy is probably about the existence of so-called 'guilt cultures' and 'shame cultures'. The assumption was that in shame cultures people's behaviour was controlled by feelings of honour and shame. Good and bad were determined by what others knew about them. In guilt cultures, on the contrary, what is good and what wrong was dictated by a person's personal conscience. These dichotomist concepts became popular after anthropologist Ruth Benedict's (1946) *The Chrysanthemum and the Sword*, in which she depicted Japan as a shame culture versus the US as a guilt culture. This distinction is now widely rejected as ethnocentrism of Western Christian scholars towards 'Non-Western' societies and a naïve underestimation of shame in their own society. Without mentioning the word 'privacy', those supporting the two concepts were in fact suggesting that the need and desire for privacy – in the sense of acting outside of the public eye – was more prominent in so-called Non-Western societies than in their home country. It should be pointed out however that the claims about guilt and shame in these debates were rarely made on the basis of intensive anthropological research as proposed in the introduction.

11.4.2 Elias' civilizing process

The German sociologist Norbert Elias who has influenced anthropology as well, also touches on this discussion about shame and conscience in his classic *The Civilizing Process* (2000[1939]). He discerns a gradual advance of human values and practices from community-imposed to personal choice. This process can be seen in many domains of public and private life, from political organization, and the state's monopolization of violence to table manners, the regulation of emotions and the shift from external restrain to self-restrain; in other words: from shame to guilt. Through all of this runs a growing emphasis on personal intimacy and individual privacy. The distinction and separation between the public and the private sphere is a crucial element in Elias' concept of the civilizing process.

Elias and his 'school' have been criticized for their evolutionist and ethnocentric view of civilization, but others have argued that the civilizing process is not a unilinear one-directional development. Counter movements and contradictory ideas occur as well. A good example of 'inconsistencies' in the history of civilization can be found in the work of the Dutch sociologist Cas Wouters.

Inspired by Elias' examination of books on manners during the past four centuries to trace processes of civilization, Wouters (1977; 2017) looked at etiquette books in various societies including Germany, UK, US, and the Netherlands. He discovered that instructions on good manners in the nineteenth and early twentieth centuries emphasized social distance and respect for privacy, particularly in the higher classes. This respect for privacy helped to sustain class differences, in Wouters' words: 'the right to privacy functions to facilitate the avoidance of lower-class people' (1977, 66). Towards the end of the twentieth century manners had become much less formal, which could be observed in practices like using the informal pronoun (*du, tu, je*), calling relative strangers by their first name, and in ways of greeting (social kissing). These developments signalled an 'increasing acceptance of greater social and psychic proximity' (1977, 69) and a blurring of the sharp boundary between public and private. As we will see in the course of this chapter, privacy ideas and practices are packed with contradictions and ambiguities.

11.4.3 Witchcraft

A 'classic' seemingly perpetual anthropological interest related to privacy is witchcraft. From long before 1937 when the British anthropologist E. Evans-Pritchard published his study of Azande witchcraft until today, witchcraft has fascinated anthropologists as both an exotic and 'irrational' phenomenon and – in another sense – as a familiar experience. Witchcraft is associated with traditional rural life as well as with modern urban society. The evil that is brought about through witchcraft is believed to come from close by, in particular from relatives who know the victim well and with whom they spend the day. Evans-Pritchard (1937, 37) wrote: 'The farther removed a man's homestead from his neighbours the safer he is from witchcraft'. And: 'a witch can injure the more severely the nearer he is to his victim'.

Nearly a century later Peter Geschiere (2013, xviii) writes: 'witchcraft is [perceived as] a form of aggression that is most dangerous because it comes from inside'.

Witchcraft has been discussed from many different perspectives, for example as a problem of social exclusion (in particular of women), as an explanatory model, and as a system of social levelling. Here it is presented as a phenomenon that disrupts or annuls the safe privacy of the home.

During my own research in Ghana I noticed that suspicions of witchcraft typically circulated among relatives (Bleek 1975). My research focused on conflicts within families. One – most hidden and fearful – conflict consisted of witchcraft suspicions and accusations between family members. Out of 27 members of two generations, only two (one dead, one alive) were not in any way involved in a case of witchcraft, either as accuser, accused, or assumed victim. The most malicious aspect of these witchcraft accusations was, however, not their high frequency, but the fact that they tended to occur between relatives living closely together. Strangely enough, witchcraft accusations did not necessarily originate from conflicts. Their occurrence was more obscure. Actual conflicts could pass without any allusion to witchcraft while outwardly peaceful relationships could be riddled with witchcraft suspicions. Hidden jealousy was usually given as an explanation for this. The enemy was hiding within the family. Privacy at home was not a secure and safe condition; 'home' and therefore 'privacy' proved precarious and ambiguous concepts. Dangers lurk everywhere. Respectability and safety were not secured in relations with intimate connections.[1]

It would be too simple to regard witchcraft as an exotic superstition that will eventually disappear as did the witch craze in Europe several

1 The Kenyan philosopher John Mbiti wrote that life in an African village made every member of the community dangerously naked in the sight of the other members. "It is paradoxically the centre of love and hatred, of friendship and enmity, of trust and suspicion, of joy and sorrow, of generous tenderness and bitter jealousies. It is paradoxically the heart of security and insecurity, of building and destroying the individual and community." (Mbiti, 1989: 204).

centuries ago. It also reveals a widespread awareness of the ambiguity of close proximity as exists within families. Those near you are also the ones who can harm you most severely. The Ghanaian proverb, which refers to witchcraft, also applies to the private sphere in European families: 'It is the insect in your cloth that bites you'. Familiarity does not only breed contempt; it can also destroy you.

11.4.4 Lack of privacy at home: HIV/AIDS

The recent problems around HIV/AIDS and its stigmatization illustrate this insecurity of the home. Let me quote somewhat freely from a paper I wrote with two Ghanaian colleagues who carried out research among people living with HIV/AIDS (Van der Geest et al. n.d.). One worked in the community, the other in the hospital where the patients went for treatment and medicines. Almost all HIV-positive persons whom the researcher in the community met had kept their status hidden. Thanks to the absence of overt symptoms or progression of the disease, little change occurred in their everyday lives after testing positive. They were therefore not compelled to reveal their status to those in their environment.

Many of those who were married or in a sexual relationship, especially women, did not even reveal their status to their partner. They knew what the consequences could be if their partner were to find out: breakdown of the relationship and divorce, loss of financial support, and possibly disclosure to others. More than 80% of the HIV positive people who were followed in the community had not disclosed their status to their family or friends. For those whose status had been disclosed to relatives, two consequences were most likely to happen: exclusion or collective concealment by the family, to prevent what Goffman (1968, 44) has called 'courtesy stigma': stigmatization by association. A severe example of exclusion and collective concealment was the case of a very sick woman whose relatives refused to spend an extra penny on her when they found out that she had tested HIV positive. 'She was going to die anyway and the money would go waste', a nurse explained. About three weeks after the researcher met her, he saw her obituary notice all around the community. A grand funeral was held for her. The family had rejected the live body but celebrated the dead one to avoid the shame of AIDS. In a clear act of collective impression management, the funeral was the family's strategy to keep the real cause of the woman's death 'private', even though many in the community were probably aware of the fact that she had died of AIDS.

Though people living with HIV were usually more likely to trust people in their own household than others, it was found that they were not inclined to inform them if they were receiving treatment for HIV/AIDS. Some avoided the nearby hospital and looked for treatment further away where they would not encounter acquaintances who could ask questions about the reason for the hospital visit. At home they kept all medical records in their possession – hospital cards, prescription forms, and even their medicines – away from prying eyes. One woman explained that she hid her medicines in a locked suitcase; one man hid his pills under the family sofa. After a hospital visit, some patients disposed of the ARV packages and leaflets even before they got out of the hospital. Another strategy was to scratch off the writing on the containers or put the medicines into a different box altogether.

The researcher in the hospital remarked that in spite of dangers lurking in the hospital, people with HIV/AIDS found there a safe haven once they were inside and met companions in misfortune and caring nurses and peers who were involved in the treatment and education on a voluntary basis. All these people knew the secret of the patients and thus formed a safe audience for their stories and problems. Stigmatized individuals viewed those who shared their particular stigma as their 'own'; they belonged to the same in-group, in contrast to those who were ignorant if not hostile towards HIV/AIDS. The shared experience of stigma created a strong sense of solidarity among the clients, and health workers sympathized with them and supported them in that situation. Health workers were adopted as 'parents', nurses became 'mothers', who helped them to take decisions on treatment and marital problems, while peer voluntary workers became the 'uncles' and 'aunts' who advised and assisted where needed. Clients, finally, shared with their fellow patients as 'siblings' their worries on a wide range of issues.

The ambivalence toward and hidden insecurity of the family home should be taken into account if we try to understand that people in certain situations avoid the privacy and intimacy of their own family and rather seek help from outsiders whom they trust more than their relatives. The hospital as a 'home' illustrates this. The rejection of family care derives from this long existing ambiguity and tension in the heart of Ghanaian families. Privacy is at risk in the place where it is widely believed to be most secured.

11.4.5 Physical and social privacy

In contrast to this example of the home not securing privacy and safety, we now turn to anthropological discussions of houses and living conditions that do offer privacy in spite of the fact that physical privacy seems entirely missing.

There seems to be almost general agreement that a longing for some kind of privacy is universal (e.g. Westin 1967; Altman 1975, 1977; Moore Jr

Fig. 11.2: Longhouse (Patterson & Chiswick 1981, 133)

1984; Lindstrom 2015), also when the physical and social conditions that make privacy possible (or nearly impossible) vary widely. This opinion is not based on a worldwide statistically tested investigation but on the few anthropological observations in widely varying cultural settings that they could find. Where living conditions hardly provide physical privacy, rules of proper behaviour and keeping social distance create an imaginary wall that protects mutual privacy. Patterson and Chiswick (1981) described such a situation for people in Kalimantan, Indonesia, who lived in a 'long house' that was shared by 150 individuals, comprising 22 families:

> The longhouse (…) appears to offer little in the way of physical environmental mechanisms useful in privacy regulation. The density is relatively high, a large portion of the family's space (the gallery and deck) is semi-public and open to communal view (and often communal use), and the apartments are separated only by insubstantial walls. The walls are ineffective as a sound barrier, and family noises can be clearly heard in neighboring apartments and in the gallery. Further, there is frequent and easy access from apartment to apartment. The apartment itself offers little within-family privacy, with large families eating and sleeping in the same room. (p. 135)

But the families had social mechanisms that provided the privacy they wanted, such as rules about who could enter the house, restricted movements in the night, and working patterns that excluded other families.

Hirschauer (2005) presents a comparable yet completely different example of people in a North American elevator, 'a place where strangers come together' in a small cabin for only a few seconds. Standing order and techniques of not speaking and avoiding eye contact prevent this unusual proximity of unknown bodies from being experienced as an intrusion of privacy. These are techniques of 'civil inattention', again a term coined by Goffman (1963, 84); they are 'A display of disinterestedness without disregard' (Hirschauer 2005, 41). (cf. what Westin and Altman write about 'reserve').

Van Hekken, a Dutch anthropologist, carried out fieldwork in a rural community in Tanzania, where neighbours can hear almost everything that happens next door. The rule of safeguarding privacy is buttressed by the belief that 'making noise' can cause a sickness called *ikigune* (curse) or *mbe sya bandu* (people's breath). When the neighbours hear a father and son shout at one another in the house next door, they start talking about it and may ask the father what happened. This talking and thinking about what happens in the neighbours' house can eventually lead to the sickness of a person in the house where the quarrel has taken place (Van Hekken 1986, 70).

In the extremely poor Malawian village where Janneke Verheijen did her research, people knew almost everything about each other. They tried to hide food and small luxuries such as soap or batteries from their neighbours to prevent jealousy and evade the social obligation of sharing. George Foster (1972) in a classic article on envy writes that hiding your 'riches' is one of the most effective ways of preventing other people's envy and – as a consequence – witchcraft. But in the Malawian community hiding was nearly impossible. Hiding batteries, for example, would mean not using them. If people would hear sounds of music, they would conclude that their neighbour had got some money to buy batteries. They would wonder how she managed to get it. From trading? Or perhaps from a man, a secret lover? Batteries would thus ignite gossip that entered the private intimate sphere of a woman and her household. In a footnote, Verheijen refers to a remark by Vaughan (1987, 34) about the survival strategies of rural Malawians during the severe famine of 1949. She writes that 'the food that could be found was brought to the household at night so that neighbours would not see it, and eating was done indoors instead of outside as usual' (Verheijen 2013, 211).

The severe poverty in this Malawian village and the lack of physical distance are colossal factors leading to the anxiety about privacy among its inhabitants. Such conditions are entirely absent in most Western societies. Privacy is indeed a privilege of the more well to do. The extreme caution to keep certain things hidden from neighbours is not necessary in my Dutch neighbourhood. Of course, there is a lot that I decide not to share with my neighbours, but it does not require any effort to do so. Technological and architectural facilities take care of this. Moreover, the neighbours have less reason to be curious about my private activities and possessions and they don't need my help to survive.

11.4.6 Gossip

In the previous two examples from Tanzania and Malawi, gossip was mentioned as a threat to privacy. Gossip has been studied and discussed by generations of anthropologists. The irony is that anthropologists themselves are invaders of people's privacy and eagerly engage in gossip to achieve their goals, not only because it provides much desired information but also because it confirms that the researcher has been accepted by the community.[2] André Köbben (1947, 42) overheard his two Surinamese co-researchers saying that anthropology is simply 'collecting gossip'. 'Not at all a bad definition of the bizarre activities of the ethnographer', he added. Indeed, for an anthropologist there is hardly better proof of success than becoming part of the local gossip network. But if that success is not reached, the researcher can himself initiate the gossip by luring or pressing someone into divulging private information about others. I must confess: it is a method that I have employed frequently. I always promised the person however that I would keep the information confidential, which is exactly what usually happens during gossip: 'I will tell you something but don't tell anyone else'.

The interesting thing about gossip is that it is private conversation about private matters. In other words, gossip remains 'private', because it is exchanged in secret, even if that secrecy is continuously broken.

11.4.7 Secrecy

Secrets – like gossip – have always fascinated anthropologists, partly for the same reason that people in general are attracted to them: what is hidden causes curiosity. But secrets also relate to 'cultural constructions of personhood, identity construction, and the dynamics of interpersonal relationships' (Lindstrom 2017, 374). When I asked an old man in Ghana what friendship is, he responded without thinking for a moment: 'A friend is someone with whom you share secrets'. The sharing of secrets 'fits' in with

2 Interestingly, the philosopher Aristotle used the term *anthropologos* (lit. 'talk about people') in the sense of 'gossip'. It had nothing to do with present-day anthropology, but the coincidence is amusing (Bok 1989, 90).

what Koops et al. (2017) call the intimate zone: the privacy that is enjoyed by a small group of people who trust one another. Those intimates are a blessing for the one who carries the secret. A secret that cannot be shared misses its main attraction. The joy of having or knowing a secret lies in telling someone about it. The secret is like glue that binds people. Secret societies, common male groupings throughout Africa that exclude women, are based on sharing a secret (or an assumed secret). The secret of secret societies may even be that there is no secret.

If sharing secrets with another person is a common way to establish and strengthen friendship, that friendship can again be effectively destroyed 'by spilling these secrets to third parties. Personal secrets are a social currency that people invest in their relationships' (Lindstrom 2015, 377). Sharing secrets is a telling example of what Simmel called privacy as a means to include others.

11.4.8 Lying in defence of privacy

The threat to privacy mainly comes from two sources: from concrete human persons (usually those who are close to the individual) and from advanced technology (behind which distant human beings hide). The technological threat is warded off by counter-technology; the more direct human threat is countered by age-old 'social techniques' of concealment, lying for example.

To follow up on pressing interlocutors to divulge intimate information – about themselves or about others – people may resort to lying. One could argue that someone during conversation with a researcher could simply decline to answer questions that he considers too invasive and personal, but in actual practice a refusal to answer is likely to be interpreted as an implicit confession. So an explicit lie becomes the only effective option in order to keep the intruder at bay. In a context of social inequality, the resort to lying seems particularly necessary because an outright refusal ('This is none of your business') may be considered rude and disrespectful. It should be taken into account however that lying is easily detected in a conversation-like face-to-face interview and could lead to further questions and props that eventually bring about the 'true' information.[3] As a matter of fact, lies usually show the way to matters that are most relevant and point to what is really at stake. Anthropologists, therefore, take a keen interest in lies since they are the sentries to the private domain.

3 Lying and other forms of concealment to protect one's identity is fairly common and accepted on the Internet (see further below, and: Hancock 2007; Hancock et al. 2007).

Ethical guidelines stipulate that researchers should respect the privacy of their interlocutors and not cause any harm to them, for example by revealing their identity. In qualitative research – as is the case in anthropology – individual interlocutors usually play a prominent role (through case histories, narratives, life histories, and anecdotes). Anonymizing interlocutors may pose difficult dilemmas for the writer. A common practice to conceal the identity of interlocutors without impoverishing the richness of the data is giving interlocutors fictitious names and changing some insignificant details of their identity. That has also been my strategy: I did not keep silent about the confidential things I recorded (on the contrary, I collected them in order to spread them through publications), but I made sure that those who gave me the confidential information could not be traced, nor the identity of those about whom they spoke.[4]

11.4.9 Privacy of the anthropologist

When anthropologists, especially those working outside their own cultural setting, discuss privacy, it mainly is their own privacy. Lack of privacy (in combination – paradoxically – with loneliness) is a common complaint of ethnographers who practice participant observation and try to live closely with the people among whom they carry out research. Malinowski, widely crowned as the pioneer of anthropological fieldwork, preferred to live in a tent at a safe distance from Trobriand families. This enabled him to write and read (and sleep) without being disturbed by the villagers. Jean Briggs (1970) spent almost two years with an Inuit family, including two Arctic winters where she stayed with the family in an igloo. She wrote a candid reflection on that long period in which she describes her moments of loneliness and longing for more privacy. But she realized that her longing clashed with local obligations of hospitality and sociability towards her.

To add one more example out of a myriad of anthropological 'confessions', Paul Spencer (1992, 53) recalled how – after a first period of fieldwork in a Maasai community in Kenya, he needed a break and 'a dose of English culture to be able to relax with others in my own language, and to indulge in some privacy'. Such yearning for privacy in the safety of one's own culture is a well-known experience among ethnographic researchers, not only in the past, the pre-Internet era, but also today with its numerous options for communication with people at home (see for example the list of fieldwork frustrations in Pollard 2009).

4 For a more nuanced account, see Van der Geest 2003.

11.4.10 Undesired intimacy

'Undesired intimacy' (*ongewenste intimiteit*) is the most common, explicit and effective qualification to express anger and disgust about the transgression of boundaries of human integrity in the Dutch language. The closest equivalent in English is probably 'sexual harassment', but 'undesired intimacy' is more to the point as it names the deeper source of discomfort and disgust: unbearable violation of one's private sphere. Moreover, the expression can also refer to other forms of privacy invasion than sexual insolence or aggression. One obvious example is the privacy surrounding defecation. Defecation is a normal daily activity which is not morally wrong. Yet in almost all societies defecation is done in a private location. It is normal to lock oneself up in a small apartment to be alone, not because you are doing a shameful thing but because it would be shameful to defecate for others to be seen. If people have no access to a private toilet and must defecate in a public place or in the open field, the same rules of observing privacy apply. One should not stare at defecating people. One should as it were pretend that there is no one relieving himself (another kind of civic inattention).

The discomfort is mutual: the one confronted with another person's act or substance of defecation is as embarrassed and disgusted as the person who is 'caught' in the act. But there are exceptions that reveal the deeper cause of the discomfort. The average mother is not disgusted while changing her baby's nappy and the baby cares even less. To a lesser extent, lovers and close friends are not worried by each other's faeces, especially not at a young age. When people grow older and acquire a more prominent own identity, the (mutual) invasion of one's intimate sphere begins to be felt more strongly. The body substances of others become dirtier and unpleasant. Those of more distant others, relatives, neighbours, colleagues with whom one has a more superficial relationship are considered equally dirty and disturbing since one does not want to share intimacy with them. Hall's (1975) grades of proximity apply here, not in metric measurements but in terms of emotional and psychological distance.

Mary Douglas' (1966) concept of dirt as 'matter out of place' or 'disorder' clearly fits here as well. Sharing intimacy with a non-intimate person, whether it is bodily waste, or sex, or secret information or anything personal, upsets and causes revulsion. It is out of place, improper. But, by a remarkable twist of human experience, the amount of discomfort caused by sharing intimacy with a non-intimate person may be less disturbing if that person is a complete stranger. The stranger does not have a clear identity that invades mine. I may never see him/her again.

The example of the use of a public toilet illustrates this. When I realize that people at the airport are waiting for me to leave the toilet after I have defecated, I feel uneasy, but I can escape untainted since I will never again see the person who will enter my smell. But if the same situation presents itself at the anthropological institute where I work, I will be much more worried. Every time the student or colleague after me meets me s/he may remember the incident and feel equally uneasy. It is unintentional but nevertheless a moment of light undesired intimacy (cf. Lea 2000). It may seem a long jump, but in the same way the divulging of private information to a researcher who will disappear in a few months' time may be easier and safer than telling the same things to a relative or neighbour.

11.4.11 Thoughts

Authors agree that thoughts are the ultimate bastion of privacy. The SF fantasy of a 'thought police' is indeed the most frightening spectre of a future world that some believe to be on its way, referring to the growing power of big data technology.

Some years ago, during guest lectures in Vienna I gave the students an assignment to write about one page on what they considered most private and why. One female student wrote the following:

> I think this is an inappropriate assignment. With all respect: you are my teacher and I am your student. I think it is not important for a teacher to know this about his students. There should be a limit between the teacher-student relationship.

I accepted the critique but continued giving similar assignments during lectures on privacy at my own university. I never received the same severe rebuke, but in a more shrouded way I was told the same thing occasionally. The most frequent response, next to defecation and nakedness, was: thoughts. Three quotes from one assignment:

> What do you consider most private in your life? The first thing that came to my mind when I was asked this question was my thoughts. Even when I am physically unable to withdraw myself from public view, I can still exclude others from what I think (...).
> What if all aspects of my life, except for my thoughts, were open to the public? Everyone would know what I do every moment of every day, and thus everyone would get a pretty accurate picture of who I am, but I still

> have the soothing possibility to withdraw my thoughts from the public realm, thus I still have some privacy. Nevertheless, this kind of privacy is, in practice, totally useless. If I cannot share my thoughts with a select group of people, what is the use of having private thoughts at all? (...)
> In conclusion, I think intimate relationships are the most valuable thing in private life, because privacy works both ways. If there is no possibility to share my thoughts with a select group or individual, there is no point in having privacy. Privacy is often defined as a freedom, and my choice to share certain thoughts or experiences with only those who I choose, gives me that freedom.

The author escapes from the dilemma s/he formulated in the second paragraph: privacy is the freedom to share and not to share. Paradoxically private thoughts are not necessarily shared with soul mates. As we saw before, some thoughts may rather be kept hidden precisely from friends, partners, or children and shared with a passer-by or a distant acquaintance or a researcher. Privacy lies in the possibility – the freedom – to share or not to share, to open the door or keep it closed.

The possibility of keeping private thoughts secret may be in danger if we believe some pessimists and writers of dystopic fiction, as I just mentioned, but for the time being there is more reason for optimism. Thoughts represent the hard core of privacy. The old German song 'Die Gedanken sind frei' comes to mind; even where no freedom exists in the popular sense of the word, and where nearly permanent surveillance takes place, as in captivity, there is the freedom of thoughts, privacy.

> Die Gedanken sind frei, wer kann sie erraten?
> Sie fliehen vorbei wie nächtliche Schatten.
> Kein Mensch kann sie wissen, kein Jäger erschießen
> mit Pulver und Blei: Die Gedanken sind frei![5]

11.5 New challenges and topical discussions

This section will discuss two recent issues I can think of that raise questions about privacy: first the globalization through Internet and social media which according to many has deeply changed the experience and meaning of

5 Thoughts are free, who can guess them? They fly by like nocturnal shadows. No man can know them, no hunter can shoot them, with powder and lead: Thoughts are free!

privacy; the second is the reality of living longer and the failing control over privacy it entails for the elderly. Both phenomena are frequently discussed in political debates and public media.

11.5.1 Internet and social media

In an editorial to a special issue on privacy and the Internet Jacquelyn Burkell (2008) sketches how the landscape of privacy has changed:

> Frequent flyer plans archive our travel histories, debit cards track our purchases, cell phones announce our location, online registrations for Web sites collect our identifying information, social networking profiles reflect our personal lives, blogs display to any who choose to look details about our attitudes, preferences, and desires. And that, of course, is only the start. When digitized, information held by government such as health records or income tax records becomes (at least potentially) part of the mix. Our digital shadows grow ever more complex, ever more revealing, and ever more interesting to those with a desire to know who we are, what we do, and what we think.

There is no doubt that the Internet has enormously affected and expanded the threats to our privacy but authors disagree about the question if it has indeed changed our sense of privacy and dealing with it, as Burkell seems to suggest. Some rather argue that the Internet has provided us with new potentials to secure our privacy (as we will see further below).

Anthropology is only hesitatingly engaging with the Internet as a research topic but the public debate about the loss of privacy through the Internet and social media *has* triggered a growing interest in privacy among anthropologists in contrast to their earlier negligence (see the previous section). The use of Facebook in particular has received ample attention from a group of researchers around Daniel Miller, well known for his work on material culture. Miller has now turned to 'digital anthropology'. In 2000, his first ethnography of Facebook use in Trinidad appeared, co-authored with Don Slater. They rejected the idea that the Internet constitutes a different reality and emphasized that it should be studied a part of the 'real' social world:

> We need to treat Internet media as continuous with and embedded in other social spaces, that they happen within mundane social structures and relations that they may transform but that they cannot escape into

> a self-enclosed cyberian apartness (Miller and Slater 2000, 4; see also Dourish and Bell 2011, 59).

In 2008, Miller and Heather Horst edited a volume on digital anthropology including contributions on media technologies in everyday life, 'Geomedia' (location-tracking technologies), disability, personal communication, social networking, and gaming. Privacy concerns are discussed in most of the chapters. In 2011, Miller published another Facebook ethnography on Trinidad, in which he followed twelve different people who all used Facebook for very different purposes to demonstrate the wide variety of meanings and goals that people attach to this social medium.

A most fascinating and efficient project on social media resulted in eleven monographs by Miller and colleagues. Nine are ethnographies about the use of social media in different locations: Brazil, Chile, China (rural and urban), England, India, Italy, Trinidad, and Turkey; one is a comparative overview all nine ethnographies, and one contrasts the visuals that people post on Facebook in England and Trinidad. The publisher UCL Press set an example of how Internet can facilitate the study of Internet by placing the entire series of eleven studies as open access on the Internet.

Let me highlight the most relevant observations and claims that have been made by the authors of this series. First of all: the traditional anthropological approach of lengthy participatory fieldwork is stressed:

> Everything we do and encounter is related as part of our lives, so our approach to people's experience needs to be holistic. The primary method of anthropology is empathy: the attempt to understand social media from the perspective of its users (…) this project was always collaborative and comparative, from conception to execution to dissemination (Miller et al. 2017, xi).

The authors of the nine ethnographies had built relationships with their interlocutors over a long period and were thus able to place their Internet practices in the context of other aspects of their lives and to look at these practices from the users' point of view.

A recurring finding is that Internet users are not helpless people that fall victim to the machinations of the Internet and lose grip on their private lives. Privacy is a process of optimal management of disclosure and withdrawal. Most users of Facebook were well aware of what they could share with whom and what not. They wanted to be 'seen' (cf. Tufecki 2008), but also knew how to hide themselves if needed. Young people in rural

China used avatars and user aliases and shared passwords with peers to conceal their identity from strangers while allowing friends to read their messages (McDonald 2016, 184-185). Costa (2016, 125) describes the case of a young woman in Turkey who strategically manipulated her presentation on Facebook:

> she made public those images in which she appeared more modest and decorous, but kept completely private those photos that could have damaged her reputation in Hasan's [her boyfriend] circle. She did her best to appear beautiful and be appreciated by her boyfriend and his family. She was well aware of all Facebook privacy settings, and she accurately changed them in every different circumstance. Her intricate uses of Facebook's privacy settings were probably much more elaborate than those envisaged in Palo Alto in California.

Marwick and boyd (2014, 1051) writing about network privacy quote a young man:

> Every teenager wants privacy. Every single last one of them, whether they tell you or not, wants privacy (…) Just because teenagers use internet sites to connect to other people doesn't mean they don't care about their privacy (…) So to go ahead and say that teenagers don't like privacy is pretty ignorant and inconsiderate honestly, I believe, on the adult's part. (1051-1052) (see also boyd 2014)

But in an earlier statement boyd (2006) had been less optimistic. She accused Facebook of making complete openness the default which had led users into unintended public exposure (especially children; see also Livingstone 2008) (quoted by Broadbent 2012, 149). Nicolescu (2016, 102) reported that 48% of respondents in a household survey in Southeast Italy declared 'they had never changed their Facebook profiles to private. Most of them did not know there was such a possibility'. Not only ignorance causes privacy risks on the Internet however, ingenuity can also be a threat. Costa (2016, 1130), for example, describes clever ways of young people in Turkey to circumvent privacy locks and leak confidential information to outsiders, often in the case of (broken) love affairs. The overall conclusion of the digital ethnographers however is that – contrary to public opinion – Internet visitors are reasonably competent to secure their privacy if they want to. But a 100% success rate in the protection of privacy is never possible, neither in 'real' life nor online.

Another notable conclusion of the researchers is that the Internet does not present a totally different reality than in ordinary life but is rather a continuation of existing living conditions and views. In his description of a rural English community Miller (2016, 5) discovered that English people exploited social media to do what they had always been doing: calibrating 'the precise distance they desire for a given social relationship – neither too cold nor too hot but 'just right''. Another observation by Miller is that Internet enables the English people to stay in contact with old friends, relatives, and colleagues who moved away from the village in this time of increased mobility:

> Facebook had helped them to return to the older experience of when this was a community, not just a shared workspace. Similarly there are many examples of WhatsApp groups that form around family members now living in different places. There are also attempts to retain the community of the school class when people drift to dispersed colleges and work. In all such cases Facebook seems a bulwark against the potential loss of community (Miller 2016, 185).

Similar observations of digital ways of continuing and enhancing existing emotions and experiences have been reported from Trinidad (Miller and Slater 2000; Sinanan 2017), Turkey (Costa 2016), and Italy (Nicolescu 2016) to mention only a few.

Social media may also change local traditions and views related to privacy. Costa (2016, 52) described how in Turkey, where everything taking place in the house was private, ordinary family events such as meals lost much of their strictly private character due to images posted on Facebook. The same applied to expressions of affection and body presentation as a result of Internet 'images of engaged and married couples holding hands or hugging each other and photos portraying the bodies and faces of women'. Girls may add strangers to their Facebook profile and have private conversations with them, which is considered morally reprehensible. 'they smartly change the privacy settings to avoid being seen by other friends and relatives or they create fake profiles', because being seen to be in touch with strangers would be condemned even if it did not have any romantic intention (p. 100).

One of the most remarkable 'discoveries' was that in certain circumstances Facebook and other social media provide a privacy that does not exist in the home; they offer an escape from the privacy-less conditions of daily life. MacDonald (2016, 186) noted that Facebook users in China were sometimes more concerned about how their peers and family would react to their online behaviour than the administrative powers. Horst (2012, 66) writes that young people 'turn to sites like Facebook because they feel that what they can do and express in these spaces is more private than their physical homes'. She further notes that bedrooms are important spaces 'where young people feel relatively free to develop or express their sense of self or identity'. Privacy in this case is a greater problem in the house than online.

The rapidly growing literature on privacy-related repercussions of Internet use is more extensive than what I could present here. Future anthropological research should also focus on the use of personal information for commercial or political purposes and the storage of big data that contain the most diverse information on our personal habits, preferences, and movements. Excesses such as sharing private pictures and messages to harm a person, fraud, extortion, and blackmail also need more attention from a social and cultural perspective.

11.5.2 Ageing, care and privacy

Finally, I want to draw attention to one of the most pressing challenges of present-day society, *ageing*, through the lens of privacy. Much has been written about the demographic transition taking place as a result of the spectacular increase of life expectancy during the past century and its prognoses for the future. The economic burden, the shortage of professional and informal caregivers, the implications for medical facilities, the impact on family life, and many other aspects of this transition have been extensively discussed in various media. Studies of what growing old means to older people themselves in 'Western' society mainly focus on health problems, in particular (fear of) dementia, loneliness, and growing dependency (next to optimistic accounts of vital ageing and active retirement). How fragile old age affects the security and comfort of privacy is however little being studied.

When growing old leads to decreasing physical and mental well-being, it will unavoidably also affect conditions of privacy. This loss of privacy arrives in two ways. One is the way of 'normal' development. Giving up – bit by bit – parts of privacy is a natural necessity linked to the fact that older persons need the help of others to carry out activities that have become too difficult for them. The other route is that they are 'robbed' of their privacy, especially in care institutions. To start with the latter, violation of older people's privacy after they have moved from their own house to an institution is a rather common topic in anthropological studies. Mary

Applegate and Janice Morse (1994) start their account of care in a nursing home in the US with a complaint of an older person:

> What a disgrace to be seen crying by that fat Doris. The door of my room has no lock. They say it is because I might be taken ill in the night, and then how could they get in to tend me (*tend*-as though I were a crop, a cash crop). So they may enter my room any time they choose. Privacy is a privilege not granted to the aged or the young. (p. 413)

Rules for privacy are part of the institution's policy but are frequently disregarded due to the heavy workload. The objectives of the caregiver come first and privacy was invaded if care activities required it. The autonomy of the resident was thus jeopardized leading to a loss of self-worth and dignity. Patients were reduced to objects in the eyes of the researchers:

> Many times, residents were ignored as if they were invisible. Things were done to residents without consideration for their feelings, including respect for their privacy (...) staff made no attempt to knock when they entered the bathroom. One nurse was observed changing an incontinent patient's pants in the corridor. Staff did not consider the patient by requesting permission to enter their lockers. (Applegate and Morse 1994, 427)

In a similar vein Eleanor Schuster (1976) explored the experiences of older people in an institution in the US. She observed that problems arise 'when the person's ability to control the degree and form of distancing is impaired or impinged upon in some way. Often, such dissonance is seen by the individual as 'invasion of privacy'' (p. 246). Two Indian anthropologists who carried out research in a Dutch home for older people and a nursing home were fascinated by the strong desire for privacy among the residents in those institutions. They described various strategies they employed to defend their privacy, both against co-residents and staff members (Chowdhury 1990; Chatterji 2016).[6] Undesired invasion of privacy in these and other studies of ageing is indicated by terms like 'dehumanization', 'objectification', 'lack of respect', and 'loss of dignity'. Infringement of their privacy is experienced as violation of their personhood.

But there is also a loss of privacy that is a natural and unavoidable fact of life in old age. It is a loss that is necessary to grow old successfully and gracefully. The freedom and independence of the younger years allowed a

6 For an overview of privacy studies in the context of nursing, see Leino-Kilpi et al. 2001.

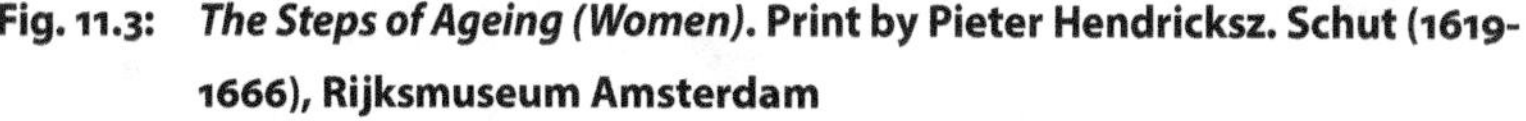

Fig. 11.3: *The Steps of Ageing (Women)*. Print by Pieter Hendricksz. Schut (1619-1666), Rijksmuseum Amsterdam

high degree of privacy but that privacy must be gradually given up when old age necessitates the older person to ask for assistance. Loss of independence – implies loss of control over privacy: the intimacy of the ageing body, private matters like bank accounts and handling money, and taking highly personal decisions about housing, hospital, and family. Growing old requires a continuous adjusting of one's life and accepting the 'interferences' of others. In this respect it represents a return to early childhood, to the position of a child that has no privacy. Popular prints from the seventeenth century illustrate this cyclical movement of life. The bed-ridden older person on the right finds herself as devoid of privacy as the new-born baby on the left. The art of growing old is to accept this circularity without turning bitter or resentful (cf. Von Faber et al. 2001). This reconciliation with the loss of one of life's most cherished values (or the failure to accept this loss) has been insufficiently studied from an anthropological perspective.

Looking at ageing as a process of losing control over privacy reveals that respect for privacy and privacy management is a challenge for the elderly as well as for the people around them, both at home and in institutions.

11.6 Conclusion

This chapter started with a brief introduction to the anthropological perspective: its contextualizing approach and its small-scale participatory style of doing research with a special interest in social and cultural differences and personal experience. The following section argued that the old definitions and concepts by Westin (1967) and Altman (1975) of privacy still provided fruitful starting points for the exploration of meanings and experiences around privacy in varying social and cultural settings. Altman's most important insight is that 'privacy' is not a more-or-less static condition but a dynamic process of having control over what one wants to share with selected others and what not.

The section on classic texts and traditional debates revealed the relative neglect of privacy by anthropologists. A surprising discovery since working in other cultures and living closely with their interlocutors confronted them with striking differences in local managements and experiences of privacy. Observations about privacy remained however largely implicit in their ethnographic work. Indirect allusions to privacy can be found in debates about shame, social manners, witchcraft, family life, stigmatization (HIV/AIDS), gossip, secrets, lying, and disgust. Privacy was given more explicit attention in discussions about social versus physical privacy, in reflections on the researcher's own privacy, and ethical accounts about confidentiality regarding interlocutors. The section ended with a few remarks about thoughts as the ultimate privacy control.

The digital age we are living in poses an important new challenge for the anthropological study of privacy. Concerns about privacy in the face of the overwhelming presence of the Internet and social media fill the chapters of this handbook. Privacy has become highly political. It is one of the hottest topics in public debates in almost every domain of society. This explosion of national and international disputes has also affected anthropologists. The past decade anthropology has devoted more attention to privacy as a central value of personhood and social living than in the entire one and a half century of its existence. The chapter ends with a plea for more ethnographic and theoretical exploration of societal processes using privacy as its lens, in particular with regard to the digitalization of our environment and the challenges of ageing.

Finally, this chapter on privacy ethnography suggests surprising similarities between privacy in the present digital era and in the pre-digital *Gemeinschaft*-type community where nearly anyone was known to anyone. Where the baker was acquainted with the family and the character of the woman he employed in his shop and neighbours were aware of each other's peculiarities. Where families knew the family of the boy with whom their daughter had fallen in love. Where the grocer knew what his client wanted to buy before she had spoken a word. And so forth. Life in small-scale communities was not so different from Bentham's (and Foucault's) Panopticon and the present situation of increasing digitalization of information for political, commercial, and security purposes. The paradoxes and tensions in present-day navigating between privacy and the public eye (cf. Koops 2017) differ mainly in size from what past generations always have been doing and coping with. I admit the difference in size is significant, but the similarities are no less significant. Without accepting some intrusions of privacy society cannot exist.

Further reading

Altman, I. (1977). 'Privacy Regulation: Culturally Universal or Culturally Specific?' *Journal of Social Issues* 33, 66-84.

Applegate, M. and J.M. Morse. (1994). 'Personal Privacy and Interactional Patterns in a Nursing Home'. *Journal of Aging Studies* 8(4), 413-434.

Magi, T.J. (2011). Fourteen Reasons Privacy Matters: A Multi-disciplinary Review of Scholarly Literature. *Library Quarterly* 81 (2): 187-209.

Miller, D. (2016). *Social Media in an English Village: Or How to Keep People at Just the Right Distance*. London: UCL Press.

Smith, J.M. (2004). 'Personal Privacy: Cultural Concerns' in N.J. Smelser, and P.B. Baltes (eds.), *International Encyclopedia of the Social and Behavioral Sciences* volume 11. Amsterdam: Elsevier Science, 250-254.

Van der Geest, S. (2018). 'Lying in Defence of Privacy: Anthropological and Methodological Observations'. *International Journal of Social Research Methodology* Open access. doi.org/10.1080/13645579.2018.1447866

References

Altman, I. (1975). *The Environment and Social Behaviour: Privacy, Personal Space, Territory, Crowding*. Monterey: Brooks/Cole Publishing.

Altman, I. (1977). 'Privacy Regulation: Culturally Universal or Culturally Specific?' *Journal of Social Issues* 33, 66-84.

Applegate, M. and J.M. Morse. (1994). 'Personal Privacy and Interactional Patterns in a Nursing Home'. *Journal of Aging Studies* 8(4), 413-434.

Benedict, R. (1946). *The Chrysanthemum and the Sword: Patterns of Japanese Culture*. Boston: Houghton Mifflin Company.

Bleek, W. (1975). *Marriage, Inheritance and Witchcraft: a Case Study of a Rural Ghanaian Family*. Leiden: Africa Study Centre.

Bok, S. (1989). *Secrets: on the Ethics of Concealment and Revelation*. New York: Vintage Books.

boyd, d. (2006). 'Facebook's 'privacy trainwreck': Exposure, Invasion, and Drama'. *Apophenia Blog*, 8 September. www.danah.org/papers/FacebookAndPrivacy.html.

boyd, d. (2014). *It's Complicated: the Social Lives of Networked Teens*. New Haven: Yale University Press.

Briggs, J.L. (1970). *Never in Anger: Portrait of an Eskimo Family*. Cambridge, MA: Harvard University Press.

Broadbent, S. (2012). 'Approaches to Personal Communication' in H.A. Horst and D. Miller (eds.), *Digital Anthropology*. Oxford: Berg, 127-145.

Chatterji, R. (2006). 'Thinking about Dementia' in A. Leibing and L. Cohen (eds.), *Culture, Loss, and the Anthropology of Senility*. New Brunswick: Rutgers University Press, 218-239.

Chowdhury, S.D. (1990). 'Privacy, Space and the Person in a Home for the Aged'. *Etnofoor* 3(2): 32-47.

Costa, E. (2016). *Social Media in Southeast Turkey: Love, Kinship and Politics*. London: UCL Press.

DeCew, J. (2013). 'Privacy'. Stanford Encyclopedia of Philosophy. http://plato.stanford.edu/entries/privacy/ 18 July 2015.

Desjarlais, R.R. (1991). 'Poetic Transformations of 'Yolmo sadness''. *Culture, Medicine and Psychiatry* 15, 387-420.

Dourish, P. and G. Bell. (2011). *Divining a Digital Future: Mess and Mythology in Ubiquitous Computing*. Cambridge, MA/London: MIT Press.

Elias, N. (2000[1939]). *The Civilizing Process. Sociogenetic and Psychogenetic Investigations*. Cambridge, MA: Blackwell.

Foster, G. (1972). 'The Anatomy of Envy: A Study in Symbolic Behavior'. *Current Anthropology* 13(2), 165-202.

Goffman, E. (1963). *Behavior in Public Places. Notes on the Social Organization of Gatherings*. New York: Free Press.

Goffman, E. (1968). *Stigma: Notes on the Management of Spoiled Identity*. Harmondsworth: Penguin.

Hancock, J.T. (2007). 'Digital Deception: When, Where, and How People Lie Online' in K. McKenna et al. (eds.), *Oxford Handbook of Internet Psychology*. Oxford: Oxford University Press, 287-301.

Hancock, J.T., C. Toma, and N. Ellison. (2007). 'TheTruth about Lying in Online Dating Profiles' in *CHI 2007: Proceedings of the ACM Conference on Human Factors in Computing Systems*. New York: Association for Computing Machinery, 449-452.

Hirschauer, S. (2005). On doing being a stranger: The practical constitution of civil inattention. *Journal for the Theory of Social Behaviour* 35 (1): 41-67.

Horst, H.A. and D. Miller (eds.). (2012). *Digital Anthropology*. Oxford: Berg.

Köbben, A.J.F. (1967). 'Participation and Quantification: Fieldwork among the Djuka' i: D.G. Jongmans and P.C.W. Gutkind (eds.), *Anthropologists in the Field*. New York: Humanities Press, 35-55.

Koops, J.-B. (2017). 'Bert-Jaap Koops over huisrecht 2.0: Rechtsbescherming op straat voldoet niet meer door ICT'. https://www.tilburguniversity.edu/nl/thema/veiligheid/audio-video/koops-over-huisrecht-en-ict/

Koops, J.-B. et al. (2017). 'A Typology of Privacy'. *University of Pennsylvania Journal of International Law* 3(2), 483-575.

Lea, R. (2000). *The Performance of Control and the Control of Performance: Towards a Social Anthropology of Defecation*. PhD dissertation, Brunel University London (unpublished).

Leino-Kilpi, H. et al. (2001). 'Privacy: a Review of the Literature'. *International Journal of Nursing Studies* 38, 663-671.

Lindstrom, L. (2015). 'Anthropology of Secrecy' in *International Encyclopedia of the Behavioral and Social Sciences*. Elsevier vol. 21, 374-379.

Livingstone, S. (2008). 'Taking Risky Opportunities in Youthful Content Creation: Teenagers' Use of Social Networking Sites for Intimacy, Privacy, and Self-expression'. *New Media and Society* 10(3), 393-411.

Mbiti, J. (1989). *African Religions and Philosophy.* Oxford: Heinemann Publishers.
Miller, D. (2011). *Tales from Facebook: Culture Online in Trinidad.* Cambridge: Polity Press.
Miller, D. (2016). *Social Media in an English Village: Or How to Keep People at Just the Right Distance.* London: UCL Press.
Miller, D. and D. Slater. (2000). *The Internet: an Ethnographic Approach.* New York: New York University Press.
Miller, D. et al. (2017). *How the World Changed Social Media.* London: UCL Press.
Moore Jr, B. (1984). *Privacy: Studies in Social and Cultural History.* London: M.E. Sharpe.
Nicolescu, R. (2016). *Social Media in Southeast Italy: Crafting Ideals.* London: UCL Press.
Patterson, A.H. and N.R. Chiswick. (1981). 'The Role of the Social and Physical Environment in Privacy Maintenance among the Iban of Borneo'. *Journal of Environmental Psychology* 1, 131-139.
Pollard, A. (2009). 'Field of Screams: Difficulty and Ethnographic Fieldwork'. *Anthropology Matters* 11(2), 1-24.
Schuster, E. (1976). 'Privacy, the Patient and Hospitalization'. *Social Science and Medicine* 10, 245-248.
Simmel, A. (1971). 'Privacy is Not an Isolated Freedom' in J.R. Pennock and J.W. Chapman (eds.), *Privacy.* New York: Atherton Press, 71-88.
Smith, J.M. (2004). 'Personal Privacy: Cultural Concerns' in N.J. Smelser, and P.B. Baltes (eds.), *International Encyclopedia of the Social and Behavioral Sciences* volume 11. Amsterdam: Elsevier Science, 250-254.
Spencer, P. (1992). 'Automythologies and the Reconstruction of Ageing' in J. Okely and H. Callaway (eds.), *Anthropology and Autobiography.* London/New York: Routledge, 49-63.
Tufecki, Z. (2008) 'Can You See Me Now? Audience and Disclosure Regulation in Online Social Network Sites'. *Bulletin of Science, Technology & Society* 28(1), 20-36.
Van der Geest, S. (1998). 'Participant Observation in Demographic Research: Fieldwork Experiences in a Ghanaian Community' in A.M. Basu and P. Aaby (eds.), *The Methods and Uses of Anthropological Demography.* Liège: IUSSP, 39-56.
Van der Geest, S. (2003). 'Confidentiality and Pseudonyms: a Fieldwork Dilemma from Ghana'. *Anthropology Today* 19(1), 14-18.
Van der Geest, S. (2007). 'The Social Life of Faeces: System in the Dirt' in R. van Ginkel and A. Strating (eds.), *Wildness and Sensation: an Anthropology of Sinister and Sensuous Realms.* Amsterdam: Het Spinhuis, 381-397.
Van der Geest, S. (2018). 'Lying in Defence of Privacy: Anthropological and Methodological Observations'. *International Journal of Social Research Methodology.* Open access. doi.org/10.1080/13645579.2018.1447866.
Van der Geest, S., J.M. Dapaah, and B.J. Kwansa (n.d.) 'Avoided Family Care, Diverted Intimacy: How People with HIV Find New Kinship in Two Ghanaian Hospitals'. Submitted.
Van Hekken, P.M. (1986). *Leven en werken in een Nyakyusa dorp* [Life and work in a Nyakyusa village]. Leiden: African Studies Centre.
Verheijen, J. (2011). *Balancing Men, Morals and Money: Women's Agency between HIV and Security in a Malawi Village.* Leiden: African Studies Centre.
Von Faber, M. et al. (2001). 'Successful Aging in the Oldest Old: Who Can Be Characterized as Successfully Aged?' *Archives of Internal Medicine* 161, Dec 10/24, 2694-2700.
Westin, A. (1967). *Privacy and Freedom.* New York: Atheneum.
Wouters, C. (1977). 'Informalization and the Civilizing Process' in P.R. Gleichman (ed.), *Human Figurations: Essays for Norbert Elias. Amsterdams Sociologisch Tijdschrift,* 437-453.
Wouters, C. (2007) *Informalization: Manners and Emotions since 1890.* London: Sage.

About the Authors

W.J.M. Aerdts LLM MA is a lecturer employed by the department of Intelligence & Security of the Institute of Security and Global Affairs (Leiden University, the Netherlands). Before, she used to work for the University of Amsterdam where she taught courses for the minor Intelligence Studies. During her time at the Dutch Ministry of Foreign Affairs, Willemijn conducted research on the Western Balkan. Nowadays, her research focusses on intelligence methodology, analysis techniques and oversight. Willemijn is a Worldconnector and a board member of the Netherlands Intelligence Security Association (NISA).

Anita L. Allen is an expert on privacy law, the philosophy of privacy, bioethics, and contemporary values, and is recognized for scholarship about legal philosophy, women's rights, and race relations. She is a graduate of Harvard Law School and received her Ph.D. in Philosophy from the University of Michigan. She was the first African American woman to hold both a PhD in philosophy and a law degree. She was an Associate Attorney with Cravath, Swaine and Moore At Penn she is the Vice Provost for Faculty and the Henry R. Silverman Professor of Law and Professor of Philosophy. She was elected to the National Academy of Medicine in 2016. In 2010 she was appointed by President Obama to the Presidential Commission for the Study of Bioethical Issues.

Her books include *Unpopular Privacy: What Must We Hide* (Oxford, 2011); *The New Ethics: A Guided Tour of the 21st Century Moral Landscape* (Miramax/Hyperion, 2004); *Why Privacy Isn't Everything: Feminist Reflections on Personal Accountability* (Rowman and Littlefield, 2003); and *Uneasy Access: Privacy for Women in a Free Society* (Rowman and Littlefield, 1988), the first monograph on privacy written by an American philosopher. Her textbooks include: *Privacy Law and Society* (Thomson/West, 2016), the most comprehensive textbook on the US law of privacy and data protection on the market, with chapters on the common law, constitutional law, federal statutory law, surveillance law and international standards.

Allen, who has published more than a hundred scholarly articles, book chapters and essays, has also contributed to popular magazines, newspapers and blogs, and has frequently appeared on nationally broadcast television and radio programs. Allen is active as a member of editorial, advisory, and charity boards, and in professional organizations relating to her expertise in law, philosophy and health care. Elected President of the American Philosophical Association (Eastern) in 2018, she was a member of the NIH Precision Medicine IRB, and has served on the boards of the Hastings Center, the Bazelon Center for Mental Health Law, and the Electronic Privacy Information Center.

Kenneth A. Bamberger is The Rosalinde and Arthur Gilbert Foundation Professor of Law at the University of California, Berkeley. He is Faculty co-Director of the Berkeley Center for Law and Technology (BCLT) and of the Berkeley Institute for Jewish Law and Israel Studies, and is a core faculty member of the Berkeley Center for Law and Business (BCLB).

Prof. Bamberger is an expert on government regulation, corporate compliance, and technology in both the United States and Europe. At Berkeley, he teaches Administrative Law; the First Amendment (Speech and Religion); Corporate Compliance; the Law and Technology Workshop; and Jewish Law. For his recent book, *Privacy on the Ground: Driving Corporate Behavior in the United States and Europe* (MIT Press), Bamberger and his co-author, Berkeley I-School Prof. Deirdre Mulligan, were awarded the 2016 Privacy Leadership Award from the International Association of Privacy Professionals. His articles on "*Platform Market Power*," and "*Saving Governance-by-Design*," will be published in 2018.

Bamberger graduated from Harvard Law School, where he was President of the *Harvard Law Review*. Before coming to Berkeley Law, he clerked for federal appeals court Judge Amalya L. Kearse and U.S. Supreme Court Justice David H. Souter, served as a Bristow Fellow in the Office of the United States Solicitor General, and was an associate, and then counsel, at the Wilmer Hale firm in Washington DC.

Outside the law school, Prof. Bamberger has served on the advisory boards of the Future of Privacy Forum, the Israel Institute, and The Taube Foundation for Jewish Life; on the Program Committee for the European Privacy Law Scholars Conference (PLSC); and on the ADL's Anti-Cyberhate Working Group. In the fall of 2017, he was selected for the U.S. Department of Commerce-European Commission list of arbitrators developed as part of the EU-U.S. Privacy Shield Framework.

From 2013-2015, Bamberger was the first co-chair of UC Berkeley's Center for Jewish Studies. He has been a Visiting Professor at Tel Aviv University

Law School, a Visiting Fellow at the Shalom Hartman Institute in Jerusalem, and he is a founding board member of the U.S.-Israel Tech Policy Institute.

Amitai Etzioni is a University Professor and Professor of International Relations at The George Washington University. He served as a Senior Advisor at the Carter White House, taught at Columbia University, Harvard University, and the University of California at Berkeley, and served as president of the American Sociological Association (ASA). A study by Richard Posner ranked him among the top 100 American intellectuals. He is the author of numerous op-eds and his voice is frequently heard in the media. He is the author of many books, including *The Active Society, The Moral Dimension, The New Golden Rule, My Brother's Keeper*—and most recently *Happiness is the Wrong Metric*. And two books on privacy, *The Limits of Privacy* (1999) and *Privacy in a Cyber Age: Policy and Practice* (2015).

Sjaak van der Geest is emeritus professor of Medical Anthropology at the University of Amsterdam. He has done fieldwork in Ghana and Cameroon on a variety of subjects including sexual relationships and birth control, the use and distribution of medicines, popular song texts, meanings of growing old, and concepts of dirt and hygiene. His interest in privacy is of a more recent date and enables him to reflect on fieldwork experiences throughout his anthropological career. He was the founder and editor-in-chief of the journal *Medische Antropologie* (now: *Medicine Anthropology Theory – MAT*) and assistant editor of several other journals in the field of medical anthropology. Website: www.sjaakvandergeest.nl

Aviva de Groot is a PhD researcher at the Tilburg Institute for Law, Technology, and Society. With her research, she aims to identify explanatory benchmarks and modalities for providing rights relevant understanding of data driven technologies and their applications to laymen users. In particular, she focusses on automated decision making processes. With privacy at the core, her interests more broadly concern humans and technology, their mutual shaping and how this effects our understanding of human rights protections. She has professional and research experience in fields where technology supports human interaction and where humans interact with machines. As concerns the former, in dumber times, the transformation of citizens' reality to their administrative truth already proved to be a tricky process. Years of professional legal experience in a specialized field of administration law, dealing with phantom vehicle registrations, taught her how the complex interplay of law, policy and technology should discourage a narrow a focus on either element. Her Information Law master's thesis research project dealt with privacy concerns arising from the use of social robots in health care settings. She focused in particular on levels of transparency and understanding of human-robot interaction.

Mikko Hypponen is a worldwide authority on computer security and the Chief Research Officer of F-Secure. He has written on his research for the New York Times, Wired and Scientific American and lectured at the universities of Oxford, Stanford and Cambridge. He sits in the advisory boards of EUROPOL and the Monetary Authority of Singapore.

Sjoerd Keulen is an assistant professor of public administration at the Erasmus University, Rotterdam. Sjoerd read both History and International Relations at the University of Groningen. Sjoerd wrote his PhD-thesis at the University of Amsterdam on the post-war history of changing ideal types of policy making in the Netherlands. Thereafter he was a researcher for the parliamentary enquiry on housing associations of the Netherlands House of Representatives and a senior advisor at the Inspectorate of the Budget of the Netherlands Ministry of Finance. Sjoerd's research focuses on qualitative analyses of accountability, policy evaluation and on connecting history to financial management and budgetary policy. Together with Ronald Kroeze Sjoerd published on management history and leadership in management journals and history journals.

Dr. **Wouter Koelewijn** wrote his dissertation at the eLaw, Center for and Digital Technologies of the University of Leiden, on electronical exchanges of sensitive personal data in the policing domain. He worked as a lawyer specialized in health care governance & finance, and as advisor and senior manager for PWC in the health care domain. He currently works as health care lawyer and partner of Van Benthem & Keulen and also as associate professor Health Law at Amsterdam University. Wouter also teaches postacademic courses on Privacy & Data Protection.

Matthijs Koot is employed as IT security specialist at Secura BV, and is guest researcher at the System and Network Engineering laboratory of the University of Amsterdam. Matthijs holds a MSc in System & Network Engineering (2005-2006) and a PhD in data anonymity (2007-2011) from the University of Amsterdam. In his work for Secura (previously known as Madison Gurkha) he has over six years of experience in, among others, testing security of computer systems and networks, social engineering, red teaming, code reviews, design reviews, and supporting privacy-by-design and security-by-design practices. In 2018 he joined as board member of the Netherlands Intelligence Studies Association (NISA).

Ronald Kroeze is an Assistant Professor of History at the Vrije Universiteit Amsterdam. He teaches courses on European history, democracy, corruption and global justice. In his research and publications, he focuses on the history of governance, transparency, privacy, democracy and (anti)corruption as well as the history of management and leadership. A recent publication is Ronald Kroeze, Guy Geltner and André Vitoria (eds.), *A History of Anticorruption. From Antiquity until the Modern Era.* Oxford: Oxford University Press, 2018. On his research he has also published in *Journal for Modern European History, BMGN-Low Countries Historical Review, Management & Organizational History* and *Business History.* Kroeze was a postdoctoral researcher at the University of Amsterdam and a research fellow at Humboldt University, Warwick University and University of Avignon. He is a member of the Amsterdam Centre for Political Thought and the European Research Network 'Politics and corruption: history and sociology', funded by the CNRS.

Prof.dr.ir. **Cees de Laat** chairs the System and Network Engineering (SNE) laboratory at the Faculty of Science at University of Amsterdam. The SNE lab conducts research on leading-edge computer systems of all scales, ranging from global-scale systems and networks to embedded devices. Across these multiple scales our particular interest is on extra-functional properties of systems, such as performance, programmability, productivity, security, trust, sustainability and, last but not least, the societal impact of emerging systems-related technologies. Cyber Infrastructure is rapidly evolving from relatively simple fixed components to programmable and virtualized objects with many degrees of freedom, owned and operated by different entities in multiple administrative connected domains on the Internet. Harnessing this complexity in a transparent trust-able way for safe and secure data processing is a major research topic that nowadays defines the focus of his research. For current activities and projects see: http://delaat.net/.

Viktor Mayer-Schönberger is Professor of Internet Governance and Regulation at the University of Oxford. In addition to his recent book "Reinventing Capitalism in the Age of Big Data" with Thomas Ramge) and the best-selling "Big Data" (with Kenneth Cukier), Mayer-Schönberger has published ten books, including the awards-winning "Delete: The Virtue of Forgetting in the Digital Age" and is the author of over a hundred articles and book chapters on the information economy. After successes in the International Physics Olympics and the Austrian Young Programmers Contest, Mayer-Schönberger studied in Salzburg, Harvard and at the London School of Economics. In 1986 he founded Ikarus Software, a company focusing on data security and developed the Virus Utilities, which became a best-selling Austrian software product. He advises governments, businesses and NGOs on new economy and information society issues.

Deirdre K. Mulligan is an Associate Professor in the School of Information at UC Berkeley, and a faculty Director of the Berkeley Center for Law & Technology. Mulligan's research explores legal and technical means of protecting values such as privacy, freedom of expression, and fairness in emerging technical systems. Her book, Privacy on the Ground: Driving Corporate Behavior in the United States and Europe, a study of privacy practices in large corporations in five countries, conducted with UC Berkeley Law Prof. Kenneth Bamberger was recently published by MIT Press. Mulligan and Bamberger received the 2016 International Association of Privacy Professionals Leadership Award for their research contributions to the field of privacy protection. Mulligan is currently serving on the Defense Advanced Research Projects Agency's Information Science and Technology Advisory Board, the Board of Directors of the Center for Democracy and Technology, a leading advocacy organization protecting global online civil liberties and human rights, and the Board of the Partnership for AI. Prior to joining the School of Information. Mulligan began her academic career as a Clinical Professor of Law, the founding Director of the Samuelson Law, Technology & Public Policy Clinic, and Director of Clinical Programs at the UC Berkeley School of Law. Prior to Berkeley, she served as staff counsel at the Center for Democracy & Technology in Washington, D.C. Mulligan holds a JD from Georgetown University Law Center and a BA from Smith College.

Sandra Petronio is Founding Director of Communication Privacy Management Center at IUPUI, Professor, Department of Communication Studies in the Indiana University and IU School of Medicine, Senior Affiliate Faculty in Fairbanks Center for Medical Ethics. Served as Founding Director Translating Research into Practice Center. Received B.A State University of New York at Stony Brook, received M.A. and Ph.D. University of Michigan. Studies privacy, disclosure, and confidentiality and developed the evidenced-based "*Communication Privacy Management*" (CPM) theory examining how people manage their private information. Her 2002 book on CPM theory, "Boundaries of Privacy: Dialectics of Disclosure" won National Communication Association Gerald R. Miller Award and from the International Association of Relationship Research. Petronio published five books, numerous scholarly articles. Received National Communication Association's Bernard J. Brommel Lifetime Award for Excellence in scholarship, received National Communication Association Mark Knapp Scholar's Award for research in Interpersonal, Western States Communication Association Distinguished Scholar Award, and Founding Fellow International Association of Relationship Research. In 2018, received National Communication Association Distinguished Scholar. Email address is: petronio@iupui.edu and www.cpmcenter.iupui.edu

Robin Pierce, JD, PhD, is associate professor at the Tilburg Institute for Law, Technology, and Society (TILT, Tilburg University, The Netherlands). She obtained her law degree (Juris Doctor) from University of California, Berkeley and a PhD from Harvard University where her work focused on genetic privacy. She has taught courses in Data Protection and Privacy, Ethical and Legal Issues in Biotechnology, and Social Issues in Biology. She currently teaches a course on Regulation for the LLM Law & Technology at Tilburg Law School. Her current research addresses the themes of data protection law and health data, as well as legal, policy and ethical issues regarding AI in healthcare, and regulation of novel biomedical technologies.

Jo Pierson, Ph.D., is Associate Professor in the Department of Media and Communication Studies at the Vrije Universiteit Brussel (VUB) in Belgium (Faculty of Social Sciences & Solvay Business School). He is also Senior Researcher and Unit Leader at the research centre SMIT (Studies on Media, Innovation and Technology) since 1996. In this position he is in charge of the research unit 'Privacy, Ethics & Literacy', in cooperation with imec (R&D and innovation hub in nanoelectronics and digital technology). He lectures undergraduate and postgraduate courses at Vrije Universiteit Brussel, Hasselt University and University of Amsterdam, covering socio-technical issues of digital media design and use. Drawing upon media and communication studies, in combination with science and technology studies, his interdisciplinary research focus is on data, privacy, public values and user empowerment in online platforms. He is also elected member of the International Council of the International Association for Media and Communication Research (IAMCR).

Charles Raab is Professorial Fellow in Politics and International Relations, School of Social and Political Science, University of Edinburgh; Director, Centre for Research into Information, Surveillance and Privacy (CRISP); co-Chair, Independent Digital Ethics Panel for Policing (IDEPP); Fellow, Alan Turing Institute (ATI) and member, ATI Data Ethics Group. Research on privacy, data protection, surveillance, 'smart' environments, security, democracy, identity. Publications include (with C. Bennett), *The Governance*

of Privacy: Policy Instruments in Global Perspective (2003; 2006); (with B. Goold), *Protecting Information Privacy* (2011); (with Surveillance Studies Network), *A Report on the Surveillance Society* (2006); (with W. Webster *et al*, eds.), *Video Surveillance: Practices and Policies in Europe* (2012). Evidence to UK parliamentary committees (e.g., Intelligence and Security Committee, 2014); Specialist Adviser, House of Lords Constitution Committee for inquiry, *Surveillance: Citizens and the State*, HL Paper 18, Session 2008-09. Fellow, Academy of Social Sciences; Fellow, Royal Society of Arts.

Dr. **Priscilla Regan** is a Professor in the Schar School of Policy and Government at George Mason University. Prior to joining that faculty in 1989, she was a Senior Analyst in the Congressional Office of Technology Assessment (1984-1989) and an Assistant Professor of Politics and Government at the University of Puget Sound (1979-1984). From 2005 to 2007, she served as a Program Officer for the Science, Technology and Society Program at the National Science Foundation. Since the mid-1970s, Dr. Regan's primary research interests have focused on both the analysis of the social, policy, and legal implications of organizational use of new information and communications technologies, and also on the emergence and implementation of electronic government initiatives by federal agencies.

Dr. Regan has published over forty articles or book chapters, as well as Legislating Privacy: Technology, Social Values, and Public Policy (University of North Carolina Press, 1995). As a recognized researcher in this area, Dr. Regan has testified before Congress and participated in meetings held by the Department of Commerce, Federal Trade Commission, Social Security Administration, and Census Bureau. She has received grants from the National Science Foundation. She was a member of the National Academy of Sciences, Computer Science and Telecommunications Board, Committee on Authentication Technologies and their Privacy Implications. Dr. Regan received her PhD in Government from Cornell University and her BA from Mount Holyoke College.

Beate Roessler is Professor of Philosophy at the University of Amsterdam. She formerly taught philosophy at the Free University, Berlin, Germany, and at the University of Bremen, Germany. She was a fellow at the *Institute for Advanced Study* (Wissenschaftskolleg) in Berlin, at the *Center for Agency, Value, and Ethics* at Macquarie University, Sydney, a two-month fellow at the University of Melbourne, Law School and a visiting professor at the New York University. She is a co-editor of the *European Journal of Philosophy* and a member of various advisory boards.

Her publications include *Social Dimensions of Privacy. Interdisciplinary Perspectives* (ed. with D. Mokrosinska), Cambridge UP 2015; "Meaningful Work: Arguments from Autonomy", in: *Journal of Political Philosophy* 2012; *The Value of Privacy*, Polity Press 2005; *Privacies. Philosophical Evaluations*, (ed.), Stanford University Press, 2004. Her book *Autonomie: ein Versuch über das gelungene Leben*, was published in 2017 and will be published in English in 2019.

Edo Roos Lindgreen is professor of Data Science in Auditing at the University of Amsterdam. He is one of the co-founders of the Amsterdam Platform for Privacy Research and member of the organizing committee of the Amsterdam Privacy Conference. He serves as the program director for the Executive Programme of Digital Auditing and the Executive MSc of Internal Auditing at the Amsterdam Business School. Until 2018, he was a partner at KPMG Advisory N.V.

Marijn Sax is a PhD candidate at the Institute for Information Law and Department of Philosophy of the University of Amsterdam. He has a background in Political Science (BSc.) and Philosophy (BA., MA., both *cum laude*) and is mainly interested in questions concerning ethics, privacy and technology. Marijn's research focuses on health apps, and more specifically on the ethical dimensions of this new phenomenon and how ethical considerations can inform legal regulation. Marijn's research is part of the Personalised Communication project.

Dr. **Tjeerd Schiphof** is assistant professor at the University of Amsterdam. Since long he has specialized in Information Law, more in particular with regard to art, media and cultural heritage. The study of privacy concerns in the archival sector is one of his research objects. He is affiliated to the Bachelor programme Media and Information, and the Master programme Archival Studies, both being offered by the UvA department of Media Studies.

Bart van der Sloot specializes in Privacy and Big Data. He also publishes regularly on the liability of Internet Intermediaries, data protection and internet regulation. Bart has studied philosophy and law in the Netherlands and Italy and has also successfully completed the Honours Programme of the Radboud University.

He currently works at the Tilburg Institute for Law, Technology, and Society of the University of Tilburg, Netherlands. Bart formerly worked for the Institute for Information Law, University of Amsterdam, and for the Scientific Council for Government Policy (WRR) (part of the Prime Minister's Office of the Netherlands) to co-author a report on the regulation of Big Data in relation to security and privacy.

Bart van der Sloot is the general editor of the international journal *European Data Protection Law Review* and board member of the European Human Rights Cases. Bart also is the scientific director of the Privacy & Identity Lab.

Finally, Bart van der Sloot is the coordinator of the Amsterdam Platform for Privacy Research (APPR), which consists of about 70 employees at the University of Amsterdam who in their daily teaching and research activities focus on privacy-related issues. In that position, he also founded the minor Privacy Studies, an interdisciplinary minor at bachelor level about privacy and new technological developments. Previously, he was the conference coordinator of the Amsterdam Privacy Conference 2012, 2015 and 2018. Personal website: www.bartvandersloot.com

Cass R. Sunstein is currently the Robert Walmsley University Professor at Harvard. From 2009 to 2012, he was Administrator of the White House Office of Information and Regulatory Affairs. He is the founder and director of the Program on Behavioral Economics and Public Policy at Harvard Law School. Mr. Sunstein has testified before congressional committees on many subjects, and he has been involved in constitution-making and law reform activities in a number of nations.

Mr. Sunstein is author of many articles and books, including Republic.com (2001), Risk and Reason (2002), Why Societies Need Dissent (2003), The Second Bill of Rights (2004), Laws of Fear: Beyond the Precautionary Principle (2005), Worst-Case Scenarios (2001), Nudge: Improving Decisions about Health, Wealth, and Happiness (with Richard H. Thaler, 2008), Simpler:

The Future of Government (2013) and most recently Why Nudge? (2014) and Conspiracy Theories and Other Dangerous Ideas (2014). He is now working on group decisionmaking and various projects on the idea of liberty.

Giliam de Valk is, since 2016, an assistant professor at the Institute of Security and Global Affairs (ISGA) of Leiden University. He lectures and researches on intelligence, analysis and methodology. He wrote his dissertation (2005) at the Law Faculty of University of Groningen on the quality of intelligence analyses. Before, he has worked at the University of Amsterdam and the Netherlands Defense Academy in Breda, and at both institutes he initiated and coordinated a minor on intelligence studies. Besides on intelligence, he has lectured on issues as strategy, humanitarian intervention, counterinsurgency, terrorism and counter-terrorism. From 2004-2012, he had been the secretary of the Netherlands Intelligence Security Association (NISA), of which he is still a board member.

Ine van Zeeland MA is a PhD researcher at the Studies in Media, Innovation and Technology (SMIT) research centre at the Vrije Universiteit Brussel, which is affiliated with R&D hub imec. She mostly works on research projects around privacy and data protection, with occasional excursions into algorithmic accountability and digital literacy. Her main focus is on the day-to-day tactics of data protection professionals, as well as organisational efforts to comply with the European Union's General Data Protection Regulation (GDPR).

For Product Safety Concerns and Information please contact our EU
representative GPSR@taylorandfrancis.com
Taylor & Francis Verlag GmbH, Kaufingerstraße 24, 80331 München, Germany

www.ingramcontent.com/pod-product-compliance
Lightning Source LLC
Chambersburg PA
CBHW050556270326
41926CB00012B/2074

* 9 7 8 9 4 6 2 9 8 8 0 9 5 *